Cosmological Inflation, Dark Matter and Dark Energy

Cosmological Inflation, Dark Matter and Dark Energy

Special Issue Editor

Kazuharu Bamba

MDPI • Basel • Beijing • Wuhan • Barcelona • Belgrade

Special Issue Editor
Kazuharu Bamba
Fukushima University
Japan

Editorial Office
MDPI
St. Alban-Anlage 66
4052 Basel, Switzerland

This is a reprint of articles from the Special Issue published online in the open access journal *Symmetry* (ISSN 2073-8994) from 2018 to 2019 (available at: https://www.mdpi.com/journal/symmetry/special_issues/Cosmological_Inflation_Dark_Matter_Dark_Energy).

For citation purposes, cite each article independently as indicated on the article page online and as indicated below:

LastName, A.A.; LastName, B.B.; LastName, C.C. Article Title. *Journal Name* **Year**, *Article Number*, Page Range.

ISBN 978-3-03921-764-9 (Pbk)
ISBN 978-3-03921-765-6 (PDF)

Contents

About the Special Issue Editor

Kazuharu Bamba received his Ph.D. from Osaka University in 2006 and served as JSPS Postdoctoral Research Fellow at Yukawa Institute for Theoretical Physics (YITP), Kyoto University. He also worked at Kinki University as a Postdoctoral Research Fellow and then he moved to National Tsing Hua University, Taiwan, in 2008. In 2010, he became Assistant Professor at Kobayashi-Maskawa Institute for the Origin of Particles and the Universe (KMI), Nagoya University, and worked on dark energy problems and modified gravity theories. In 2014, he was appointed Lecturer at Ochanomizu University. In 2015, he began as Associate Professor at Fukushima University in Japan, where he teaches physics and cosmology. He has published more than 110 papers in international refereed journals. Since 2014, he has been the Associate Editor-in-Chief of *Universe* (MDPI), and also an Editor of *Entropy* since 2014 and *Symmetry* since 2018.

Preface to "Cosmological Inflation, Dark Matter and Dark Energy"

Based on recent cosmological observations, such as type Ia supernovae, cosmic microwave background (CMB) radiation large-scale structure, baryon acoustic oscillations (BAO), and weak lensing, the universe has experienced an accelerated phase of its expansion, not only regarding the early universe but also in present times. The former is called "inflation" and the latter is called "the late-time cosmic acceleration". It is also well known that the three energy components of the universe are dark energy (about 68%), dark matter (about 27%), and baryon (about 5%).

A number of studies have been executed for the origins of the field to realize inflation, dark matter, and dark energy. The future detection of primordial gravitational waves is strongly expected in order to know the energy scale of the inflationary phase. Moreover, there are two possibilities for the origin of dark matter, namely, new particles in particle theory models beyond the standard model and astrophysical objects. Furthermore, two representative studies have been proposed for the true character of dark energy. One is the introduction of some unknown matter called dark energy with the negative pressure in the framework of general relativity. The other is the modification of gravity at large scales, leading to the so-called geometrical dark energy.

The main aim of this book is to understand various cosmological aspects, including the origins of inflation, dark matter, and dark energy. It is one of the most significant and fundamental issues in modern physics and cosmology. In addition to phenomenological approaches, more fundamental studies are considered from higher-dimensional theories of gravity, quantum gravity, and quantum cosmology, physics in the early universe, quantum field theories, and gauge field theories in curved spacetime as well as strings, branes, and the holographic principle.

This book consists of the 13 peer-reviewed articles published in the Special Issue "Cosmological Inflation, Dark Matter and Dark Energy" in Symmetry. As Guest Editor of this Special Issue, I have invited the authors to write original articles to the Issue and rearranged the contents for this book.

The organization of this book is as follows. The first part (2 articles) concerns the origin and nature of dark matter. The second (7 articles) details the mechanisms for the cosmic accelerations of dark energy as well as inflation. The third (4 articles) covers gravity theories and their quantum aspects.

I would like to sincerely acknowledge MDPI and am greatly appreciative of the Managing Editor, Ms. Dalia Su, for her very kind support and warm assistance during this project. Moreover, I am highly grateful to the Editor-in-Chief Professor Dr. Sergei D. Odintsov for giving me the chance to serve as Guest Editor of this Special Issue. Furthermore, since this is my first memorial editorial book, I would like to express my sincere gratitude to my supervisors Professor Dr. Jun'ichi Yokoyama, Professor Dr. Misao Sasaki, Professor Dr. Motohiko Yoshimura, Professor Dr. Fumio Takahara; my professors Professor Dr. Shin'ichi Nojiri, Professor Dr. Akio Sugamoto, Professor Dr. Chao-Qiang Geng, Professor Dr. Nobuyoshi Ohta; my important collaborators on the topics in this book, Professor Dr. Salvatore Capozziello, Professor Dr. Emmanuel N. Saridakis, Professor Dr. Shinji Tsujikawa; as well as to all of my collaborators. I would also like to thank all the authors for the submission of their articles to this Special Issue of *Symmetry*.

Kazuharu Bamba
Special Issue Editor

symmetry

MDPI

Article

Dark Matter as a Non-Relativistic Bose–Einstein Condensate with Massive Gravitons

Emma Kun [1], Zoltán Keresztes [1,*], Saurya Das [2] and László Á. Gergely [1]

1 Institute of Physics, University of Szeged, Dóm tér 9, H-6720 Szeged, Hungary; kun@titan.physx.u-szeged.hu (E.K.); gergely@physx.u-szeged.hu (L.Á.G.)

2 Theoretical Physics Group and Quantum Alberta, Department of Physics and Astronomy, University of Lethbridge, 4401 University Drive, Lethbridge, AB T1K 3M4, Canada; saurya.das@uleth.ca

* Correspondence: zkeresztes@titan.physx.u-szeged.hu

Received: 15 September 2018; Accepted: 15 October 2018; Published: 17 October 2018

Abstract: We confront a non-relativistic Bose–Einstein Condensate (BEC) model of light bosons interacting gravitationally either through a Newtonian or a Yukawa potential with the observed rotational curves of 12 dwarf galaxies. The baryonic component is modeled as an axisymmetric exponential disk and its characteristics are derived from the surface luminosity profile of the galaxies. The purely baryonic fit is unsatisfactory, hence a dark matter component is clearly needed. The rotational curves of five galaxies could be explained with high confidence level by the BEC model. For these galaxies, we derive: (i) upper limits for the allowed graviton mass; and (ii) constraints on a velocity-type and a density-type quantity characterizing the BEC, both being expressed in terms of the BEC particle mass, scattering length and chemical potential. The upper limit for the graviton mass is of the order of 10^{-26} eV/c^2, three orders of magnitude stronger than the limit derived from recent gravitational wave detections.

Keywords: dark matter; galactic rotation curve

---

1. Introduction

The universe is homogeneous and isotropic at scales greater than about 300 Mpc. It is also spatially flat and expanding at an accelerating rate, following the laws of general relativity. The spatial flatness and accelerated expansion are most easily explained by assuming that the universe is almost entirely filled with just three constituents, namely visible matter, Dark Matter (DM) and Dark Energy (DE), with densities ρ_{vis}, ρ_{DM} and ρ_{DE}, respectively, such that $\rho_{vis} + \rho_{DM} + \rho_{DE} = \rho_{crit} \equiv 3H_0^2/8\pi G \approx 10^{-26}\ kg/m^3$ (where H_0 is the current value of the Hubble parameter and G the Newton's constant), the so-called critical density, and $\rho_{vis}/\rho_{crit} = 0.05, \rho_{DM}/\rho_{crit} = 0.25$ and $\rho_{DM}/\rho_{crit} = 0.70$ [1,2]. It is the large amount of DE which causes the accelerated expansion. In other words, 95% of its constituents is invisible. Furthermore, the true nature of DM and DE remains to be understood. There has been a number of promising candidates for DM, including weakly interacting massive particles (WIMPs), sterile neutrinos, solitons, massive compact (halo) objects, primordial black holes, gravitons, etc., but none of them have been detected by dedicated experiments and some of them fail to accurately reproduce the rotation curves near galaxy centers [3,4]. Similarly, there has been a number of promising DE candidates as well, the most popular being a small cosmological constant, but any computation of the vacuum energy of quantum fields as a source of this constant gives incredibly large (and incorrect) estimates; another popular candidate is a dynamical scalar field [5,6]. Two scalar fields are also able to model both DM and DE [7]. Extra-dimensional modifications through a variable brane tension and five-dimensional Weyl curvature could also simulate the effects of DM and DE [8]. In other theories, dark energy is the thermodynamic energy of the internal motions of a polytropic DM fluid [9,10].

Therefore, what exactly are DM and DE remain as two of the most important open questions in theoretical physics and cosmology.

Given that DM pervades all universe, has mass and energy, gravitates and is cold (as otherwise it would not clump near galaxy centers), it was examined recently whether a Bose–Einstein condensate (BEC) of gravitons, axions or a Higgs type scalar can account for the DM content of our universe [11,12]. While this proposal is not new, and in fact BEC and superfluids as DM have been considered by various authors [13–34], the novelty of the new proposal was twofold: (i) for the first time, it computed the quantum potential associated with the BEC; and (ii) it showed that this potential can in principle account for the DE content of our universe as well. It was also argued in the above papers that, if the BEC is accounting for DE gravitons, then their mass would be tightly restricted to about 10^{-32} eV/c^2. Any higher, and the corresponding Yukawa potential would be such that gravity would be shorter ranged than the current Hubble radius, about 10^{26} m, thereby contradicting cosmological observations. Any lower and unitarity in a quantum field theory with gravitons would be lost [35].

In this paper, we discuss the possibility of a BEC formed by scalar particles, interacting gravitationally through either the Newton or Yukawa potential. Such a BEC, interacting only through massless gravitons has been previously tested as a viable DM candidate by confronting with galactic rotation curves [30,36].

In this paper, we solve the time-dependent Scrödinger equation for the macroscopic wavefunction of a spherically symmetric BEC, where in place of the potential we plug-in a sum of the external gravitational potential and local density of the condensate, proportional to the absolute square of the wavefunction itself, times the self-interaction strength. The resultant non-linear Schrödinger equation is known as the Gross–Pitaevskii equation. For the self-interaction, we assume a two-body δ-function type interaction (the Thomas–Fermi approximation), while we assume that the external potential being massive-gravitational in nature, satisfying the Poisson equation with a mass term. The BEC-forming bosons could be ultra-light, raising the question of why we use the non-relativistic Schrödinger equation. This is because, once in the condensate, they are in their ground states with little or no velocity, and hence non-relativistic for all practical purposes. Solving these coupled set of equations, we obtain the density function, the potential outside the condensate and also the velocity profiles of the rotational curves. We then compare these analytical results with observational curves for 12 dwarf galaxies and show that they agree with a high degree of confidence for five of them. For the remaining galaxies, no definitive conclusion can be drawn with a high confidence level. Nevertheless, our work provides the necessary groundwork and motivation to study the problem further to provide strong evidence for or against our model.

This paper is organized as follows. In the next section, we set the stage by summarizing the coupled differential equations that govern the BEC wavefunction and gravitational potential and find the BEC density profiles. In Section 3, we construct the corresponding analytical rotation curves. In Section 4, we compare these and the rotational curves due to baryonic matter with the observational curves for galaxies. In Section 5, we find most probable bounds on the graviton mass, as well as derive limits for a velocity-type and a density-type quantity characterizing the BEC.

2. Self-Gravitating, Spherically Symmetric Bec Distribution in the Thomas-Fermi Approximation

A non-relativistic Bose–Einstein condensate in the mean-field approximation is characterized by the wave function $\psi(\mathbf{r},t)$ obeying

$$i\hbar\frac{\partial}{\partial t}\psi(\mathbf{r},t) = \left[-\frac{\hbar^2}{2m}\Delta + mV_{ext}(\mathbf{r}) + \lambda\rho(\mathbf{r},t)\right]\psi(\mathbf{r},t)\,, \tag{1}$$

known as the Gross–Pitaevskii equation [37–39]. Here, $\hbar$ is the reduced Planck constant, $\mathbf{r}$ is the position vector; t is the time; Δ is the Laplacian; m is the boson mass;

$$\rho\left(\mathbf{r},t\right)=\left|\psi(\mathbf{r},t)\right|^{2} \tag{2}$$

is the probability density; the parameter $\lambda > 0$ measures the atomic interactions and is also related to the scattering length [40], characterizing the two-body interatomic potential energy:

$$V_{self}=\lambda\delta\left(\mathbf{r}-\mathbf{r}'\right); \tag{3}$$

and finally $V_{ext}\left(\mathbf{r}\right)$ is an external potential. For a stationary state,

$$\psi(\mathbf{r},t)=\sqrt{\rho\left(\mathbf{r}\right)}\exp\left(\frac{i\mu}{\hbar}t\right) \tag{4}$$

where μ is a chemical potential energy [40,41]. When μ is constant, Equation (1) reduces to present works [22,30]

$$mV_{ext}+V_{Q}+\lambda\rho=\mu\,, \tag{5}$$

where V_Q is the quantum correction potential energy:

$$V_{Q}=-\frac{\hbar^{2}}{2m}\frac{\Delta\sqrt{\rho}}{\sqrt{\rho}}\,. \tag{6}$$

We mention that Equation (5) is valid in the domain where $\rho\left(\mathbf{r}\right)\neq 0$.

The quantum correction V_Q has significant contribution only close to the BEC boundary [21], therefore it can be neglected in comparison to the self-interaction term $\lambda\rho$. This *Thomas–Fermi approximation* becomes increasingly accurate with an increasing number of particles [42].

We assume $V_{ext}\left(\mathbf{r}\right)$ to be the gravitational potential created by the condensate. In the case of massive gravitons, it is described by the Yukawa-potential in the non-relativistic limit:

$$V_{ext}=U_{Y}\left(\mathbf{r}\right)=-\int\frac{G\rho_{BEC}\left(\mathbf{r}'\right)}{\left|\mathbf{r}-\mathbf{r}'\right|}e^{-\frac{\left|\mathbf{r}-\mathbf{r}'\right|}{R_{g}}}d^{3}\mathbf{r}'\,, \tag{7}$$

with $\rho_{BEC}=m\rho$, gravitational constant G, and characteristic range of the force R_g carried by the gravitons with mass m_g. The relation between R_g and m_g is $R_g=\hbar/\left(m_g c\right)$, where c is the speed of light and $\hbar$ is the reduced Planck constant. The Yukawa potential obeys the following equation:

$$\Delta U_{Y}-\frac{U_{Y}}{R_{g}^{2}}=4\pi G\rho_{BEC}\,. \tag{8}$$

Contrary to Equation (5), Equation (8) is also valid in the domain where $\rho\left(\mathbf{r}\right)=0$. In the massless graviton limit, we recover Newtonian gravity, in particular Equations (7) and (8) reduce to the Newtonian potential and Poisson equation.

2.1. *Mass Density and the Gravitational Potential inside the Condensate*

The Laplacian of Equation (5) using Equation (8) gives

$$\Delta\rho_{BEC}+\frac{4\pi Gm^{2}}{\lambda}\rho_{BEC}=-\frac{m^{2}}{\lambda R_{g}^{2}}U_{Y}\,. \tag{9}$$

For a spherical symmetric matter distribution, Equations (8) and (9) become

$$\frac{d^2(rU_Y)}{dr^2} - \frac{1}{R_g^2}(rU_Y) = 4\pi G(r\rho_{BEC}) , \tag{10}$$

$$\frac{d^2(r\rho_{BEC})}{dr^2} + \frac{1}{R_*^2}(r\rho_{BEC}) = -\frac{m^2}{\lambda R_g^2}(rU_Y) . \tag{11}$$

where we introduced the notation

$$\frac{1}{R_*^2} = \frac{4\pi G m^2}{\lambda} . \tag{12}$$

This system gives the following fourth order, homogeneous, linear differential equation for $r\rho_{BEC}$:

$$\frac{d^4(r\rho_{BEC})}{dr^4} + \Lambda^2 \frac{d^2(r\rho_{BEC})}{dr^2} = 0 , \tag{13}$$

with

$$\Lambda = \sqrt{\frac{1}{R_*^2} - \frac{1}{R_g^2}} . \tag{14}$$

In the case of massless gravitons, πR_* gives the radius of the BEC halo [30]. To have a real Λ, $R_g > R_*$ should hold, constraining the graviton mass from above. Typical dark matter halos have πR_* of the order of 1 kpc which gives the following upper bound for the graviton mass: $m_g c^2 < 4 \times 10^{-26}$ eV. Then, the general solution of Equation (13) is

$$r\rho_{BEC} = A_1 \sin(\Lambda r) + B_1 \cos(\Lambda r) + C_1 r + D_1 . \tag{15}$$

with integration constants A_1, B_1, C_1 and D_1. This is why we impose the reality of Λ. For the imaginary case the general solution would contain runaway hyperbolic functions. This is also the solution of the system in Equations (10) and (11). Requiring ρ_{BEC} to be bounded, we have $D_1 = -B_1$. Then, the core density of the condensate is

$$0 < \rho^{(c)} \equiv \rho_{BEC}(r=0) = A_1\Lambda + C_1 , \tag{16}$$

and the solution can be written as

$$\rho_{BEC}(r) = \left(\rho^{(c)} - C_1\right)\frac{\sin(\Lambda r)}{\Lambda r} + B_1\frac{\cos(\Lambda r) - 1}{r} + C_1 . \tag{17}$$

Substituting $\rho_{BEC}(r)$ in Equation (11), the gravitational potential is

$$-\frac{m^2}{\lambda R_g^2}(rU_Y) = \left(\rho^{(c)} - C_1\right)\frac{\sin(\Lambda r)}{\Lambda R_g^2} + \frac{B_1}{R_g^2}\cos(\Lambda r) - \frac{B_1}{R_*^2} + \frac{C_1}{R_*^2}r . \tag{18}$$

Being related to the mass density by Equation (5) gives

$$B_1 = 0 , \; C_1 = -\frac{m\mu}{\lambda R_g^2 \Lambda^2} . \tag{19}$$

The BEC mass distribution ends at some radial distance R_{BEC} (above which we set ρ_{BEC} to zero), allowing to express C_1 in terms of $\rho^{(c)}$, R_{BEC} and Λ as

$$C_1 = \rho^{(c)}\frac{\sin(\Lambda R_{BEC})}{\Lambda R_{BEC}}\left(\frac{\sin(\Lambda R_{BEC})}{\Lambda R_{BEC}} - 1\right)^{-1} . \tag{20}$$

Finally, we consider the massless graviton limiting case $m_g \to 0$. Then, $R_g \to \infty$ implies $\Lambda = \sqrt{4\pi G m^2/\lambda} = 1/R_*$ and $C_1 = 0$ (by Equation (19)). Then, $\rho_{BEC}(r)$ coincides with Equation (40) [22].

2.2. Gravitational Potential Outside the Condensate

The potential U is determined up to an arbitrary constant A_2, i.e.,

$$U^{out} = U_Y^{out} + A_2 \,. \tag{21}$$

Here, U_Y^{out} satisfies Equation (8) with $\rho_{BEC} = 0$. The solution for U_Y^{out} is

$$U_Y^{out} = B_2 \frac{e^{-\frac{r}{R_g}}}{r} + C_2 \frac{e^{\frac{r}{R_g}}}{r} \,. \tag{22}$$

Since an exponentially growing gravitational potential is non-physical, $C_2 = 0$ and

$$U^{out} = A_2 + B_2 \frac{e^{-\frac{r}{R_g}}}{r} \,. \tag{23}$$

The constants A_2 and B_2 are determined from the junction conditions: the potential is both continuous and continuously differentiable at $r = R_{BEC}$:

$$\begin{aligned} A_2 \;=\; & \frac{4\pi G \rho^{(c)}}{1+\frac{R_{BEC}}{R_g}} \frac{R_*^2 R_g^2}{1-\frac{\sin(\Lambda R_{BEC})}{\Lambda R_{BEC}}} \left[\frac{\Lambda}{R_g} \sin\left(\Lambda R_{BEC}\right) \right. \\ & \left. \frac{1}{R_*^2} \frac{\sin\left(\Lambda R_{BEC}\right)}{\Lambda R_{BEC}} - \frac{\cos\left(\Lambda R_{BEC}\right)}{R_g^2} \right] , \end{aligned} \tag{24}$$

$$B_2 = \frac{4\pi G \rho^{(c)}}{\frac{1}{R_{BEC}} + \frac{1}{R_g}} \frac{R_*^2}{1-\frac{\sin(\Lambda R_{BEC})}{\Lambda R_{BEC}}} \left[\cos\left(\Lambda R_{BEC}\right) - \frac{\sin\left(\Lambda R_{BEC}\right)}{\Lambda R_{BEC}} \right] e^{\frac{R_{BEC}}{R_g}} \,. \tag{25}$$

In the next section, we see that the continuous differentiability of the gravitational potential coincides with the continuity of the rotation curves.

3. Rotation Curves in Case of Massive Gravitons

Newton's equation of motions give the velocity squared of stars in circular orbit in the plane of the galaxy as

$$v^2(R) = R \frac{\partial U}{\partial R} \,. \tag{26}$$

Here, R is the radial coordinate in the galaxy's plane and U is the gravitational potential. In the case of massive gravitons, U is given by $U = U_Y + A$, where U_Y satisfies the Yukawa-equation with the relevant mass density and A is a constant.

The contribution of the condensate to the circular velocity is

$$v_{BEC}^2(R) = \frac{4\pi G \rho^{(c)} R_*^2}{1-\frac{\sin(\Lambda R_{BEC})}{\Lambda R_{BEC}}} \left[\frac{\sin\left(\Lambda R\right)}{\Lambda R} - \cos\left(\Lambda R\right) \right] \tag{27}$$

for $r \leq R_{BEC}$ and

$$v_{BEC}^2(R) = -B_2 \left(\frac{1}{R} + \frac{1}{R_g} \right) e^{-\frac{R}{R_g}} \tag{28}$$

for $r \geq R_{BEC}$.

In the relevant situations, the stars orbit inside the halo and their rotation curves are determined by the parameters: $\rho^{(c)} R_*^2$, R_{BEC} and Λ. In the limit $m_g \to 0$, the v^2 of the BEC with massless gravitons is recovered, given as Böhmer proposed [22]

$$v_{BEC}^2(R) = 4\pi G \rho^{(c)} R_*^2 \left[\frac{\sin(R_*^{-1} R)}{R_*^{-1} R} - \cos(R_*^{-1} R) \right] \tag{29}$$

for $r \leq R_{BEC}$ and

$$v_{BEC}^2\,(R) = 4G\rho^{(c)} \frac{R_*}{R} \tag{30}$$

for $r \geq R_{BEC}$.

4. Best-Fit Rotational Curves

4.1. Contribution of the Baryonic Matter in Newtonian and in Yukawa Gravitation

The baryonic rotational curves are derived from the distribution of the luminous matter, given by the surface brightness $S = F/\Delta\Omega$ (radiative flux F per solid angle $\Delta\Omega$ measured in radian squared of the image) of the galaxy. The observed S depends on the redshift as $1/(1+z)^4$, on the orientation of the galaxy rotational axis with respect to the line of sight of the observer, but independent from the curvature index of Friedmann universe. Since we investigate dwarf galaxies at small redshift ($z < 0.002$), the z-dependence of S is negligible. Instead of S given in units of solar luminosity $L_\odot$ per square kiloparsec ($L_\odot/kpc^2$), the quantity μ given in units of $mag/arcsec^2$ can be employed, defined through

$$S(R) = 4.255 \times 10^{14} \times 10^{0.4(\mathcal{M}_\odot - \mu(R))}, \tag{31}$$

where R is the distance measured the center of the galaxy in the galaxy plane and $\mathcal{M}_\odot$ is the absolute brightness of the Sun in units of mag. The absolute magnitude gives the luminosity of an object, on a logarithmic scale. It is defined to be equal to the apparent magnitude appearing from a distance of 10 parsecs. The bolometric absolute magnitude of a celestial object $\mathcal{M}_\star$, which takes into account the electromagnetic radiation on all wavelengths, is defined as $\mathcal{M}_\star - \mathcal{M}_\odot = -2.5 \log(L_\star/L_\odot)$, where $L_\star$ and $L_\odot$ are the luminosity of the object and of the Sun, respectively.

The brightness profile of the galaxies $\mu(R)$ was derived in some works [43–45] from isophotal fits, employing the orientation parameters of the galaxies (center, inclination angle and ellipticity). This analysis leads to $\mu(R)$ which would be seen if the galaxy rotational axis was parallel to the line-of-sight. We used this $\mu(R)$ to generate $S(R)$.

The surface photometry of the dwarf galaxies are consistent with modeling their baryonic component as an axisymmetric exponential disk with surface brightness [46]:

$$S(R) = S_0 \exp[-R/b] \tag{32}$$

where b is the scale length of the exponential disk, and S_0 is the central surface brightness. To convert this to mass density profiles, we fitted the mass-to-light ratio ($\Upsilon = M/L$) of the galaxies.

In Newtonian gravity, the rotational velocity squared of an exponential disk emerges as Freeman proposed [46]:

$$v^2(R) = \pi G S_0 \Upsilon b \left(\frac{R}{b}\right)^2 (I_0 K_0 - I_1 K_1), \tag{33}$$

with I and K the modified Bessel functions, evaluated at $R/2b$. In Yukawa gravity, a more cumbersome expression has been given in the work of De Araujo and Miranda [47] as

$$\begin{aligned} v^2(R) &= 2\pi G S_0 \Upsilon R \times \left[\int_{b/\lambda}^{\infty} \frac{\sqrt{x^2 - b^2/\lambda^2}}{(1+x^2)^{3/2}} J_1 \left(\frac{R}{b} \sqrt{x^2 - b^2/\lambda^2} \right) dx \right. \\ &\left. - \int_0^{b/\lambda} \frac{\sqrt{b^2/\lambda^2 - x^2}}{(1+x^2)^{3/2}} I_1 \left(\frac{R}{b} \sqrt{b^2/\lambda^2 - x^2} \right) dx \right], \end{aligned} \tag{34}$$

where $\lambda = h/m_g/c = 2\pi R_g$ is the Compton wavelength. For $b/\lambda \ll 1$, the Newtonian limit is recovered.

4.2. Testing Pure Baryonic and Baryonic + Dark Matter Models

We chose 12 late-type dwarf galaxies from the Westerbork HI survey of spiral and irregular galaxies [43–45] to test rotation curve models. The selection criterion was that these disk-like galaxies have the longest R-band surface photometry profiles and rotation curves. For the absolute R-magnitude of the Sun, $\mathcal{M}_{\odot,R} = 4.42^m$ [48] was adopted. Then, we fitted Equation (32) to the surface luminosity profile of the galaxies, calculated with Equation (31) from $\mu(R)$. The best-fit parameters describing the photometric profile of the dwarf galaxies are given in Table 1.

We derived the pure baryonic rotational curves by fitting the square root of Equation (33) to the observed rotational curves allowing for variable M/L. The pure baryonic model leads to best-fit model-rotation curves above 5σ significance level for all galaxies (the χ^2-s are presented in the first group of columns in Table 1), hence a dark matter component is clearly required.

Then, we fitted theoretical rotation curves with contributions of baryonic matter and BEC-type dark matter with massless gravitons to the observed rotational curves in Newtonian gravity. The model–rotational velocity of the galaxies in this case is given by the square root of the sum of velocity squares given by Equations (29) and (33) with free parameters Υ, $\rho^{(c)}$ and R_*. The best-fit parameters are given in the second group of columns of Table 1. Adding the contribution of a BEC-type dark matter component with zero-mass gravitons to rotational velocity significantly improves the χ^2 for all galaxies, as well as results in smaller values of M/L. The fits are within 1σ significance level in five cases (UGC3851, UGC6446, UGC7125, UGC7278, and UGC12060), between 1σ and 2σ in three cases (UGC3711, UGC4499, and UGC7603), between 2σ and 3σ in one case (UGC8490), between 3σ and 4σ in one case (UGC5986) and above 5σ in two cases (UGC1281 and UGC5721). We note that the bumpy characteristic of the BEC model results in the limitation of the model in some cases, the decreasing branch of the theoretical rotation curve of the BEC component being unable to follow the observed plateau of the galaxies (UGC5721, UGC5986, and UGC8490). The theoretical rotation curves composed of a baryonic component plus BEC-type dark matter component with massless gravitons are presented on Figure 1.

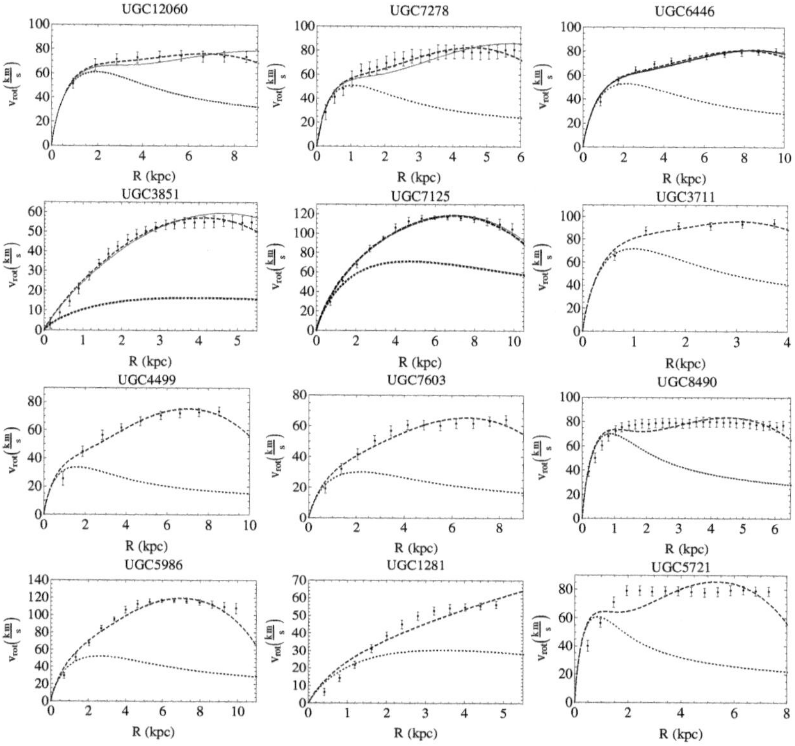

Figure 1. Theoretical rotational curves of the dwarf galaxy sample. The dots with error-bars denote archive rotational velocity curves. The model rotation curves are denoted as follows: pure baryonic in Newtonian gravitation with dotted line, baryonic + BEC with massless gravitons in Newtonian gravitation with dashed line, and baryonic + BEC with the upper limit on m_g in Yukawa gravitation with continuous line.

Table 1. Parameters describing the theoretical rotational curve models of the 12 dwarf galaxies. Best-fit parameters of the pure baryonic model in the first group of columns: central surface brightness S_0, scale parameter b, M/L ratio Υ, along with the χ^2 of the fit. This model results in best-fit model-rotation curves above 5σ significance level for all galaxies. Best-fit parameters of the baryonic matter + BEC with massless gravitons appear in the second group of columns: M/L ratio Υ, characteristic density $\rho^{(c)}$, distance parameter R_*, along with the χ^2 of the fit and the respective significance levels. Constraints on the parameter m^2/λ are also derived. In five cases, the fits χ^2 are within 1σ and marked as boldface. The fits are between 1σ and 2σ in three cases, between 2σ and 3σ in one case, between 3σ and 4σ in one case and above 5σ in two cases. Best-fit parameters of the baryonic matter + BEC with massive gravitons are given in the third group of columns only for the well-fitting galaxies: the range for R_{BEC} and the upper limit on m_g are those for which the fit remains within 1σ. Corresponding constraints on the parameter m/μ are also derived.

	Pure Baryonic				**Baryonic + BEC with $m_g = 0$**						**Baryonic + BEC with $m_g > 0$**			
ID	S_0 $10^8 \frac{L_\odot}{kpc^2}$	b kpc	Υ	χ^2	Υ	$\rho^{(c)}$ $10^7 \frac{M_\odot}{kpc^3}$	R_* kpc	$\frac{m^2}{\lambda}$ $10^{-31} \frac{kg s^2}{m^5}$	χ^2	sign. lev.	R_{BEC} kpc	m_g $10^{-26} \frac{eV}{c^2}$	$\frac{m}{\mu}$ $10^{-10} \frac{s^2}{m^2}$	sign. lev.
UGC12060	0.7	0.90	11.23	155	5.50 ± 0.33	1.07 ± 0.11	2.650 ± 0.118	1.78 ± 0.16	**1.69**	$1\sigma = 5.89$	$[7.3 \div 10.6]$	< 0.95	< 7.02	$1\sigma = 7.08$
UGC7278	6.1	0.49	2.59	499	0.81 ± 0.06	3.53 ± 0.23	1.702 ± 0.048	4.32 ± 0.24	**7.91**	$1\sigma = 21.36$	$[4.6 \div 6.8]$	< 1.40	< 5.46	$1\sigma = 22.44$
UGC6446	1.9	1.00	3.89	809	1.37 ± 0.11	1.02 ± 0.09	3.040 ± 0.128	1.36 ± 0.11	**7.91**	$1\sigma = 8.18$	$[9.2 \div 10]$	< 0.42	< 4.27	$1\sigma = 9.86$
UGC3851	0.5	1.80	2.74	86	0.74 ± 0.18	1.91 ± 0.22	1.509 ± 0.038	5.50 ± 0.28	**11.30**	$1\sigma = 20.28$	$[4.3 \div 5.5]$	< 1.26	< 11.4	$1\sigma = 21.36$
UGC7125	1.2	2.20	4.50	285	1.78 ± 0.18	2.26 ± 0.21	2.670 ± 0.071	1.76 ± 0.93	**11.82**	$1\sigma = 12.64$	$[8.2 \div 8.6]$	< 0.31	< 2.44	$1\sigma = 13.74$
UGC3711	5.2	0.46	4.40	232	2.00	8.06	1.212	-	5.11	$2\sigma = 6.18$	-	-	-	-
UGC4499	1.4	0.75	6.30	603	1.00	1.34	2.590	-	8.51	$2\sigma = 11.31$	-	-	-	-
UGC7603	2.1	1.00	1.88	462	0.40	1.07	2.470	-	13.46	$2\sigma = 15.78$	-	-	-	-
UGC8490	2.8	0.40	9.52	1350	4.06	3.35	1.715	-	40.27	$3\sigma = 50.55$	-	-	-	-
UGC5986	4.4	1.20	3.95	1682	0.48	3.17	2.620	-	32.12	$4\sigma = 38.54$	-	-	-	-
UGC1281	1.0	1.60	1.33	231	0.53	0.75	3.70	-	48.74	$5\sigma = 43.98$	-	-	-	-
UGC5721	4.9	0.40	5.79	1388	1.75	2.84	1.982	-	88.56	$5\sigma = 50.21$	-	-	-	-

We attempted to distinguish among galaxies to be included in well-fitting or less well-fitting classes based on their baryonic matter distribution. Several factors affect the goodness of the fits, as follows. The best-fit falls outside the 1σ significance level in the case of seven galaxies. Among these galaxies, UGC8490 and UGC5721 have (a_1) steeply rising rotational curve due to their centralized baryonic matter distribution ($b < 0.5$ kpc, $v_{max} > 50$ km s^{-1}) with (a_2) long, approximately constant height observed plateau. Joint fulfilment of these criteria does not occur for the well-fitting galaxies, as $b \gtrsim 0.5$ kpc for them. The rest of the galaxies with best-fits falling outside the 1σ significance level have (b_1) slowly rising rotational curve due to their less centralized baryonic matter distribution ($b > 0.5$ kpc, $v_{max} < 50$ km s^{-1}) with (b_2) short, variable height observed plateau, holding relatively small number of observational points ($N \leq 15$, a small N lowers the 1σ significance level). The well-fitting galaxies do not belong to this group, as either they hold more observational points, or have a longer, approximately constant height observed plateau. We expect that for the galaxies not falling in the classes with baryonic and observational characteristics summarized by either properties (a_1)–(a_2) or (b_1)–(b_2) the BEC dark matter model represents a good fit. Finally, we note the galaxy UGC3711 represents a special case due to the lack of sufficient observational data. Although the shape of its rotational curve is very similar to that of the best-fitting galaxy, UGC12060, it is based on just six observational points, lowering the 1σ level. Its points also have smaller error bars, which increases the χ^2. This results in the best-fit rotational curve of UGC3711 falling outside out the 1σ significance level.

Finally, we fitted the theoretical rotational curves given by both a baryonic component and a non-relativistic BEC component with massive gravitons, employing Yukawa gravity. The parameters Υ, $\rho^{(c)}$ and R_* were kept from the best-fit galaxy models composed of baryonic matter + BEC with massless gravitons. The model–rotational velocity of the galaxies arises as the square root of the sum of velocity squares given by Equations (27) and (34) with free parameters R_{BEC} and R_g. Adding mass to the gravitons in the BEC model leads to similar performances of the fits.

5. Discussion and Concluding Remarks

We estimated the upper limit on the graviton mass, employing first the theoretical condition of the existence of the constant Λ, then analyzing the modelfit results of those five dwarf galaxies for which the fit of the BEC model with massive gravitons to data was within 1σ significance level.

Keeping the best-fit parameters $\rho^{(c)}$, R_*, we varied the value of R_{BEC} and R_g and calculated the χ^2 between model and data. The upper limit on the graviton mass m_g has been estimated from the values of R_g, for which $\chi^2 = 1\sigma$ has been reached. The results are given in Table 1. We plotted the theoretical rotation curves given by a baryonic plus a non-relativistic BEC component with massive gravitons with limiting mass in Figure 1. As shown in Table 2, the fit with the rotation curve data has improved the limit on the graviton mass in all cases.

Table 2. Constraints for both the upper limit for the mass of the graviton (first from the existence of Λ, second from the rotation curves) and for the velocity-type and density-type BEC parameters (related to the mass of the BEC particle, scattering length and chemical potential) in the case of the five well-fitting galaxies.

ID	$m_g (\Lambda \in \mathbb{R})$ $10^{-26} \frac{eV}{c^2}$	m_g $10^{-26} \frac{eV}{c^2}$	$\bar{v}_{BEC}$ $\frac{m}{s}$	$\bar{\rho}_{BEC}$ $10^6 \frac{M_\odot}{kpc^3}$
UGC12060	< 1.51	< 0.95	37,724	3.75
UGC7278	< 2.35	< 1.40	42,800	11.69
UGC6446	< 1.32	< 0.42	48,383	4.68
UGC3851	< 2.65	< 1.26	29,571	7.1
UGC7125	< 1.5	< 0.31	63,964	10.61

Comparing the theoretical rotation curves derived in our model with the observational ones, we found the upper limit to the graviton mass to be of the order of 10^{-26} eV/c^2 . We also note that the constraint on the graviton mass imposed from the dispersion relations tested by the first three observations of gravitational waves, 7.7×10^{-23} eV/c^2 [49], is still weaker than the present one.

For the BEC, we could derive two accompanying limits: (i) first m^2/λ has been constrained from the corresponding values of R_* arising from the fit with the massless gravity model; and then (ii) m/μ has been constrained from the constraints derived for the graviton mass and our previous fits through Equations (19) and (20). These are related to the bosonic mass, chemical potential and scattering length, but only two combinations of them, a velocity-type quantity

$$\bar{v}_{BEC} = \sqrt{\frac{\mu}{m}} \tag{35}$$

and a density-type quantity

$$\bar{\rho}_{BEC} = \frac{m^2}{\lambda} \bar{v}_{BEC}^2 \tag{36}$$

were restricted, both characterizing the BEC. Their values are also given in Table 2 for the set of five well-fitting galaxies.

If the BEC consists of massive gravitons with the limiting masses $m = m_g$ determined in Table 2, the chemical potential μ and the constant characterizing the interparticle interaction λ can be determined as presented in Table 3.

Table 3. Constraints on μ and λ assuming $m = m_g$ in case of the five well fitting galaxies.

ID	$\mu(m = m_g)$ $10^{-53} \frac{m^2}{s^2}$ **kg**	$\lambda(m = m_g)$ $10^{-94} \frac{m^5}{s^2}$ **kg**
UGC12060	< 2.41	< 16.08
UGC7278	< 4.57	< 14.40
UGC6446	< 1.75	< 4.14
UGC3851	< 1.96	< 9.17
UGC7125	< 2.26	< 1.74

With this, we established observational constraints for both the upper limit for the mass of the graviton and for the BEC.

Author Contributions: Conceptualization, L.Á.G. and S.D.; Data curation, E.K.; Formal analysis, L.Á.G., E.K. and Z.K.; Funding acquisition, L.Á.G., Z.K. and S.D.; Investigation, E.K.; Methodology, L.Á.G., E.K. and Z.K.; Software, E.K.; Supervision, L.Á.G. and Z.K.; Validation, L.Á.G., Z.K. and S.D.; Visualization, E.K.; Writing—original draft, L.Á.G., E.K., Z.K. and S.D.; Writing—review & editing, L.Á.G.

Funding: This work was supported by the Hungarian National Research Development and Innovation Office (NKFIH) in the form of the grant 123996 and by the Natural Sciences and Engineering Research Council of Canada and based upon work from the COST action CA15117 (CANTATA), supported by COST (European Cooperation in Science and Technology). The work of Z.K. was also supported by the János Bolyai Research Scholarship of the Hungarian Academy of Sciences and by the UNKP-18-4 New National Excellence Program of the Ministry of Human Capacities.

Acknowledgments: This work was supported by the Hungarian National Research Development and Innovation Office (NKFIH) in the form of the grant 123996 and by the Natural Sciences and Engineering Research Council of Canada and based upon work from the COST action CA15117 (CANTATA), supported by COST (European Cooperation in Science and Technology). The work of Z.K. was also supported by the János Bolyai Research Scholarship of the Hungarian Academy of Sciences and by the UNKP-18-4 New National Excellence Program of the Ministry of Human Capacities.

Conflicts of Interest: The founding sponsors had no role in the design of the study; in the collection, analyses, or interpretation of data; in the writing of the manuscript, or in the decision to publish the results.

References

1. Perlmutter, S.; Aldering, G.; Goldhaber, G.; Knop, R.A.; Nugent, P.; Castro, P.G.; Deustua, S.; Fabbro, S.; Goobar, A.; Groom, D.E.; et al. Measurements of Ω and Λ from 42 high-redshift supernovae. *Astrophys. J.* **1999**, *517*, 565–586. [CrossRef]
2. Riess, A.G.; Filippenko, A.V.; Challis, P.; Clocchiatti, A.; Diercks, A.; Garnavich, P.M.; Gilliland, R.L.; Hogan, C.J.; Jha, S.; Kirshner, R.P.; et al. Observational evidence from supernovae for an accelerating universe and a cosmological constant. *Astrophys. J.* **1998**, *116*, 1009–1038. [CrossRef]
3. Young, B.L. A survey of dark matter and related topics in cosmology. *Front. Phys.* **2017**, *12*, 121201. [CrossRef]
4. Plehn, T. Yet another introduction to Dark Matter. *arXiv* **2017**, arXiv:1705.01987.
5. Copeland, E.J.; Sami, M.; Tsujikawa, S. Dynamics of Dark Energy. *Int. J. Mod. Phys.* **2006**, *15*, 1753–1935. [CrossRef]
6. Frieman, J.A.; Turner, M.S.; Huterer, D. Dark Energy and the accelerating universe. *ARAA* **2008**, *46*, 385–432. [CrossRef]
7. Gergely, L. Á.; Tsujikawa, S. Effective field theory of modified gravity with two scalar fields: Dark energy and dark matter. *Phys. Rev. Lett.* **2014**, *89*, 064059. [CrossRef]
8. Gergely, L.Á. Friedmann branes with variable tension. *Phys. Rev. Lett.* **2008**, *78*, 0084006. [CrossRef]
9. Kleidis, K.; Spyrou, N.K. Polytropic dark matter flows illuminate dark energy and accelerated expansion. *A&A* **2015**, *576*, 23.
10. Kleidis, K.; Spyrou, N. Dark Energy: The Shadowy Reflection of Dark Matter? *Entropy* **2016**, *18*, 94, [CrossRef]
11. Das, S.; Bhaduri, R.K. Dark matter and dark energy from a Bose-Einstein condensate. *Class. Quantum Gravity* **2015**, *32*, 105003. [CrossRef]
12. Das, S.; Bhaduri, R.K. Bose-Einstein condensate in cosmology. *arXiv* **2018**, arXiv:1808.10505.
13. Hu, W.; Barkana, R.; Gruzinov, A. Fuzzy cold dark matter: the wave properties of ultralight particles. *Phys. Rev. Lett.* **2000**, *85*, 1158–1161, [CrossRef] [PubMed]
14. Ureña-López, L.A. Bose-Einstein condensation of relativistic Scalar Field Dark Matter. *J. Cosmol. Astropart. Phys.* **2009**, *1*, 014. [CrossRef]
15. Sinha, K.P.; Sivaram, C.; Sudarshan, E.C.G. Aether as a superfluid state of particle-antiparticle pairs. *Found. Phys.* **1976**, *6*, 65–70. [CrossRef]
16. Sinha, K.P.; Sivaram, C.; Sudarshan, E.C.G. The superfluid vacuum state, time-varying cosmological constant, and nonsingular cosmological models. *Found. Phys.* **1976**, *6*, 717–726. [CrossRef]
17. Bohua Li, M.A. Cosmology with Bose-Einstein-Condensed Scalar Field Dark Matter. Master's Thesis, The University of Texas at Austin, Austin, TX, USA, 2013.
18. Morikawa, M. Structure Formation through Cosmic Bose Einstein Condensation Unified View of Dark Matter and Energy. In Proceedings of the 22nd Texas Symposium on Relativistic Astrophysics at Stanford University, Stanford, CA, USA, 13–17 December 2004.
19. Fukuyama, T.; Morikawa, M. The relativistic gross-pitaevskii equation and cosmological bose-einstein condensation quantum structure in the universe. *Prog. Theor. Phys.* **2006**, *115*, 1047–1068. [CrossRef]
20. Moffat, J.W. Spectrum of cosmic microwave fluctuations and the formation of galaxies in a modified gravity theory. *arXiv* **2016**, arXiv:astro-ph/0602607.
21. Wang, X.Z. Cold Bose stars: Self-gravitating Bose-Einstein condensates. *Phys. Rev. D* **2001**, *64*, 124009. [CrossRef]
22. Böhmer, C.G.; Harko, T. Can dark matter be a Bose Einstein condensate? *J. Cosmol. Astropart. Phys.* **2007**, *6*, 025. [CrossRef]
23. Harko, T.; Mocanu, G. Cosmological evolution of finite temperature Bose-Einstein condensate dark matter. *Phys. Rev. D* **2012**, *85*, 084012. [CrossRef]
24. Sikivie, P. Dark Matter Axions. *Int. J. Mod. Phys. A* **2010**, *25*, 554–563. [CrossRef]
25. Dvali, G.; Gomez, C. Black Hole's Quantum N-Portrait. *arXiv* **2011**, arXiv:1112.3359.
26. Chavanis, P.H. Growth of perturbations in an expanding universe with Bose-Einstein condensate dark matter. *A&A* **2012**, *537*, A127.

27. Kain, B.; Ling, H.Y. Cosmological inhomogeneities with Bose-Einstein condensate dark matter. *Phys. Rev. D* **2012**, *85*, 023527. [CrossRef]
28. Suárez, A.; Robles, V.H.; Matos, T. A review on the scalar field/Bose-Einstein condensate dark matter model. *Accel. Cosmic Expans.* **2014**, *38*, 107-142.
29. Ebadi, Z.; Mirza, B.; Mohammadzadeh, H. Infinite statistics condensate as a model of dark matter. *J. Cosmol. Astropart. Phys.* **2013**, *11*, 057. [CrossRef]
30. Dwornik, M.; Keresztes, Z.; Gergely, L.A. Rotation curves in Bose-Einstein Condensate Dark Matter Halos, In *Recent Development in Dark Matter Research*; Kinjo, N., Nakajima, A., Eds.; Nova Science Publishers: New York, NY, USA, 2014; pp. 195–219.
31. Bettoni, D.; Colombo, M.; Liberati, S. Dark matter as a Bose-Einstein Condensate: The relativistic non-minimally coupled case. *J. Cosmol. Astropart. Phys.* **2014**, *2*, 004. [CrossRef]
32. Gielen, S. Quantum cosmology of (loop) quantum gravity condensates: An example. *Classi. Quantum Gravity* **2014**, *31*, 155009. [CrossRef]
33. Schive, H.Y.; Chiueh, T.; Broadhurst, T. Cosmic structure as the quantum interference of a coherent dark wave. *Nat. Phys.* **2014**, *10*, 496–499. [CrossRef]
34. Davidson, S. Axions: Bose Einstein condensate or classical field? *Astropart. Phys.* **2015**, *65*, 101–107. [CrossRef]
35. Ali, A.F.; Das, S. Stringent theoretical and experimental bounds on graviton mass. *Int. J. Mod. Phys. D* **2016**, *25*, 1644001. [CrossRef]
36. Dwornik, M.; Keresztes, Z.; Kun, E.; Gergely, L.A. Bose-Einstein condensate Dark Matter halos confronted with galactic rotation curves. *Adv. High Energy Phys.* **2017**, *4025386*, 14. [CrossRef]
37. Gross, E.P. Structure of a quantized vortex in boson systems. *Nuovo Cim.* **1961**, *20*, 454. [CrossRef]
38. Gross, E.P. Hydrodynamics of a superfluid condensate. *J. Math. Phys.* **1963**, *4*, 195. [CrossRef]
39. Pitaevskii, L.P. Vortex lines in an imperfect bose gas. *Sov. Phys. JETP* **1961**, *13*, 451.
40. Rogel-Salazar, J. The Gross-Pitaevskii equation and Bose-Einstein condensates. *Eur. J. Phys.* **2013**, *34*, 247. [CrossRef]
41. Giorgini, S.; Pitaevskii, L.P.; Stringari, S. Thermodynamics of a trapped Bose-condensed gas. *J. Low Temp. Phys.* **1997**, *109*, 309. [CrossRef]
42. Lieb, E.H.; Seiringer, R.; Yngvason, J. Bosons in a trap: A rigorous derivation of the Gross-Pitaevskii energy functional. *arXiv* **2000**, arXiv:math-ph/9908027.
43. Swaters, R.A. Dark Matter in Late-type Dwarf Galaxies. Ph.D. Thesis, University of Groningen, Groningen, The Netherlands, 1999.
44. Swaters, R.A.; Balcells, M. The Westerbork HI survey of spiral and irregular galaxies. II. R-band surface photometry of late-type dwarf galaxies. *A&A* **2002**, *390*, 863–878, arXiv:astro-ph/0204526.
45. Swaters, R.A.; Sancisi, R.; van Albada, T.S.; van der Hulst, J.M. The rotation curves shapes of late-type dwarf galaxies. *A&A* **2009**, *493*, 871–892, arXiv:0901.4222.
46. Freeman, K.C. On the Disks of Spiral and S0 Galaxies. *Astrophys. J.* **1970**, *160*, 811. doi:10.1086/150474. [CrossRef]
47. De Araujo, J.C.N.; Miranda, O.D. A solution for galactic disks with Yukawian gravitational potential. *Gen. Relativ. Gravit.* **2007**, *39*, 777–784. [CrossRef]
48. Binney, J.; Merrifield, M. *Galactic Astronomy*; Princeton University Press: Princeton, NJ, USA, 1998.
49. Abbott, B.P.; Abbott, R.; Abbott, T.D.; Acernese, F.; Ackley, K.; Adams, C.; Adams, T.; Addesso, P.; Adhikari, R.X.; Adya, V.B.; et al. GW170104: Observation of a 50-Solar-Mass Binary Black Hole Coalescence at Redshift 0.2. *Phys. Rev. Lett.* **2017**, *118*, 221101. [CrossRef] [PubMed]

symmetry

MDPI

Article

Hadronic and Hadron-Like Physics of Dark Matter

Vitaly Beylin [1], Maxim Yu. Khlopov [1,2,3,*], Vladimir Kuksa [1] and Nikolay Volchanskiy [1,4]

1 Institute of Physics, Southern Federal University, Stachki 194, Rostov on Don 344090, Russia; vitbeylin@gmail.com (V.B.); vkuksa47@mail.ru (V.K.); nikolay.volchanskiy@gmail.com (N.V.)
2 APC Laboratory 10, rue Alice Domon et Léonie Duquet, CEDEX 13, 75205 Paris, France
3 National Research Nuclear University "MEPHI" (Moscow State Engineering Physics Institute), 31 Kashirskoe Chaussee, Moscow 115409, Russia
4 Bogoliubov Laboratory of Theoretical Physics, Joint Institute for Nuclear Research, Joliot-Curie 6, Dubna 141980, Russia
* Correspondence: maxim51khl@yahoo.com

Received: 29 March 2019; Accepted: 17 April 2019; Published: 23 April 2019

Abstract: The problems of simple elementary weakly interacting massive particles (WIMPs) appeal to extend the physical basis for nonbaryonic dark matter. Such extension involves more sophisticated dark matter candidates from physics beyond the Standard Model (BSM) of elementary particles. We discuss several models of dark matter, predicting new colored, hyper-colored or techni-colored particles and their accelerator and non-accelerator probes. The nontrivial properties of the proposed dark matter candidates can shed new light on the dark matter physics. They provide interesting solutions for the puzzles of direct and indirect dark matter search.

Keywords: cosmology; particle physics; cosmo–particle physics; QCD; hyper-color; dark atoms; composite dark matter

1. Introduction

The nature of dark matter is inevitably linked to beyond the Standard Model (BSM) physics of elementary particles. In the lack of direct experimental evidences for this physics, methods of cosmo–particle physics [1–6] are needed for its study involving proper combination of cosmological, astrophysical and experimental physical probes.

The most popular simplest dark matter candidate—elementary weakly interacting massive particles (WIMPs)—finds support neither in direct dark matter searches, nor in searches for supersymmetric (SUSY) particles at the Large Hadron Collider (LHC). The latter removes strong theoretical motivation for WIMPs as the lightest supersymmetric particles and opens the room for wider class of BSM models and corresponding dark matter candidates [7–12].

In this paper, we turn to a possibility of hadronic, hyperhadronic, and composite dark matter candidates. In particular, in the scenario with hadronic dark matter it is suggested that such candidates consist of a new heavy quark and a light standard one.

It can be shown that the effects of new physics are related to the new massive stable quarks, which fit into the limits imposed by both the electroweak and cosmo–particle data. The bound states of these heavy fermions with light QCD quarks can be considered as (pseudoscalar) neutral dark matter candidates. Masses of these particles as the lifetime of the charged component are estimated. Besides, we study a low-energy asymptotics of the potential for their interactions with nucleons and with each other. Remind, there is well-known Sommerfeld–Gamov–Sakharov enhancement when heavy particles annihilate. Here, this effect for the states considered is also discussed.

An extension of the Standard Model (SM) with an additional symplectic hypercolor gauge group is analyzed. The extension keeps the Higgs boson of SM as a fundamental field but permits the Higgs to participate in mixing with composite hyperhadrons. New heavy hyperquarks are assumed to be vector-like in the character of their interactions with the intermediate vector bosons. We also consider the properties of pseudo-Nambu–Goldstone (pNG) bosons emerging as a result of dynamical symmetry breaking $SU(2n_F) \to Sp(2n_F)$, with n_F being a number of hyperquark flavors. Some versions of the model are invariant under specific global symmetries that ensure the stability of a neutral pseudoscalar field and scalar diquark states (hyperbaryons). Possible signals of the emergence of these lightest states at colliders are also discussed on the basis of the dark matter (DM) two-component model. Consideration of the DM relic density kinetics allows us to evaluate masses of these neutral stable states and, consequently, to analyze some processes with their participation. Here, we briefly describe possible channels of the hyperhadron production at colliders having a specific signature of final state. Moreover, there occurs an interesting feature of cosmic ray scattering off the dark matter particles; the study of a diffuse photon spectrum produced by annihilation of the DM candidates results in a possible prominent manifestation of the two-component structure of dark matter. It is especially important signal because one of the DM components does not interact with vector bosons directly and is, in some sense, invisible in electroweak processes. The generalization of vector-like model symmetry to include three hyperquark flavors significantly expands the spectrum of states, leading to new additional stable hadrons. They can be observed and identified as specific features of the energy and angular spectra of photons and/or leptons recorded mainly by space telescopes. Virtually all additional new hadronic states are quite massive, which prevents their production and study at colliders.

The dark atom scenario assumes existence of stable multiple charged particles that can be predicted in some non-supersymmetric BSM models. It involves minimal number of parameters of new physics—the mass of the new charged particles. Particles with charge -2 bind with primordial helium nuclei in a neutral OHe atom. The nuclear interactions of its α-particle shell dominantly determine cosmological evolution and astrophysical effects of these atoms. However, the nontrivial structure of the OHe atom with the radius of the Bohr orbit equal to the size of helium nucleus and strongly interacting atomic shell make impossible to apply usual approximations of atomic physics to its analysis. Qualitatively this approach can shed light on the puzzles of direct dark matter searches. It can give explanation to the observed excess in radiation in positronium annihilation line in the galactic bulge. It can explain the excess of high energy positrons in cosmic rays as indirect effects of composite dark matter. Stable multiple charged constituents of dark atoms are a challenge for their direct search at the LHC and the such searches acquire the meaning of the direct probe of dark atom cosmology.

The paper is organized as follows. In the framework of the hadronic scenario we consider new stable hadrons, in which a new stable heavy quark is bound by the standard QCD interaction with ordinary light quarks (Section 2). In hyperhadronic models, new heavy quarks are bound by hypercolor strong interactions (Section 3). We also consider the scenario with dark atoms, in which the ordinary Coulomb interaction binds new stable -2 charged particle with primordial helium nucleus (Section 4). We discuss the physical motivation for these extensions of the Standard model and their experimental and observational signatures. We conclude (Section 5) by the discussion of the cross-disciplinary test of these BSM models in the context of cosmo–particle physics.

2. New Stable Hadrons

In this Section, we consider theoretical and experimental motivations for hadronic dark matter. As a rule, DM candidates are interpreted as stable heavy particles which interact with standard particles through the weak vector bosons (WIMP). The last rigid experimental restrictions on the value of WIMP-nucleon interaction cross section [13] expel some variants of WIMPs as candidates for dark matter. Therefore, alternative variants are considered in literature, namely the model with fermions from fourth generation, hypercolor models, dark atoms in composite DM, and so on (see the review [14] and references therein). It was shown in refs. [15–22] that the existence of hadronic DM candidates, which consist of a new heavy stable quark and a light standard one, is not excluded by cosmological data. In particular, such possibility was carefully considered in the chiral-symmetric extension of SM [22].

We consider the scenarios where the strong interaction of new heavy quarks with light standard quarks, which is described by $SU_C(3)$ symmetry, forms new stable heavy states. This possibility was analyzed in the extensions of SM with fourth generation [15–19], in the framework of mirror and chiral-symmetric models [22,23], and in the models with a singlet quark [24–29]. The simplest variant of the chiral-symmetric model was realized in ref. [22], where quark content and quantum numbers of new heavy mesons and fermions were represented, and the low-energy phenomenology of new heavy pseudoscalar mesons was described. It is shown in this work that the existence of new hadrons does not contradict to cosmo–chemical data and precision restrictions on new electroweak effects. Here, we should note that the chiral-symmetric scenario may encounter experimental and theoretical difficulties, which was not analyzed in Ref. [22]. The scenario with fourth generation and its phenomenology was also considered in literature, although there are strong restrictions from invisible Z-decay channel, unitarity of quark-mixing matrix, flavor-changing neutral currents, and others. The principal problem of the extension with fourth generation is contained in new quarks contributions into the Higgs decay channels [30]. It was shown that new quarks contribution to vector gauge boson coupling can be compensated by heavy (with the mass around 50 GeV) neutrino contribution [15,31,32]. Then Higgs should have the dominant channel of decay to the fourth neutrino, which is excluded by experimental measurements at the LHC. The proposed solution is that the fourth family gets masses from additional heavy Higgses and the standard Higgs (125 GeV) has suppressed couplings to the fourth family. In this section, we consider the hypothesis of hadronic DM candidates which can be built in the framework of the chiral-symmetrical and singlet quark extensions of SM.

2.1. Gauge Structure of Chiral-Symmetrical Model with New Quarks

The chiral-symmetrical extension of the standard set of fermionic fields was considered in ref. [22], where the phenomenology of new heavy hadrons is described. In this subsection, we consider the group structure of new quark sector and analyze corresponding gauge interactions of quarks with vector bosons. This aspect was not analyzed in ref. [22] but has significance in electroweak precision test of the model. In the model under consideration, new multiplets of the up and down quarks has chiral-symmetric structure with respect to the standard set of quarks:

$$Q = \{Q_R = \begin{pmatrix} U \\ D \end{pmatrix}_R ; \; U_L, \; D_L\}. \tag{1}$$

Thus, in contrast to the Standard Model structure, the right-hand components of new quarks are doublets and left-hand ones are singlets. The structure of covariant derivatives follows from this definition in the standard way:

$$\begin{aligned} D_\mu Q_R =& (\partial_\mu - ig_1 Y_Q V_\mu - \frac{ig_2}{2}\tau_a V_\mu^a - ig_3 t_i G_\mu^i) Q_R; \\ D_\mu U_L =& (\partial_\mu - ig_1 Y_U V_\mu - ig_3 t_i G_\mu^i) U_L, \\ D_\mu D_L =& (\partial_\mu - ig_1 Y_D V_\mu - ig_3 t_i G_\mu^i) D_L. \end{aligned} \tag{2}$$

In Equation (2), the values Y_A, $A = Q, U, D$, are the hypercharges of quark multiplets (doublets and singlets), t_i are generators of $SU_C(3)$ group, which describes the standard color (strong) interaction. We should note that the coupling constants g_1 and g_2 are equal to the standard ones at the energy scale, where the chiral symmetry is restored. At low energy they depend on the details of symmetry violation scenario. Here, the gauge boson fields V_μ^a are chiral partner of the standard gauge bosons which are expected to be superheavy. The status of the abelian gauge field V_μ depends on its physical interpretation in low-energy electromagnetic processes. If we interpret abelian gauge field V_μ as standard one, then the mixing of V_μ and V^3_μ in the standard way is forbidden, because it leads to contradictions with precision electroweak measurements. Moreover, a direct interpretation of the field V_μ and the weak hypercharge $Y_Q = \bar{q}$, where $\bar{q}$ is an average electric charge of a quark multiplet (in our case, doublet and singlets), leads to wrong a $V - A$ structure of photon interaction with fermions. To escape these obstacles, we have two options:

- to interpret the field V_μ as a new non-standard abelian field which mixes with V^3_μ in analogy with the standard procedure,
- to assume that V_μ is standard abelian field which does not mix with V^3_μ.

The first option leads to a mirror world with exotic "electroweak" vector bosons (in particular, mirror or dark photons) and standard QCD-like strong interactions. In the framework of this scenario, it is difficult to build coupled heavy-light states of type (qQ) and satisfy cosmo–chemical restrictions which were considered in ref. [22]. So, we consider the second option and analyze the interaction of V_μ with additional heavy quark sector. This scenario needs a redefinition of the hypercharge operator $\hat{Y}_Q$, which we consider below. The standard definition of the hypercharge operator $\hat{Y}_Q = (\hat{q} - \hat{t})$ in the case of the doublet Q can be realized with the help of 2×2 matrices of charge $\hat{q}$ and isospin $\hat{t} = \tau_3/2$. This operator in matrix representation acts on the standard left quark doublet as follows:

$$\hat{Y}_Q Q_L = \begin{pmatrix} \hat{q}_u - 1/2 & 0 \\ 0 & \hat{q}_d + 1/2 \end{pmatrix} \begin{pmatrix} u \\ d \end{pmatrix}_L . \quad (3)$$

The operator of charge $\hat{q}$ in (3) is defined by equalities $\hat{q}_u u = 2/3\, u$ and $\hat{q}_d d = -1/3\, d$. So, it follows from (3) that

$$\hat{Y}_Q Q_L = \frac{1}{6} \begin{pmatrix} 1 & 0 \\ 0 & 1 \end{pmatrix} \begin{pmatrix} u \\ d \end{pmatrix}_L = \frac{1}{6} \cdot Q_L. \quad (4)$$

Thus, the operator action reduces to multiplication by the coefficient $1/6$ (one half of the average charge of the doublet) and the standard V-structure of γ- and $V - A$ structure of Z-interactions arise as a result of the mixing of A_μ and W^3_μ. From this simple analysis, it follows that the presence of the isospin operator in definition of the hypercharge operator is connected with the presence of singlet-triplet mixing. So, the hypercharge operator in the absence of mixing has the form $\hat{Y}_Q = \hat{q}$ (without $\hat{t}$):

$$\hat{Y}_Q Q_R = \begin{pmatrix} \hat{q}_u & 0 \\ 0 & \hat{q}_d \end{pmatrix} \begin{pmatrix} U \\ D \end{pmatrix}_R . \quad (5)$$

Taking into account equalities $\hat{q}_u \cdot U_R = \hat{Y}_U \cdot U_L = 2/3 \cdot U$ and $\hat{q}_d \cdot D_R = \hat{Y}_D \cdot D_L = -1/3 \cdot D$, it follows from the Equation (2) that vector-like interactions of physical fields γ and Z with the new quarks $U = U_R + U_L$ and $D = D_R + D_L$:

$$\mathscr{L}^{int}_Q = g_1 V_\mu \bar{Q} \gamma^\mu \hat{q} Q = g_1 (c_w A_\mu - s_w Z_\mu)(q_U \bar{U} \gamma^\mu U - q_D \bar{D} \gamma^\mu D), \quad (6)$$

where we omit the strong interaction term which has the standard structure. In the expression (6), abelian field V_μ is the standard mixture of the physical fields A_μ and Z_μ, $c_w = \cos\theta_w$, $s_w = \sin\theta_w$, $g_1 c_w = e$ and θ_w is the Weinberg angle. Indirect limits for the new quarks follow from electroweak measurements of FCNC processes and the value of polarizations. Since the new fermions are stable, there are no FCNC processes in our scenario. The constraints which are caused by the vector-boson polarization measurements will be considered below.

2.2. The Extension of Standard Quark Sector with Stable Singlet Quark

There are many scenarios of SM extensions with singlet (isosinglet) quarks which are considered in literature. The singlet quark (SQ) is usually defined as a Dirac fermion with the quark quantum numbers having the standard $U_Y(1)$ and $SU_C(3)$ gauge interactions. In contrast to a standard quark, it is a singlet with respect to $SU_W(2)$ transformations, that is, it does not interact with the non-abelian weak charged boson W. The high-energy origin and low-energy phenomenology of SQ were discussed in literature (see the recent works [20,33–35], and references therein). As a rule, SQ is supposed to be an unstable particle, which is caused by the mixing of SQ with ordinary quarks. Such mixing leads to FCNC appearing at the tree level, which is absent in SM. This results in an additional contributions to rare lepton decays and $M^0 - \bar{M}^0$ oscillations. There are strong restrictions on the value of singlet-ordinary quark mixing. Here, we suggest an alternative variant with a stable SQ, namely, the scenario with the absence of such mixing. Further, we analyze this variant and apply it to the description of possible DM candidate. Because the SQ together with ordinary quarks are in confinement they form the bound states of type (Sq), (Sqq), (SSq), and more complicated. Here, we consider the main properties of (Sq)-states and describe the lightest state $M^0 = (\bar{S}q)$ (which should be stable) as DM particle.

Further, we analyze the scenario with SQ, which in general case can be up, U, or down, D, type ($q = 2/3$ or $q = -1/3$ respectively). According to the above definitions, the minimal Lagrangian describing interactions of the singlet quark S with the gauge bosons is as follows:

$$\mathscr{L}_S = i\bar{S}\gamma^\mu(\partial_\mu - ig_1 qV_\mu - ig_s t_a G^a_\mu)S - M_S\bar{S}S. \tag{7}$$

In (7), the hypercharge $Y/2 = q$ is a charge of the S, $t_a = \lambda_a/2$ are generators of $SU_C(3)$ group, and M_S is a phenomenological mass of the S. It can not get mass by the Higgs mechanism because the corresponding term is forbidden by SU(2) symmetry. However, the mass term in expression (7) is allowed by the symmetry of the model. The abelian part in (7) contains the interactions of SQ with photon and Z-boson:

$$\mathscr{L}_S^{int} = g_1 qV_\mu\bar{S}\gamma^\mu S = qg_1(c_w A_\mu - s_w Z_\mu)\bar{S}\gamma^\mu S. \tag{8}$$

In expression (8), the values $c_w = \cos\theta_w$, $s_w = \sin\theta_w$, $g_1 c_w = e$ and θ_w is the Weinberg angle. We should note that the interaction of SQ with the vector bosons has a vector-like form, so SQ is usually called a vector quark [34,35].

Now, we take into account the restrictions on the processes with SQ participation, which follow from the experimental data. New sequential quarks are excluded by LHC data on Higgs properties [33]. Because SQ does not interact with the Higgs doublet, it is not excluded by data on Higgs physics. The limits on new quarks for colored factors $n_{eff} = 2, 3, 6$ are about 200 GeV, 300 GeV, and 400 GeV respectively [36]. As we show further, these limits are much less than our estimations with the assumption that the new heavy quark is a DM particle. The scenario with the long-lived heavy quarks, which takes place when SQ slightly mixes with an ordinary quark, was discussed in the review [20].

2.3. Constraints on the New Quarks Following from the Precision Electroweak Measurements

The constraints on the new heavy quarks follow from the electroweak measurements of FCNC and vector boson polarization. As was noted earlier, there is no mixing of the new quarks with the standard ones and the new quarks do not contribute to rare processes. The contributions of the new quarks to polarization tensors of the vector bosons are described by the Peskin–Takeuchi (PT) parameters [37]. From Equations (6) and (8), it follows that interactions of the new quarks with the vector bosons in both scenarios (the chiral-symmetrical extension and the model with SQ) have the same structure and their contributions into the PT parameters can be described by general expressions. In the models under consideration, the new heavy quarks contribute into polarization tensors of γ and Z-bosons, namely $\Pi_{\gamma\gamma}$, $\Pi_{\gamma Z}$, Π_{ZZ}. Note, because the W-boson does not interact with the new quarks, $\Pi_{WW} = 0$. To extract the transverse part $\Pi(p^2)$ of the polarization tensor $\Pi_{\mu\nu}(p^2)$, we have used the definition $\Pi_{\mu\nu}(p^2) = p_\mu p_\nu P(p^2) + g_{\mu\nu}\Pi(p^2)$. In our case, $\Pi_{ab}(0) = 0$, $a, b = \gamma, Z$ and the PT parameters, which we take from [38], can be represented by the following expressions:

$$S = \frac{4s_w^2 c_w^2}{\alpha}\left[\frac{\Pi_{ZZ}(M_Z^2, M_Q^2)}{M_Z^2} - \frac{c_w^2 - s_w^2}{s_w c_w}\Pi'_{\gamma Z}(0, M_Q^2) - \Pi'_{\gamma\gamma}(0, M_Q^2)\right]; \quad T = -\frac{\Pi_{ZZ}(0, M_Q^2)}{\alpha M_Z^2} = 0;$$
$$U = -\frac{4s_w^2}{\alpha}\left[c_w^2\frac{\Pi_{ZZ}(M_Z^2, M_Q^2)}{M_Z^2} + 2s_w c_w \Pi'_{\gamma Z}(0, M_Q^2) + s_w^2 \Pi'_{\gamma\gamma}(0, M_Q^2)\right]. \tag{9}$$

In Equation (9), $\alpha = e^2/4\pi$, M_Q is a mass of the new heavy quark and $\Pi_{ab}(p^2)$ are defined at $p^2 = M_Z^2$ and $p^2 = 0$. In (9), the functions $\Pi_{ab}(p^2, M_Q^2)$, $a, b = \gamma, Z$, can be represented in a simple form:

$$\Pi_{ab}(p^2, M_Q^2) = \frac{g_1^2}{9\pi^2} k_{ab} F(p^2, M_Q^2); \quad k_{ZZ} = s_w^2,\ k_{\gamma\gamma} = c_w^2,\ k_{\gamma Z} = -s_w c_w;$$
$$F(p^2, M_Q^2) = -\frac{1}{3}p^2 + 2M_Q^2 + 2A_0(M_Q^2) + (p^2 + 2M_Q^2)B_0(p^2, M_Q^2), \tag{10}$$

where we take into account the contribution of the new quark with $q = 2/3$. The function $F(p^2, M_Q^2)$ in Equation (10) contains divergent terms in the one-point, $A_0(M_Q^2)$, and two-point, $B_0(p^2, M_Q^2)$, Veltman functions. These terms are compensated exactly in the physical parameters S, T, and U, defined by the expressions (9). In the case of the D-type quark, the contributions are four times smaller. Using the standard definitions of the functions $A_0(M_Q^2)$ and $B_0(p^2, M_Q^2)$ and the equality $B_0'(0, M_Q^2) = M_Q^2/6$, we get:

$$S = -U = \frac{k s_w^4}{9\pi}\left[-\frac{1}{3} + 2\left(1 + 2\frac{M_Q^2}{M_Z^2}\right)\left(1 - \sqrt{\beta}\arctan\frac{1}{\sqrt{\beta}}\right)\right]. \tag{11}$$

In Equation (11), $\beta = 4M_Q^2/M_Z^2 - 1$, $k = 16(4)$ at $q = 2/3(-1/3)$, and $k = 20$ in the case of the chiral-symmetric model. We have checked by direct calculations that in the limit of infinitely heavy masses of the new quarks, $M_Q^2/M_Z^2 \to \infty$, the parameters S and U go to zero as $\sim M_Z^2/M_Q^2$. From Equation (11) it follows that for the value of mass $M_Q > 500$ GeV the parameters $S, U < 10^{-2}$. Experimental limits are represented in the review [39]:

$$S = 0.00 + 0.11(-0.10), \quad U = 0.08 \pm 0.11, \quad T = 0.02 + 0.11(-0.12). \tag{12}$$

Thus, the scenarios with the new heavy quarks do not contradict to the experimental electroweak restrictions.

At the quark-gluon phase of the evolution of the Universe, the new heavy quarks strongly interact with the standard ones. So, there are strong processes of scattering and annihilation into gluons and quarks, $Q\bar{Q} \to gg$ and $Q\bar{Q} \to q\bar{q}$. Additional contributions to these processes through electroweak channels $Q\bar{Q} \to \gamma\gamma$, ZZ give small differences for cross sections in the scenarios under consideration. Cross sections of annihilation are derived from the expressions for annihilation of gluons and light quarks into heavy quarks $gg \to Q\bar{Q}$ and $q\bar{q} \to Q\bar{Q}$ [39]. The cross section of two-quark annihilation [39]:

$$\frac{d\sigma}{d\Omega}(\bar{q}q \to \bar{Q}Q) = \frac{\alpha_s^2}{9s^3}\sqrt{1 - \frac{4M_Q^2}{s}}\left[(M_Q^2 - t)^2 + (M_Q^2 - u)^2 + 2M_Q^2 s\right], \tag{13}$$

where M_Q is a mass of the heavy quark ($M_Q \gg m_q$). At the threshold, $s \approx 4M_Q^2$, the parameters $t \approx u \approx -M_Q^2$. The cross section of the process $\bar{Q}Q \to \bar{q}q$ can be derived from (13) by reversing time. In this limit, from (13), we get

$$\frac{d\sigma}{d\Omega}(\bar{Q}Q \to \bar{q}q) = \frac{\alpha_s^2}{18v_r M_Q^2}, \tag{14}$$

where v_r is a relative velocity of the heavy quarks. The cross section of two-gluon annihilation [39]:

$$\frac{d\sigma}{d\Omega}(gg \to \bar{Q}Q) = \frac{\alpha_s^2}{32s}\sqrt{1-\frac{4M_Q^2}{s}}F(s,t,u;M_Q^2), \tag{15}$$

where exact expression for the function $F(s,t,u;M_Q^2)$ is rather complicated. In the approximation $s \approx 4M_Q^2$, we get $F \approx 7/6$, and the cross section of the reversed process $\bar{Q}Q \to gg$ is:

$$\frac{d\sigma}{d\Omega}(\bar{Q}Q \to gg) = \frac{7\alpha_s^2}{3 \cdot 128 v_r M_Q^2}. \tag{16}$$

Now, we give the expressions for the total cross section of two-gluon and two-quark annihilation. Two-gluon cross section in the low-energy limit:

$$\sigma(Q\bar{Q} \to gg)v_r = \frac{14\pi}{3}\frac{\alpha_s^2}{M_Q^2}, \tag{17}$$

where $\alpha_s = \alpha_s(M_Q)$. Two-quark cross section in the limit $m_q \to 0$:

$$\sigma(Q\bar{Q} \to q\bar{q})v_r = \frac{2\pi}{9}\frac{\alpha_s^2}{M_Q^2}. \tag{18}$$

From expressions (17) and (18), it follows that the two-gluon channel dominates. The value of cross section of $\sigma v_r \sim 1/M_Q^2$ and the remaining concentration of the heavy component may be dominant at the end of the quark-gluon phase of the evolution. At the hadronization stage, the new heavy quarks, which participate in the standard strong interactions, form coupled states with ordinary quarks. Here, we consider neutral and charged states of type (qQ). The lightest of them is stable and can be suggested as a DM candidate. Here, we should note that the dominance of the heavy component before the transition from the quark-gluon to hadronization phase may be connected with the dominance of dark matter relative to ordinary matter. The possibility of the heavy-hadron existence was analyzed in [22]. It was shown that this possibility does not contradict to cosmo–chemical data. This conclusion was drawn taking account of the repulsive strong interaction of new hadrons with nucleons. This effect will be qualitatively analyzed in the next section.

The constraints on the new heavy hadrons, which follow from the cosmo–chemical data, will be discussed in the following subsection. Such constraints on the new hadrons as strongly interacting carrier of dark matter (SIMP) follow also from astroparticle physics. The majority of restrictions refer to the mass of the new hadrons or available mass/cross section parameters space [40–48]. As a rule, the low limits on the mass value do not exceed 1–2 TeV which are an order of magnitude less then our previous estimations of DM particle mass [29] (see also the estimations in this work). Here, we should pick out the results of research in [46,47], where it was reported that the mass of new hadrons $M \gtrsim 10^2$ TeV. Further we represent almost the same estimation taking account of Sommerfeld–Gamov–Sakharov enhancement effect in annihilation cross section. An additional and more detailed information on restrictions, which follows from XQC experiments, can be found in the Refs. [49,50]. In the next section, we represent effective theory of low-energy interaction of new hadrons. From this consideration, it follows that the value of interaction of the new hadrons with nucleons is of the same order of magnitude as the hadronic one (see Ref. [22]).

2.4. Composition of New Heavy Hadrons and Long-Distance Interactions with Nucleons

Due to the strong interaction new quarks together with standard ones form the coupled meson and fermion states, the lightest of which are stable. Classification of such new heavy hadrons was considered in ref. [22], where the main processes with their participation were analyzed. In the Table 1, we represent the quantum numbers and quark content of new mesons and fermions for the case of U- and D-type of new quarks.

Table 1. Classification of new hadrons.

$J^P = 0^-$	$T = \frac{1}{2}$	$M_U = (M_U^0\, M_U^-)$	$M_U^0 = \bar{U}u,\ M_U^- = \bar{U}d$
$J^P = 0^-$	$T = \frac{1}{2}$	$M_D = (M_D^+\, M_D^0)$	$M_D^+ = \bar{D}u,\ M_D^0 = \bar{D}d$
$J = \frac{1}{2}$	$T = 1$	$B_{1U} = (B_{1U}^{++}\, B_{1U}^{+}\, B_{1U}^{0})$	$B_{1U}^{++} = Uuu, B_{1U}^{+} = Uud, B_{1U}^{0} = Udd$
$J = \frac{1}{2}$	$T = 1$	$B_{1D} = (B_{1D}^{+}\, B_{1D}^{0}\, B_{1D}^{-})$	$B_{1D}^{+} = Duu, B_{1D}^{-} = Ddd, B_{1D}^{0} = Dud$
$J = \frac{1}{2}$	$T = \frac{1}{2}$	$B_{2U} = (B_{2U}^{++}\, B_{2U}^{+})$	$B_{2U}^{++} = UUu, B_{2U}^{+} = UUd$
$J = \frac{1}{2}$	$T = \frac{1}{2}$	$B_{2D} = (B_{2D}^{0}\, B_{2D}^{-})$	$B_{2D}^{0} = DDu, B_{2D}^{-} = DDd$
$J = \frac{3}{2}$	$T = 0$	(B_{3U}^{++})	$B_{3U}^{++} = UUU$
$J = \frac{3}{2}$	$T = 0$	(B_{3D}^{-})	$B_{3D}^{-} = DDD$

Most of the two- and three-quark states, which are represented in Table 1, were considered also in Refs. [20,51]. Ref. [52] considered an alternative for the DM candidates which are electromagnetically bound states made of terafermions. Here, we propose the neutral M^0-particles as candidates for DM. Another possibility is discussed in Refs. [17–19]—new charged hadrons exist but are hidden from detection. Namely, the particles with charge $q = -2$ are bound with primordial helium. In our case, the interactions of baryons B_{1Q} and B_{2Q}, where $Q = U, D$, are similar to the nucleonic interactions. These particles may compose heavy atomic nuclei together with nucleons.

Evolution of new hadrons was qualitatively studied in [22] and, here, we briefly reproduce this analysis for both cases. Matter of stars and planets may contain stable U-type particles M_U^0, B_{1U}^+ and B_{2U}^{++} as well as B_{3U}^{++} and $\bar{B}_{3U}^{++}$. The antiparticles $\bar{M}_U^0$, $\bar{B}_{1U}^+$ and $\bar{B}_{2U}^{++}$ are burning out due to interactions with nucleons N:

$$\bar{M}_U^0 + N \to B_{1U}^+ + X,\ \ \bar{B}_{1U}^+ + N \to M_U^0 + X,\ \ \bar{B}_{1U}^{++} + N \to 2M_U^0 + X, \tag{19}$$

where X are leptons or photons in the final state. There are no Coulomb barriers for the reactions (19), so the particles $\bar{M}_U^0$, $\bar{B}_{1U}^+$ and $\bar{B}_{2U}^{++}$ burn out during the evolution of the Universe. Other stable particles participate in reactions with annihilation of new quarks:

$$M_U^0 + B_{3U}^{++} \to B_{2U}^{++} + X,\ \ M_U^0 + B_{2U}^{++} \to B_{1U}^{+} + X,\ \ M_U^0 + B_{1U}^{+} \to p + X,\ \ \bar{B}_{3U}^{++} + B_{3U}^{++} \to X. \tag{20}$$

The Coulomb barrier may appear in the last reaction in (20), however, simple evaluations which are based on the quark model show that the reaction

$$B_{3U}^{++} + N \to 3\bar{M}_U^0 + X \tag{21}$$

is energetically preferred. Thus, there are no limits for total burning out of U-quarks and, along with them, all positively charged new hadrons to the level compatible with cosmological restrictions. The rest of the antiquarks $\bar{U}$ in accordance with (21) may exist inside neutral M_U^0-particles only. Concentration of these particles in matter is determined by the baryon asymmetry in the new quarks sector.

In the case of D-type hadrons, $M_{\bar{D}}^+$, B_{1D}^0, B_{2D}^0 and B_{3D}^- particles and $\bar{B}_{3D}^-$ antiparticles may be situated in matter medium, in the inner part of stars, for example. In analogy with (19) the reactions for the case of down type hadrons can be represented in the form

$$\bar{M}_{\bar{D}}^+ + N \to B_{1D}^0 + X, \;\; \bar{B}_{1D}^0 + N \to M_{\bar{D}}^+ + X, \;\; \bar{B}_{2D}^0 + N \to 2M_{\bar{D}}^+ + X. \tag{22}$$

It should be noted that the Coulomb barriers to them are absent. Annihilation of new D-quarks goes through the following channels:

$$M_{\bar{D}}^+ + B_{3D}^- \to B_{2D}^0 + X, \;\; M_{\bar{D}}^+ + B_{2D}^0 \to B_{1D}^0 + X, \;\; M_{\bar{D}}^+ + B_{1D}^0 \to p + X, \;\; \bar{B}_{3D}^- + B_{3D}^- \to X. \tag{23}$$

From this qualitative analysis and cosmo–chemical restrictions the conclusion was done that the baryon asymmetry in new quark sector exists. This asymmetry has a sign which is opposite to the ordinary baryon asymmetry sign. This conclusion mainly follows from the ratios "anomalous/natural" hydrogen $C \leqslant 10^{-28}$ for $M_Q \lesssim 1\,\text{TeV}$ [53] and anomalous helium $C \leqslant 10^{-12} - 10^{-17}$ for $M_Q \leq 10$ TeV [54]. In the case of up hadrons, the state $B_{1U}^+ = (Uud)$ is heavy proton which can form anomalous hydrogen. The anomalous state B_{1U}^+ at hadronization phase can be formed by coupling of quarks U, u, d and as a result of reaction $\bar{M}_U^0 + N \to B_{1U}^+ + X$ (the first reaction in (19)). The antiparticles $\bar{B}_{1U}^+$ are burning out due to the reaction $\bar{B}_{1U}^+ + N \to M_U^0 + X$. The states like (pM_U^0) can also manifest itself as anomalous hydrogen. But in [22] it was shown that the interaction of p and M_U^0 has a potential barrier. So, the formation of the coupled states (pM_U^0) is strongly suppressed. Baryon symmetry of new quarks is not excluded when they are superheavy.

The hadronic interactions are usually described in the meson exchange approach with the help of an effective Lagrangian. Low-energy baryon-meson interaction was described in [55] by U(1) × SU(3) gauge theory. There, U(1)-interaction corresponds to exchange by singlet vector meson and SU(3) is group of unitary symmetry. Field contents and structure of Lagrangian, for our case, are represented in [22]. It was shown that the dominant contribution to this interaction is caused by vector meson exchange. We apply this Lagrangian for analysis of MN and MM interactions. The part of the Lagrangian which will be used further is as follows:

$$\begin{aligned} \mathscr{L}_{int}(V,N,M) &= g_\omega \omega^\mu \bar{N}\gamma_\mu N + g_\rho \bar{N}\gamma_\mu \tau^a \rho_a^\mu N + i g_{\omega M}\omega^\mu (M^+\partial_\mu M - \partial_\mu M^+ M) \\ &+ i g_{\rho M}(M^+\tau^a \rho_a^\mu \partial_\mu M - \partial_\mu M^+ \tau^a \rho_a^\mu M). \end{aligned} \tag{24}$$

In (24), $N = (p,n)$ is doublet of nucleons, $M = (M^0, M^-)$, $M^+ = (\bar{M}^0, M^+)$ are new pseudoscalar mesons. Coupling constants are the following ones [22]:

$$g_\rho = g_{\rho M} = g/2, \;\; g_\omega = \sqrt{3}g/2\cos\theta, \;\; g_{\omega M} = g/4\sqrt{3}\cos\theta, \;\; g^2/4\pi \approx 3.16, \;\; \cos\theta = 0.644. \tag{25}$$

Note, the effective strong interaction does not depend on the type of new heavy quark, so we omit subscriptions U and D in (24). We should note that one-pion exchange is absent because $MM\pi$-vertex is forbidden due to parity conservation.

The potential $V(R)$ and amplitude $f(q)$ in Born approximation are connected by the relation

$$V(\vec{r}) = -\frac{1}{4\pi^2\mu}\int f(q)\exp(i\vec{q}\vec{r})\,d^3q, \tag{26}$$

where μ is a reduced mass and we consider the case of non-polarized particles. The potential of MN-interaction was calculated in ref. [22], where the relation $f(q) = -2\pi i\mu F(q)$ was used (here, $F(Q)$ is Feynman amplitude). It was shown in [22] that scalar and two-pion exchanges are strongly suppressed. The potentials of various pairs from $M = (M^0, M^-)$ and $N = (p, n)$ are described by the following expressions:

$$V(M^0, p; r) = V(M^-, n; r) \approx V_\omega(r) + V_\rho(r), \quad V(M^0, n; r) = V(M^-, p; r) \approx V_\omega(r) - V_\rho(r). \tag{27}$$

In (27), the terms $V_\omega(r)$ and $V_\rho(r)$ are as follows:

$$V_\omega(r) = \frac{g^2 K_\omega}{16\pi \cos^2\theta} \frac{1}{r} \exp(-\frac{r}{r_\omega}), \quad V_\rho(r) = \frac{g^2 K_\rho}{16\pi} \frac{1}{r} \exp(-\frac{r}{r_\rho}), \tag{28}$$

where $K_\omega = K_\rho \approx 0.92$, $r_\omega = 1.04/m_\omega$, $r_\rho = 1.04/m_\rho$. Using these values and approximate equality $m_\omega \approx m_\rho$, we rewrite the expressions (27):

$$V(M^0, p; r) = V(M^-, n; r) \approx 2.5 \frac{1}{r} \exp(-\frac{r}{r_\rho}), \quad V(M^0, n; r) = V(M^-, p; r) \approx 1.0 \frac{1}{r} \exp(-\frac{r}{r_\rho}). \tag{29}$$

Two important phenomenological conclusions follow from the expressions (29). All pairs of particles have repulsive ($V > 0$) potential and the existence of a barrier prevents the formation of the coupled states (pM^0), that is, anomalous protons.

The potential of MM interaction can be built in full analogy with MN-case. To find the sign of the potential, we use the following non-relativistic limit:

$$\mathscr{L} = \mathscr{L}_0 + \mathscr{L}_{int} \longrightarrow W_k(m, v) - V(r, t), \quad m\dot{v} = -\frac{\partial U(r, t)}{\partial r} \tag{30}$$

where $W_k(m, v)$ is kinetic term, $V(r, t)$ is potential and we separate the spatial and temporal variables. From this definition and the expression for energy, $E = W + V$, it follows that at long distance, where the monotonically decreasing function $V(r)$ is positive, $V > 0$, this function describes repulsive potential. Here, we use a relation between $\mathscr{L}_{eff}(q)$ and amplitude $F(q)$, namely $F(q) = ik\mathscr{L}_{eff}(q)$, where $k > 0$. Then, we get equality for the sign of V and F, $signum(V) = signum(iF)$. Here, the amplitude $F(q)$ is determined by one-particle exchange diagrams for the process $M_1 M_2 \to M_1^{'} M_2^{'}$. The vertices are defined by the low-energy Lagrangian (24). Then we check that MN-interactions have a repulsive character. Note that Lagrangians of NM^0 and $N\bar{M}^0$ have opposite signs. This is caused by different signs of vertices $\omega M^0 M^0$ and $\omega \bar{M}^0 \bar{M}^0$, which give dominant contribution. This effect follows from the differential structure of (24) and operator structure of field function of the M-particles. We get the vertices $\omega(q) M^0(p) M^0(p-q)$ and $\omega(q) \bar{M}^0(p) \bar{M}^0(p-q)$ in momentum representation with opposite signs, $\mathscr{L}_{eff} = \pm g_{\omega M}(2p - q)$, respectively. This leads to the potentials of interactions through ω exchange, which is repulsive for the case of NM and attractive for $N\bar{M}$ scattering. Thus, the absence of a potential barrier gives rise to the problem of coupled states $p\bar{M}^0$. To overcome this problem, we assume the existence of asymmetry in the sector of new quarks or that the particles $\bar{M}^0$ are superheavy. Interactions of baryons B_1 and B_2 are similar to the nucleonic one. Together with nucleons, they can compose an atomic nuclei.

With the help of the simple method presented above, we have checked that the potentials of $M^0 M^0$ and $\bar{M}^0 \bar{M}^0$ interactions are attractive for the case of scalar meson exchange. It is repulsive for the case of vector meson exchange. The potentials of $M^0 \bar{M}^0$ interactions have attractive asymptotes both for scalar and vector meson exchanges. Thus, the sign of potential for the cases of $M^0 M^0$ and $\bar{M}^0 \bar{M}^0$ scattering is determined by contributions of scalar and vector mesons. In the case of $M^0 \bar{M}^0$ scattering, the total potential is attractive in all channels. This property leads to the effect of enhancement of annihilation cross section (Sommerfeld–Gamov–Sakharov effect [56–61]).

2.5. The Properties of New Heavy Particles and Hadronic Dark Matter

In this subsection, we consider the main properties of new hadrons M^0, M^- and analyze the possibility that M^0 is stable and can be considered as DM candidate. We evaluate the mass of the new quark M_Q and mass splitting of the M^- and M^0 mesons, $\Delta m = m^- - m^0$. Then, we take into account the standard electromagnetic and strong interactions of new hadrons which were described in a previous subsection. The properties of mesons $M = (M^0, M^-)$ are analogous to ones of standard heavy-light mesons. Let us consider the data on mass splitting in pairs $K = (K^0, K^\pm)$, $D = (D^0, D^\pm)$, and $B = (B^0, B^\pm)$. For the case of the mesons K and B, which contain heavy down-type quarks, the mass-splitting $\Delta m < 0$. For the case of the up-type meson D, which contains heavy charm quark, $\Delta m > 0$. The value δm for all cases is $O(\text{MeV})$ and less. We take into consideration these data and assume that for the case of up-type new mesons

$$\Delta m = m(M^-) - m(M^0) > 0, \text{ and } \Delta m = O(\text{MeV}). \tag{31}$$

This assumption means that the neutral state $M^0 = (\bar{U}u)$ is stable and can be considered as the DM candidate. The charged partner $M^- = (\bar{U}d)$ is unstable, if $\delta m > m_e$, and has only one decay channel with small phase space in the final state:

$$M^- \to M^0 (W^-)^* \to M^0 e^- \bar{\nu}_e, \tag{32}$$

where $(W^-)^*$ is a W-boson in intermediate state and e^- is an electron. In the semileptonic decay (32), stable antiquark $\bar{U}$ is considered as a spectator. The width of this decay is calculated in the form-factor approach. The expression for differential width is (see review by R. Kowalski in ref. [39])

$$\frac{d\Gamma(m, \Delta m)}{d\kappa} = \frac{G_F^2}{48\pi^3} |U_{ud}|^2 (m_- + m_0)^2 m_0^3 (\kappa^2 - 1)^{3/2} G^2(\kappa), \tag{33}$$

where $m_- \approx m_0$, $\kappa = k^0/m_0 \approx 1$ and $G(\kappa) \approx 1$ (HQS approximation, [39]). Note, the value $G(\omega)$ is equivalent to the normalized form-factor $f_+(q)$. This form-factor in the vector dominance approach is usually defined by the pole expression $f_+(q) = f_+(0)/(1 - q^2/m_v^2)$. So, the HQS approximation corresponds to the conditions $q^2 \ll m_v^2$ and $f_+(0) \approx 1$ when $\kappa = k^0/m_0 \approx 1$. The expression for the total width follows from Equation (33):

$$\Gamma(m, \Delta m) \approx \frac{G_F^2 |U_{ud}|^2 m_0^5}{12\pi^3} \int_1^{\kappa_m} (\kappa^2 - 1)^{3/2} d\kappa, \tag{34}$$

where $\kappa_m = (m_0^2 + m_-^2)/2m_0 m_-$. After the integration, the expression (34) can be written in a simple form:

$$\Gamma(\Delta m) \approx \frac{G_F^2}{60\pi^3} (\Delta m)^5. \tag{35}$$

So, the width crucially depends on the mass splitting, $\Gamma \sim (\Delta m)^5$. It does not depend on the mass of heavy meson. In the interval $\Delta m = (1 - 10)$ MeV we get following estimations:

$$\Gamma \sim (10^{-29} - 10^{-24})\,\text{GeV}; \quad \tau \sim (10^5 - 10^0)\,\text{s}. \tag{36}$$

Thus, charged particle M^- can be detected in the processes of $M^0 N$-collisions. This possibility was analyzed in ref. [22] (and references therein), where indirect experimental evidences for the presence of heavy charged metastable particles in cosmic rays were considered. Here, we should note that the scenario with a long-lived co-annihilation partner is considered in refs. [20,62].

Further, we estimate the mass of new heavy hadrons in the scenario where they are interpreted as dark matter candidates. The data on the DM relic concentration lead to the equality

$$(\sigma(M)v_r)^{exp} \approx 10^{-10}\ \mathrm{GeV}^{-2}. \tag{37}$$

In (37), M is the mass of new hadron. From this equality we estimate the mass of the meson M^0. Note, the calculations are done for the case of hadron-symmetric DM. To escape the problem with anomalous helium in this case, we expect $M_Q > 10$ TeV. Evaluation of the cross section $\sigma(M^0\bar{M}^0)$ is fulfilled in the approach $\sigma(M^0\bar{M}^0) \sim \sigma(U\bar{U})$, where U is new heavy quark and we consider the light u-quark as a spectator. So, we estimate the cross section at the level of sub-processes with participation of a heavy quark, where the main contributions follow from sub-processes $U\bar{U} \to gg$ and $U\bar{U} \to q\bar{q}$. The expressions for these cross sections were represented in the third subsection (Equations (17) and (18)) and we use their sum for evaluation of the total cross section. Thus, we estimate the mass $m(M^0) \approx M_U$ from the approximate equation

$$(\sigma(M)v_r)^{exp} \approx \frac{44\pi}{9}\frac{\alpha_s^2}{M_U^2}. \tag{38}$$

From (37) and (38) it follows that $m(M^0) = M \approx M_U \approx 20$ TeV at $\alpha_s = \alpha_s(M)$. This values are in accordance with the results in Ref. [63] for the case of heavy WIMPonium.

Attractive potential of $M^0\bar{M}^0$-interaction, as was noted in the previous subsection, can increase annihilation cross section due to the light meson exchange at long distance. This effect leads to the so-called Sommerfeld–Gamov–Sakharov (SGS) enhancement [61]:

$$\sigma(M)v_r = \sigma_0(M)v_r K(2\alpha/v_r). \tag{39}$$

Here, $\sigma_0(M)$ is the initial cross section, $\alpha = g^2/4\pi$ is strong coupling which is defined in (25). At $m \ll M \approx M_U$, where m is mass of intermediate mesons (the light force mediators), the SGS factor K can be represented in the form [61]:

$$K(2\alpha/v_r) = \frac{2\pi\alpha/v_r}{1 - \exp(-2\pi\alpha/v_r)}. \tag{40}$$

The light force carriers, in the case under consideration, are ω- and ρ-mesons and $\alpha \sim 1$, so, from Equation (40) we get the estimations $10^2 \lesssim K(2\alpha/v_r)/\pi \lesssim 10^3$ in the interval $10^{-2} > v_r > 10^{-3}$. Thus, from the Equations (38)–(40) it follows that at $v_r \sim 10^{-2}$ the mass of new quark $M_U \sim 10^2$ TeV. This value agrees with the estimations of the baryonic DM mass in [64] ($M \sim 100$ TeV). So, the value M_U falls out from the mass range of the searches for anomalous hydrogen ($M_{max} \lesssim 1$ TeV) and anomalous helium ($M_{max} \lesssim 10$ TeV). In our estimations we take into account the light mesons only, ($m \ll M$). At a short distance, near $r \sim M^{-1}$, the exchange by heavy mesons is possible. The expression (40), in this case, is not valid because $M_\chi \sim M_U$, where M_χ is the mass of heavy force mediators. For evaluation of SGS factor K in this case, we use the numerical calculations in ref. [65]. From this work it follows that $K \approx 10$ in the interval $10^{-1} > v_r > 10^{-3}$, and we get from (38) and (39) the estimation $M \approx 60$ TeV which does not crucially change the situation. Here, we should note that correct description of SGS requires taking account of bosons Z and W also. Thus, SGS effects are formed at various energies which correspond to various distances. So, this effect has a very complicated and vague nature (see also ref. [66]).

3. Hypercolor Extensions of Standard Model

In this section, we consider some particular variants of models that extend SM by introducing an additional strong sector with heavy vector-like fermions, hyperquarks (H-quarks), charged under an H-color gauge group [67–78]. Depending on H-quark quantum numbers, such models can encompass scenarios with composite Higgs doublets (see e.g., [79]) or a small mixing between fundamental Higgs fields of SM and composite hadron-like states of the new strong sector making the Higgs boson partially composite. Models of this class leave room for the existence of DM candidates whose decays are forbidden by accidental symmetries. Besides, H-color models comply well with electroweak precision constraints, since H-quarks are assumed to be vector-like.

In the rest of this section, we briefly review one of the simplest realizations of the scenario described—models with two or three vector-like H-flavors confined by strong H-color force $\mathrm{Sp}(2\chi_{\tilde{c}})$, $\chi_{\tilde{c}} \geqslant 1$. The models with H-color group SU(2) [74,80] are included as particular cases in this consideration due to isomorphism $\mathrm{SU}(2) = \mathrm{Sp}(2)$ [74,80]. The global symmetry group of the strong sector with symplectic H-color group is larger than for the special unitary case—it is the group $\mathrm{SU}(2n_F)$ broken spontaneously to $\mathrm{Sp}(2n_F)$, with n_F being a number of H-flavors. We posit that the extensions of SM under consideration preserve the elementary Higgs doublet in the set of Lagrangian field operators. This doublet mixes with H-hadrons, which makes the physical Higgs partially composite. Note also that the same coset $\mathrm{SU}(2n_F)/\mathrm{Sp}(2n_F)$ can be used to construct composite two Higgs doublet model [79] or little Higgs models [81–86].

3.1. *Lagrangian and Global Symmetry of Symplectic QCD with $n_F = 2, 3$ Hyperquark Flavors*

In this section, we consider the simplest possibilities to extend the symmetry of SM, G_{SM}, by adding a symplectic hypercolor group, i.e., the gauge group of the extension under consideration is $G = G_{\mathrm{SM}} \times \mathrm{Sp}(2\chi_{\tilde{c}})$, $\chi_{\tilde{c}} \geqslant 1$. The model is postulated to have new degrees of freedom, six hyperquarks—Weyl fermions charged under H-color group. These fermions are assumed to form two weak doublets $Q^{k\underline{k}}_{\mathrm{L}(A)}$ and two singlets $S^{\underline{k}}_{\mathrm{L}(A)}$, $A = 1, 2$. In this paper, we underscore indices that are related to the H-color group $\mathrm{Sp}(2\chi_{\tilde{c}})$; the normal Latin indices (k, a, etc.) are for the weak group $\mathrm{SU}(2)_{\mathrm{L}}$. The transformation law for the H-quarks is posited to be

$$(Q^{j\underline{j}}_{\mathrm{L}(A)})' = Q^{j\underline{j}}_{\mathrm{L}(A)} - \frac{i}{2} g_1 Y_{Q(A)} \theta Q^{j\underline{j}}_{\mathrm{L}(A)} + \frac{i}{2} g_2 \theta_a \tau_a^{jk} Q^{k\underline{j}}_{\mathrm{L}(A)} + \frac{i}{2} g_{\tilde{c}} \theta_{\underline{a}} \lambda_{\underline{a}}^{\underline{jk}} Q^{j\underline{k}}_{\mathrm{L}(A)}, \tag{41}$$

$$(S^{\underline{j}}_{\mathrm{L}(A)})' = S^{\underline{j}}_{\mathrm{L}(A)} - i g_1 Y_{S(A)} \theta S^{\underline{j}}_{\mathrm{L}(A)} + \frac{i}{2} g_{\tilde{c}} \theta_{\underline{a}} \lambda_{\underline{a}}^{\underline{jk}} S^{\underline{k}}_{\mathrm{L}(A)}. \tag{42}$$

Here, θ, θ_a, $\theta_{\underline{a}}$ are transformation parameters of $\mathrm{U}(1)_{\mathrm{Y}}$, $\mathrm{SU}(2)_{\mathrm{L}}$, and $\mathrm{Sp}(2\chi_{\tilde{c}})$ respectively; τ_a are the Pauli matrices; $\lambda_{\underline{a}}$, $\underline{a} = 1 \ldots \chi_{\tilde{c}}(2\chi_{\tilde{c}} + 1)$ are $\mathrm{Sp}(2\chi_{\tilde{c}})_{\tilde{c}}$ generators satisfying the relation

$$\lambda_{\underline{a}}^{\mathrm{T}} \omega + \omega \lambda_{\underline{a}} = 0, \tag{43}$$

where T stands for "transpose", ω is an antisymmetric $2\chi_{\tilde{c}} \times 2\chi_{\tilde{c}}$ matrix, $\omega^{\mathrm{T}} \omega = 1$. From now on, $\mathrm{SU}(2)_{\mathrm{L}}$ and $\mathrm{Sp}(2\chi_{\tilde{c}})_{\tilde{c}}$ indices are omitted if this does not lead to ambiguities. The relation (43) and the analogous one holding true for the Pauli matrices of the weak group imply that the H-quarks are pseudoreal representations of the gauge symmetry groups of the model. This allows us to write the right-handed fields exhibiting transformation properties that are similar to those of the original left-handed ones:

$$Q'_{\mathrm{R}(A)} = \varepsilon \omega {Q_{\mathrm{L}(A)}}^{\mathrm{C}} = Q_{\mathrm{R}(A)} + \frac{i}{2} g_1 Y_{Q(A)} \theta Q_{\mathrm{R}(A)} + \frac{i}{2} g_2 \theta_a \tau_a Q_{\mathrm{R}(A)} + \frac{i}{2} g_{\tilde{c}} \theta_{\underline{a}} \lambda_{\underline{a}} Q_{\mathrm{R}(A)}, \tag{44}$$

$$S'_{\mathrm{R}(A)} = \omega {S_{\mathrm{L}(A)}}^{\mathrm{C}} = S_{\mathrm{R}(A)} + i g_1 Y_{S(A)} \theta S_{\mathrm{R}(A)} + \frac{i}{2} g_{\tilde{c}} \theta_{\underline{a}} \lambda_{\underline{a}} S_{\mathrm{R}(A)}, \tag{45}$$

where $\varepsilon = i\tau_2$.

The quantum numbers of the right-handed spinors $Q_{\mathrm{R}(A)}$ and $S_{\mathrm{R}(A)}$ are the same as the ones of the left-handed H-quarks except for the opposite-sign hypercharges. Therefore, setting $Y_{Q(1)} = -Y_{Q(2)} = Y_Q$ and $Y_{S(1)} = -Y_{S(2)} = Y_S$, we obtain a doublet and a singlet of Dirac fields:

$$Q = Q_{L(1)} + Q_{R(2)}, \qquad S = S_{L(1)} + S_{R(2)}. \tag{46}$$

These relations among hypercharges are also enforced independently by requiring cancellation of gauge anomalies.

Finally, the Lagrangian of the SM extension invariant under $G = G_{SM} \times Sp(2\chi_{\tilde{c}})$ reads

$$\mathscr{L} = \mathscr{L}_{SM} - \frac{1}{4} H^{\mu\nu}_{\underline{a}} H^{\underline{a}}_{\mu\nu} + i\bar{Q}\not{D}Q - m_Q \bar{Q}Q + i\bar{S}\not{D}S - m_S \bar{S}S + \delta\mathscr{L}_Y, \tag{47}$$

$$D^\mu Q = \left[\partial^\mu + \frac{i}{2} g_1 Y_Q B^\mu - \frac{i}{2} g_2 W^\mu_a \tau_a - \frac{i}{2} g_{\tilde{c}} H^\mu_{\underline{a}} \lambda_{\underline{a}}\right] Q, \tag{48}$$

$$D^\mu S = \left[\partial^\mu + i g_1 Y_S B^\mu - \frac{i}{2} g_{\tilde{c}} H^\mu_{\underline{a}} \lambda_{\underline{a}}\right] S, \tag{49}$$

where $H^\mu_{\underline{a}}$, $\underline{a} = 1 \ldots \chi_{\tilde{c}}(2\chi_{\tilde{c}} + 1)$ are hypergluon fields and $H^{\mu\nu}_{\underline{a}}$ are their strength tensors. Contact Yukawa couplings $\delta\mathscr{L}_Y$ of the H-quarks and the SM Higgs doublet $\mathscr{H}$ are permitted in the model if the hypercharges satisfy an additional linear relation:

$$\delta\mathscr{L}_Y = y_L \left(\bar{Q}_L \mathscr{H}\right) S_R + y_R \left(\bar{Q}_R \varepsilon \mathscr{H}\right) S_L + \text{h.c.} \quad \text{for } \frac{Y_Q}{2} - Y_S = +\frac{1}{2}; \tag{50}$$

$$\delta\mathscr{L}_Y = y_L \left(\bar{Q}_L \varepsilon \mathscr{H}\right) S_R + y_R \left(\bar{Q}_R \mathscr{H}\right) S_L + \text{h.c.} \quad \text{for } \frac{Y_Q}{2} - Y_S = -\frac{1}{2}. \tag{51}$$

The model can be reconciled with the electroweak precision constraints quite easily, since H-quarks are vector-like, i.e., their electroweak interactions are chirally symmetric in this scheme. Besides, this allows us to introduce explicit gauge-invariant Dirac mass terms for H-quarks.

It is easy to prove that the kinetic terms of H-quarks Q and S in the Lagrangian (47) can be rewritten in terms of a left-handed sextet as follows:

$$\delta\mathscr{L}_{\text{H-quarks, kin}} = i\bar{P}_L \not{D} P_L, \qquad P_L = \left(Q^T_{L(1)}, Q^T_{L(2)}, S_{L(1)}, S_{L(2)}\right)^T, \tag{52}$$

$$D^\mu P_L = \left[\partial^\mu + i g_1 B^\mu \left(Y_Q \Sigma_Q + Y_S \Sigma_S\right) - \frac{i}{2} g_2 W^\mu_a \Sigma^a_W - \frac{i}{2} g_{\tilde{c}} H^\mu_{\underline{a}} \lambda_{\underline{a}}\right] P_L, \tag{53}$$

$$\Sigma_Q = \frac{1}{2}\begin{pmatrix} 1 & 0 & 0 \\ 0 & -1 & 0 \\ 0 & 0 & 0 \end{pmatrix}, \qquad \Sigma_S = \begin{pmatrix} 0 & 0 & 0 \\ 0 & 0 & 0 \\ 0 & 0 & \tau_3 \end{pmatrix}, \qquad \Sigma^a_W = \begin{pmatrix} \tau_a & 0 & 0 \\ 0 & \tau_a & 0 \\ 0 & 0 & 0 \end{pmatrix}. \tag{54}$$

In the limit of vanishing electroweak interactions, $g_1 = g_2 = 0$, this Lagrangian is invariant under a global SU(6) symmetry, which is dubbed as the Pauli–Gürsey symmetry sometimes [87,88]:

$$P_L \to U P_L, \qquad U \in SU(6). \tag{55}$$

The subgroups of the SU(6) symmetry include:

- the chiral symmetry $SU(3)_L \times SU(3)_R$,
- SU(4) subgroup corresponding to the two-flavor model without singlet H-quark S,
- two-flavor chiral group $SU(2)_L \times SU(2)_R$, which is a subgroup of both former subgroups.

The global symmetry is broken both explicitly and dynamically:

- explicitly—by the electroweak and Yukawa interactions, (50) and (53), and the H-quark masses;
- dynamically—by H-quark condensate [89,90]:

$$\langle \bar{Q}Q + \bar{S}S \rangle = \frac{1}{2}\langle \bar{P}_L M_0 P_R + \bar{P}_R M_0^\dagger P_L \rangle, \qquad P_R = \omega P_L{}^C, \qquad M_0 = \begin{pmatrix} 0 & \varepsilon & 0 \\ \varepsilon & 0 & 0 \\ 0 & 0 & \varepsilon \end{pmatrix}. \tag{56}$$

The condensate (56) is invariant under $\mathrm{Sp}(6) \subset \mathrm{SU}(6)$ transformations U that satisfy a condition

$$U^{\mathrm{T}} M_0 + M_0 U = 0, \tag{57}$$

i.e., the global SU(6) symmetry is broken dynamically to its Sp(6) subgroup. The mass terms of H-quarks in (47) could break the symmetry further to $\mathrm{Sp}(4) \times \mathrm{Sp}(2)$:

$$\delta\mathscr{L}_{\text{H-quarks, masses}} = -\frac{1}{2}\bar{P}_{\mathrm{L}} M_0' P_{\mathrm{R}} + \text{h.c.}, \qquad M_0' = -M_0'^{\mathrm{T}} = \begin{pmatrix} 0 & m_Q\varepsilon & 0 \\ m_Q\varepsilon & 0 & 0 \\ 0 & 0 & m_S\varepsilon \end{pmatrix}. \tag{58}$$

The case of a two-flavor model (without the singlet H-quark) is completely analogous to the three-flavor model but is simpler than latter one—the global SU(4) symmetry is broken dynamically to its Sp(4) subgroup by the condensate of doublet H-quarks; the Lagrangian of the model is obtained from the one given by Equations (47)–(54) by simply setting to zero all terms with the H-quark S.

3.2. *Linear Sigma Model as an Effective Field Theory of Constituent H-Quarks*

Now, we proceed to construct a linear σ-model for interactions of constituent H-quarks. The Lagrangian of the model consists of kinetic terms for the constituent fermions and the lightest (pseudo)scalar composite states, Yukawa terms for the interactions of the (pseudo)scalars with the fermions, and a potential of (pseudo)scalar self-interactions U_{scalars}. The Lagrangian reads

$$\mathscr{L}_{\mathrm{L}\sigma} = \mathscr{L}_{\text{H-quarks}} + \mathscr{L}_{\mathrm{Y}} + \mathscr{L}_{\text{scalars}}, \tag{59}$$

$$\mathscr{L}_{\text{H-quarks}} = i\bar{P}_{\mathrm{L}} \not{D} P_{\mathrm{L}}, \qquad \mathscr{L}_{\mathrm{Y}} = -\sqrt{2}\varkappa\left(\bar{P}_{\mathrm{L}} M P_{\mathrm{R}} + \bar{P}_{\mathrm{R}} M^{\dagger} P_{\mathrm{L}}\right), \tag{60}$$

$$\mathscr{L}_{\text{scalars}} = D_\mu \mathscr{H}^{\dagger} \cdot D^\mu \mathscr{H} + \mathrm{Tr}\, D_\mu M^{\dagger} \cdot D^\mu M - U_{\text{scalars}}, \tag{61}$$

Here, $\varkappa$ is a coupling constant; M is a complex antisymmetric $2n_F \times 2n_F$ matrix of (pseudo)scalar fields; the multiplets $P_{\mathrm{L,\,R}}$ correspond now to the constituent H-quarks but retain all the definitions and properties of the fundamental multiplets described in the previous section. The fields transform under the global symmetry $\mathrm{SU}(2n_F)$ as follows:

$$M \to U M U^{\mathrm{T}}, \qquad P_{\mathrm{L}} \to U P_{\mathrm{L}}, \qquad P_{\mathrm{R}} \to \bar{U} P_{\mathrm{R}}, \qquad U \in \mathrm{SU}(2n_F), \tag{62}$$

where $\bar{U}$ designates the complex conjugate of the matrix U. Note also that the model comprises of the fundamental (not composite) Higgs doublet $\mathscr{H}$ of SM.

It is postulated that the interactions of the constituent H-quarks with the gauge bosons are the same as for the fundamental H-quarks. This and the transformation laws (62) define the covariant derivative for the scalar field M. The complete set of covariant derivatives present in the Lagrangian (59) is as follows:

$$D_\mu \mathscr{H} = \left[\partial_\mu + \frac{i}{2} g_1 B_\mu - \frac{i}{2} g_2 W_\mu^a\right] \mathscr{H}, \qquad D^\mu P_{\mathrm{L}} = \left[\partial^\mu + i g_1 B^\mu \left(Y_Q \Sigma_Q + Y_S \Sigma_S\right) - \frac{i}{2} g_2 W_a^\mu \Sigma_W^a\right] P_{\mathrm{L}}, \tag{63}$$

$$D_\mu M = \partial_\mu M + i Y_Q g_1 B_\mu (\Sigma_Q M + M \Sigma_Q^{\mathrm{T}}) + i Y_S g_1 B_\mu (\Sigma_S M + M \Sigma_S^{\mathrm{T}}) - \frac{i}{2} g_2 W_\mu^a (\Sigma_W^a M + M \Sigma_W^{a\mathrm{T}}), \tag{64}$$

where the matrices Σ_Q, Σ_S, Σ_W^a, $a = 1, 2, 3$ are defined by Equation (54).

3.2.1. Interactions of the Constituent H-Quarks with H-Hadrons and the Electroweak Gauge Bosons

In the case of $n_F = 3$, the field M can be expanded in a basis of fourteen "broken" generators β_a of the global symmetry group SU(6):

$$
\begin{aligned}
M &= \left[\frac{1}{2\sqrt{n_F}}(A_0 + iB_0)I + (A_a + iB_a)\beta_a\right] M_0 \\
&= \frac{1}{2}\begin{pmatrix} \bar{A}\varepsilon & \left[\frac{1}{\sqrt{n_F}}\sigma + \frac{1}{\sqrt{2n_F}}f + \frac{1}{\sqrt{2}}a_a\tau_a\right]\varepsilon & K^\star\varepsilon \\ \left[\frac{1}{\sqrt{n_F}}\sigma + \frac{1}{\sqrt{2n_F}}f - \frac{1}{\sqrt{2}}a_a\tau_a\right]\varepsilon & A\varepsilon & \varepsilon\bar{K}^\star \\ K^{\star\dagger}\varepsilon & \varepsilon K^{\star\mathrm{T}} & \frac{1}{\sqrt{n_F}}\left(\sigma - \sqrt{2}f\right)\varepsilon \end{pmatrix} \\
&+ \frac{i}{2}\begin{pmatrix} \bar{B}\varepsilon & \left[\frac{1}{\sqrt{n_F}}\eta + \frac{1}{\sqrt{2n_F}}\eta' + \frac{1}{\sqrt{2}}\pi_a\tau_a\right]\varepsilon & K\varepsilon \\ \left[\frac{1}{\sqrt{n_F}}\eta + \frac{1}{\sqrt{2n_F}}\eta' - \frac{1}{\sqrt{2}}\pi_a\tau_a\right]\varepsilon & B\varepsilon & \varepsilon\bar{K} \\ K^{\dagger}\varepsilon & \varepsilon K^{\mathrm{T}} & \frac{1}{\sqrt{n_F}}\left(\eta - \sqrt{2}\eta'\right)\varepsilon \end{pmatrix},
\end{aligned} \tag{65}
$$

Here, I is the identity matrix and new scalar fields are defined as follows:

$$
\begin{gathered}
\sigma = A_0, \quad \eta = B_0, \quad f = A_6, \quad \eta' = B_6, \quad a_a = A_{a+2}, \quad \pi_a = B_{a+2}, \quad a = 1, 2, 3, \\
A = \frac{1}{\sqrt{2}}(A_1 + iA_2), \quad B = \frac{1}{\sqrt{2}}(B_1 + iB_2), \\
K^\star = \frac{1}{2}\left[A_{10} + iA_{14} + (A_{6+a} + iA_{10+a})\tau_a\right], \qquad K = \frac{1}{2}\left[B_{10} + iB_{14} + (B_{6+a} + iB_{10+a})\tau_a\right].
\end{gathered}
$$

The generators β_a are defined in the Appendix A. A bar over a scalar field denote the complex conjugate of the field operator. In the case of $n_F = 2$, we should substitute the identity matrix I in Equation (65) by the diagonal matrix $\mathrm{diag}(1,1,1,1,0,0)$ and take into account just the first five of generators β_a, i.e., $A_a = 0 = B_a$ for $a = 6, \ldots 14$ or, equivalently, $K = 0 = K^\star$ and $f = 0 = \eta'$. In other words, only the upper left 4×4 block of the matrix (65) remains under consideration, while all other its elements are set to zero.

Assuming that the singlet meson σ develops a v.e.v. u, $\sigma = u + \sigma'$, and inserting the representation (65) into the Lagrangian (59) of the sigma model, we arrive at the following form of the Lagrangian:

$$
\begin{aligned}
&\mathscr{L}_{\text{H-quarks}} + \mathscr{L}_{\mathrm{Y}} = i\bar{Q}\not{D}Q + i\bar{S}\not{D}S - \varkappa u\left(\bar{Q}Q + \bar{S}S\right) \\
&-\varkappa\bar{Q}\left[\sigma' + \frac{1}{\sqrt{3}}f + i\left(\eta + \frac{1}{\sqrt{3}}\eta'\right)\gamma_5 + (a_a + i\pi_a\gamma_5)\,\tau_a\right]Q - \varkappa\bar{S}\left[\sigma' - \frac{2}{\sqrt{3}}f + i\left(\eta - \frac{2}{\sqrt{3}}\eta'\right)\gamma_5\right]S \\
&-\sqrt{2}\varkappa\left[(\bar{Q}\mathscr{K}^\star)\,S + i\,(\bar{Q}\mathscr{K})\,\gamma_5 S + \text{h.c.}\right] - \sqrt{2}\varkappa\left[(\bar{Q}\mathscr{A})\,\omega S^{\mathrm{C}} + i\,(\bar{Q}\mathscr{B})\,\gamma_5\omega S^{\mathrm{C}} + \text{h.c.}\right] \\
&-\frac{\varkappa}{\sqrt{2}}\left(A\bar{Q}\varepsilon\omega Q^{\mathrm{C}} + iB\bar{Q}\gamma_5\varepsilon\omega Q^{\mathrm{C}} + \text{h.c.}\right),
\end{aligned} \tag{66}
$$

$$
D_\mu Q = \partial_\mu Q + \frac{i}{2}g_1 Y_Q B_\mu Q - \frac{i}{2}g_2 W_\mu^a \tau_a Q, \qquad D_\mu S = \partial_\mu S + ig_1 Y_S B_\mu S, \tag{67}
$$

where $\mathscr{K}^\star$, $\mathscr{K}$ and $\mathscr{A}$, $\mathscr{B}$ are $\mathrm{SU(2)_L}$ doublets of H-mesons and H-diquarks (H-baryons) respectively:

$$\mathcal{K}^{\star} = \frac{1}{\sqrt{2}}\left(R_1 + iR_2\right), \mathcal{K} = \frac{1}{\sqrt{2}}\left(S_1 + iS_2\right), \mathcal{A} = \frac{1}{\sqrt{2}}\varepsilon\left(\bar{R}_1 + i\bar{R}_2\right), \mathcal{B} = \frac{1}{\sqrt{2}}\varepsilon\left(\bar{S}_1 + i\bar{S}_2\right), \quad (68)$$

$$R_1 = \frac{1}{\sqrt{2}}\begin{pmatrix} A_{10} + iA_{13} \\ -A_{12} + iA_{11} \end{pmatrix}, \qquad R_2 = \frac{1}{\sqrt{2}}\begin{pmatrix} A_{14} - iA_9 \\ A_8 - iA_7 \end{pmatrix}, \quad (69)$$

$$S_1 = \frac{1}{\sqrt{2}}\begin{pmatrix} B_{10} + iB_{13} \\ -B_{12} + iB_{11} \end{pmatrix}, \qquad S_2 = \frac{1}{\sqrt{2}}\begin{pmatrix} B_{14} - iB_9 \\ B_8 - iB_7 \end{pmatrix}. \quad (70)$$

The Lagrangian for the case of a two-flavor model, $n_F = 2$, is obtained by simply neglecting all terms with the singlet H-quark S in Equation (66). All H-hadrons the models with $n_F = 2, 3$ describe are listed in the Table 2.

Table 2. The lightest (pseudo)scalar H-hadrons in $\mathrm{Sp}(2\chi_{\tilde{c}})$ model with two and three flavors of H-quarks (in the limit of vanishing mixings). The lower half of the table lists the states present only in the three-flavor version of the model. T is the weak isospin. $\tilde{G}$ denotes hyper-G-parity of a state (see Section 3.2.4). $\tilde{B}$ is the H-baryon number. $Q_{\rm em}$ is the electric charge (in units of the positron charge $e = |e|$). The H-quark charges are $Q^U_{\rm em} = (Y_Q + 1)/2$, $Q^D_{\rm em} = (Y_Q - 1)/2$, and $Q^S_{\rm em} = Y_S$, which is seen from (67).

State	H-Quark Current	$T^{\tilde{G}}(J^{PC})$	$\tilde{B}$	$Q_{\rm em}$
σ	$\bar{Q}Q + \bar{S}S$	$0^+(0^{++})$	0	0
η	$i(\bar{Q}\gamma_5 Q + \bar{S}\gamma_5 S)$	$0^+(0^{-+})$	0	0
a_k	$\bar{Q}\tau_k Q$	$1^-(0^{++})$	0	$\pm1, 0$
π_k	$i\bar{Q}\gamma_5\tau_k Q$	$1^-(0^{-+})$	0	$\pm1, 0$
A	$\bar{Q}^C\varepsilon\omega Q$	$0\ (0^-\ )$	1	Y_Q
B	$i\bar{Q}^C\varepsilon\omega\gamma_5 Q$	$0\ (0^+\ )$	1	Y_Q
f	$\bar{Q}Q - 2\bar{S}S$	$0^+(0^{++})$	0	0
η'	$i(\bar{Q}\gamma_5 Q - 2\bar{S}\gamma_5 S)$	$0^+(0^{-+})$	0	0
$\mathcal{K}^\star$	$\bar{S}Q$	$\frac{1}{2}\ (0^+\ )$	0	$Y_Q/2 - Y_S \pm 1/2$
$\mathcal{K}$	$i\bar{S}\gamma_5 Q$	$\frac{1}{2}\ (0^-\ )$	0	$Y_Q/2 - Y_S \pm 1/2$
$\mathcal{A}$	$\bar{S}^C\omega Q$	$\frac{1}{2}\ (0^-\ )$	1	$Y_Q/2 + Y_S \pm 1/2$
$\mathcal{B}$	$i\bar{S}^C\omega\gamma_5 Q$	$\frac{1}{2}\ (0^+\ )$	1	$Y_Q/2 + Y_S \pm 1/2$

3.2.2. Interactions of the (Pseudo)scalar Fields with the Electroweak Gauge Bosons

The kinetic terms of the lightest H-hadrons in the Lagrangian (61) can be put into the following form:

$$\mathcal{T}_{\rm scalars} = \frac{1}{2}\sum_{\varphi} D_\mu\varphi \cdot D^\mu\varphi + \sum_{\Phi}\left(D_\mu\Phi\right)^\dagger D^\mu\Phi + D_\mu\bar{A}\cdot D^\mu A + D_\mu\bar{B}\cdot D^\mu B, \quad (71)$$

where $\varphi = h,\ h_a,\ \pi_a,\ a_a,\ \sigma,\ f,\ \eta,\ \eta'$ are singlet and triplet fields, $\Phi = \mathcal{K},\ \mathcal{K}^\star,\ \mathcal{A},\ \mathcal{B}$ are doublets. The fields h and h_a, $a = 1, 2, 3$ are components of the fundamental Higgs doublet

$$\mathcal{H} = \frac{1}{\sqrt{2}}\begin{pmatrix} h_2 + ih_1 \\ h - ih_3 \end{pmatrix}. \quad (72)$$

All covariant derivatives in the Lagrangian (71) follow directly from the covariant derivatives of the fields $\mathcal{H}$ (63) and M (64):

$$D_\mu h = \partial_\mu h + \frac{1}{2}(g_1\delta_3^a B_\mu + g_2 W_\mu^a)h_a, \qquad D_\mu\phi = \partial_\mu\phi, \quad \phi = \sigma,\, f,\, \eta,\, \eta', \tag{73}$$

$$D_\mu h_a = \partial_\mu h_a - \frac{1}{2}(g_1\delta_3^a B_\mu + g_2 W_\mu^a)h - \frac{1}{2}e_{abc}(g_1\delta_3^b B_\mu - g_2 W_\mu^b)h_c, \tag{74}$$

$$D_\mu \pi_a = \partial_\mu \pi_a + g_2 e_{abc} W_\mu^b \pi_c, \qquad D_\mu a_a = \partial_\mu a_a + g_2 e_{abc} W_\mu^b a_c, \tag{75}$$

$$D_\mu A = \partial_\mu A + i g_1 Y_Q B_\mu A, \qquad D_\mu B = \partial_\mu B + i g_1 Y_Q B_\mu B, \tag{76}$$

$$D_\mu \mathscr{K} = \left[\partial_\mu + i g_1 \left(\frac{Y_Q}{2} - Y_S\right) B_\mu - \frac{i}{2} g_2 W_\mu^a \tau^a\right] \mathscr{K}, \tag{77}$$

$$D_\mu \mathscr{K}^\star = D_\mu \mathscr{K}\big|_{\mathscr{K}\to\mathscr{K}^\star}, \quad D_\mu \mathscr{A} = D_\mu \mathscr{K}\Big|_{\substack{\mathscr{K}\to\mathscr{A} \\ Y_S\to -Y_S}}, \quad D_\mu \mathscr{B} = D_\mu \mathscr{K}\Big|_{\substack{\mathscr{K}\to\mathscr{B} \\ Y_S\to -Y_S}}. \tag{78}$$

3.2.3. Self-Interactions and Masses of the (Pseudo)Scalar Fields

The potential of spin-0 fields—the Higgs boson and (pseudo)scalar H-hadrons—can be written as follows:

$$U_{\text{scalars}} = \sum_{i=0}^{4} \lambda_i I_i + \sum_{0=i\leqslant k=0}^{3} \lambda_{ik} I_i I_k. \tag{79}$$

Here, I_i, $i = 0, 1, 2, 3, 4$ are the lowest dimension invariants

$$I_0 = \mathscr{H}^\dagger \mathscr{H}, \quad I_1 = \mathrm{Tr}\left(M^\dagger M\right), \quad I_2 = \mathrm{Re}\,\mathrm{Pf}\, M, \quad I_3 = \mathrm{Im}\,\mathrm{Pf}\, M, \quad I_4 = \mathrm{Tr}\left[\left(M^\dagger M\right)^2\right]. \tag{80}$$

The Pfaffian of M is defined as

$$\mathrm{Pf}\, M = \frac{1}{2^2 2!}\varepsilon_{abcd} M_{ab} M_{cd} \quad \text{for } n_F = 2, \quad \mathrm{Pf}\, M = \frac{1}{2^3 3!}\varepsilon_{abcdef} M_{ab} M_{cd} M_{ef} \quad \text{for } n_F = 3, \tag{81}$$

where ε is the $2n_F$-dimensional Levi–Civita symbol ($\varepsilon_{12\ldots(2n_F)} = +1$). We consider only renormalizable part of the potential (79) permitted by the symmetries of the model. This implies that $\lambda_{i2} = \lambda_{i3} = 0$ for all i if $n_F = 3$. Besides, the invariant I_3 is CP odd, i.e., $\lambda_3 = 0$ as well as $\lambda_{i3} = 0$ for $i = 0, 1, 2$. In the two-flavor model, one of the terms in the potential (79) is redundant because of the identity

$$I_1^2 - 4I_2^2 - 4I_3^2 - 2I_4 = 0 \tag{82}$$

that holds for $n_F = 2$. To take this into account, we set $\lambda_{22} = 0$. (As it is mentioned above, λ_{22} is also set to zero for $n_F = 3$, since we consider only renormalizable interactions.).

In the case of vanishing Yukawa couplings $y_L = y_R = 0$, the tadpole equations for v.e.v.'s $v = \langle h\rangle \neq 0$ and $u = \langle\sigma\rangle \neq 0$ read

$$\mu_0^2 = \lambda_{00} v^2 + \frac{1}{2}\Lambda_{01} u^2, \quad \mu_1^2 = \frac{1}{2}\Lambda_{01} v^2 + \Lambda_{11} u^2 - \frac{(4-n_F)\, n_F \lambda_2 u^{n_F-2}}{2\,(2\sqrt{n_F})^{n_F}} + \frac{3}{2}\zeta\frac{\langle\bar{Q}Q + \bar{S}S\rangle}{u}, \tag{83}$$

where

$$\mu_0^2 = -\lambda_0, \qquad \mu_1^2 = -\lambda_1, \qquad \Lambda_{01} = \lambda_{01} - \frac{1}{4}\lambda_{02}, \tag{84}$$

$$\Lambda_{11} = \lambda_{11} + \frac{\lambda_4}{2n_F} - \frac{1}{4}\lambda_{12} - \frac{n_F(n_F-2)\lambda_2}{2\,(2\sqrt{n_F})^{n_F}\, u^{4-n_F}} - \frac{\zeta\langle\bar{Q}Q + \bar{S}S\rangle}{2u^3}. \tag{85}$$

The condition of vacuum stability requires that the following inequalities hold:

$$\lambda_{00} > 0, \qquad \Lambda_{11} > 0, \qquad 4\lambda_{00}\Lambda_{11} - \Lambda_{01}^2 > 0. \tag{86}$$

The effects of explicit breaking of the SU($2n_F$) global symmetry can be communicated to the effective fields by different non-invariant terms in the Lagrangian [91–93]. Here, we use the most common one which is a tadpole-like term $\mathscr{L}_{\rm SB} = -\zeta\langle\bar{Q}Q + \bar{S}S\rangle(u + \sigma')$, with the parameter ζ being proportional to the current mass m_Q of the H-quarks (see [94,95], for example).

Tree masses of the (pseudo)scalars:

$$m_{\sigma,H}^2 = \lambda_{00}v^2 + \Lambda_{11}u^2 \pm \sqrt{(\lambda_{00}v^2 - \Lambda_{11}u^2)^2 + \Lambda_{01}^2 v^2 u^2}, \tag{87}$$

$$m_{\tilde{\pi}}^2 = m_{\tilde{\eta}'}^2 = m_{\tilde{B}}^2 = m_{\mathscr{K}}^2 = m_{\mathscr{B}}^2 = -\frac{\zeta\langle\bar{Q}Q + \bar{S}S\rangle}{u}, \; m_{\tilde{\eta}}^2 = \begin{cases} \frac{1}{2}\left(\frac{1}{4}\lambda_{33} - \lambda_4\right)u^2 + m_{\tilde{a}}^2 & \text{for } n_F = 2, \\ \frac{\sqrt{3}}{24}\lambda_2 u - \frac{1}{3}\lambda_4 u^2 + m_{\tilde{a}}^2 & \text{for } n_F = 3, \end{cases} \tag{88}$$

$$m_{\tilde{f}}^2 = m_{\mathscr{K}^\star}^2 = m_{\mathscr{A}}^2 = \frac{\sqrt{3}}{12}\lambda_2 u + \frac{1}{3}\lambda_4 u^2 - \frac{\zeta\langle\bar{Q}Q + \bar{S}S\rangle}{u}, \tag{89}$$

$$m_{\tilde{a}}^2 = m_{\tilde{A}}^2 = \begin{cases} \frac{1}{2}\left(\lambda_4 + \frac{1}{2}\lambda_{12}\right)u^2 + \frac{1}{4}\lambda_{02}v^2 + \frac{1}{2}\lambda_2 - \frac{\zeta\langle\bar{Q}Q\rangle}{u} & \text{for } n_F = 2, \\ m_{\tilde{f}}^2 & \text{for } n_F = 3. \end{cases} \tag{90}$$

For all n_F, the model involves a small mixing of the fundamental Higgs and H-meson σ', which makes the Higgs partially composite:

$$h = \cos\theta_s H - \sin\theta_s \sigma, \qquad \sigma' = \sin\theta_s H + \cos\theta_s \sigma, \tag{91}$$

$$\tan 2\theta_s = \frac{\Lambda_{01}vu}{\lambda_{00}v^2 - \Lambda_{11}u^2}, \qquad \operatorname{sgn}\sin\theta_s = -\operatorname{sgn}\Lambda_{01}, \tag{92}$$

where H and σ are physical fields.

3.2.4. Accidental Symmetries

If the hypercharges of H-quarks are set to zero, the Lagrangian (47) is invariant under an additional symmetry—hyper G-parity [96,97]:

$$Q^{\tilde{G}} = \varepsilon\omega Q^C, \qquad S^{\tilde{G}} = \omega S^C. \tag{93}$$

Since H-gluons and all SM fields are left intact by (93), the lightest $\tilde{G}$-odd H-hadron becomes stable. It happens to be the neutral H-pion $\tilde{\pi}^0$.

Besides, the numbers of doublet and singlet quarks are conserved in the model (47), because of the two global U(1) symmetry groups of the Lagrangian. This makes two H-baryon states stable—the neutral singlet H-baryon B and the lightest state in doublet $\mathscr{B}$, which carries a charge of $\pm 1/2$.

3.3. *Physics and Cosmology of Hypercolor SU(4) and SU(6) Models*

So, in the simplest case of zero hypercharge, it is possible to consider some experimentally observed consequences of SU(4) minimal model [80]. As it is seen from above, even in the minimal scenario of this type of hypercolor extension, there emerges a significant number of additional degrees of freedom. These new states, such as pNG or other hyperhadrons, can be detected in reactions at the collider at sufficiently high energies.

Readers should be reminded here of several papers that concern the formulation and construction of a vector-like hypercolor scheme [67,68]. In addition to the awareness of the original ideology, which made it possible to avoid known difficulties of Technicolor, main potentially observable consequences of this type of the SM extension were analyzed qualitatively and quantitatively [69,71,72,78]. Also in these articles, the possibilities of the hypercolor models for explaining the nature of DM particles were discussed in detail. Namely, this extension of the SM offers several different options as DM candidates with specific features and predictions. Some of these scenarios will be discussed in more detail below.

The vector-like hypercolor model contains two different scalar states with zero (or small) mixing, the Higgs boson and $\tilde{\sigma}$, and possibility to analyze quantitatively an effect of this mixing tends to zero. Then, we should hope to find some New Physics signals not in channels with the Higgs boson, but from production and decays of $\tilde{\sigma}$-meson (as shown by experimental data at the LHC, almost all predictions of the SM for the Higgs boson production cross sections in different channels as well as the widths of various modes of its decay are confirmed). Interestingly, the fluke two-photon signal at 750 GeV at the LHC seemed to indicate unambiguously decay of a scalar analog of the Higgs. If this were the case, the hyperpion mass in this model would have to be sufficiently small $\sim 10^2$ GeV due to the direct connection between $\tilde{\pi}$ and $\tilde{\sigma}$ masses. The condition of small mixing of scalars H and $\tilde{\sigma}$ in the conformal limit is $m_{\tilde{\sigma}} \approx \sqrt{3} m_{\tilde{\pi}}$ [68] and it means, in fact, that $\tilde{\sigma}$ is a pNG boson of conformal symmetry. Then, it should be close in mass to other pNG states. In this case, signals of formation and decays of charged and neutral (stable!) H-pions would be observed at the collider [70,80]. Nature, however, turned out to be more sophisticated.

To consider the phenomenological manifestations, we postulate a certain hierarchy of scales for numerous degrees of freedom in the model. Namely, the pNG bosons are the lightest in the spectrum of possible hyperhadrons, and the triplet of H-pions are the lightest states of pNG. This arrangement of the scales follows from the assumption that the apparent violation of the symmetry SU(4) is a small perturbation by analogy with the violation of the dynamic symmetry in the orthodox QCD scheme. There, the chiral symmetry is broken on a scale much larger than the mass scale of light quarks.

In the absence of new physics data from the LHC, we can use an estimate obtained on the assumption that the stable states in the model are dark matter candidates. In particular, the neutral H-pion $\tilde{\pi}^0$ and neutral hyperbaryon, B^0, can be such candidates. In this case, the analysis of the relic concentration of the dark matter makes it possible to estimate the range of masses of these particles. Thus, there is a natural mutual influence and collaboration of astrophysical and collider studies. So, in this scenario of the Standard Model extension, it becomes possible to identify DM particles with two representatives of the pNG states. For quantitative analysis, however, a more accurate consideration of the mass spectrum of the H-pion triplet and mass splitting between $\tilde{\pi}^0$ and hyperbaryon B^0 is necessary.

As for the mass splitting in the H-pion triplet, this parameter is defined by purely electroweak contributions [72,98] and is as follows:

$$\begin{aligned}\Delta m_{\tilde{\pi}} = \frac{G_F M_W^4}{2\sqrt{2}\pi^2 m_{\tilde{\pi}}} \Bigg[& \ln \frac{M_Z^2}{M_W^2} - \beta_Z^2 \ln \mu_Z + \beta_W^2 \ln \mu_W \\ & - \frac{4\beta_Z^3}{\sqrt{\mu_Z}} \left(\arctan \frac{2-\mu_Z}{2\sqrt{\mu_z}\beta_Z} + \arctan \frac{\sqrt{\mu_Z}}{2\beta_Z} \right) \\ & + \frac{4\beta_W^3}{\sqrt{\mu_W}} \left(\arctan \frac{2-\mu_W}{2\sqrt{\mu_W}\beta_W} + \arctan \frac{\sqrt{\mu_W}}{2\beta_W} \right) \Bigg]. \end{aligned} \quad (94)$$

Here, $\mu_V = M_V^2/m_{\tilde{\pi}}^2$, $\beta_V = \sqrt{1-\mu_V/4}$, and G_F denotes Fermi's constant. Taking the H-pion mass in a wide range 200–1500 GeV, from (94) we found the value $\Delta m_{\tilde{\pi}} \approx 0.162$–$0.170$ GeV.

Indeed, this small, non-zero and almost constant splitting of the mass in the triplet of the hyperpions obviously violates isotopic invariance. But at the same time, HG-parity remains a conserved quantum number. The reason is that the HG-parity is associated with a discrete symmetry, and not with a continuous transformation of the H-pion states. It is important to note that the inclusion of higher order corrections cannot destabilize a neutral weakly interacting H-pion, which is the lightest state in this pseudoscalar triplet. But charged H-pion states should decay by several channels.

In the strong channel, the width of the charged H-pion decay [80] can be written as

$$\Gamma(\tilde{\pi}^{\pm} \to \tilde{\pi}^{0}\pi^{\pm}) = \frac{G_F^2}{\pi} f_{\pi}^2 |U_{ud}|^2 m_{\tilde{\pi}}^{\pm} (\Delta m_{\tilde{\pi}})^2 \bar{\lambda}(m_{\pi^{\pm}}^2, m_{\tilde{\pi}^0}^2; m_{\tilde{\pi}^{\pm}}^2). \tag{95}$$

Here, $f_{\pi} = 132\,\mathrm{MeV}$ and $\pi^{\pm}$ denotes a standard pion. The reduced triangle function is defined as

$$\bar{\lambda}(a, b; c) = \left[1 - 2\frac{a+b}{c} + \frac{(a-b)^2}{c^2}\right]^{1/2}. \tag{96}$$

For the decay in the lepton channel we get:

$$\Gamma(\tilde{\pi}^{\pm} \to \tilde{\pi}^{0} l^{\pm} \nu_l) = \frac{G_F^2 m_{\tilde{\pi}^{\pm}}^3}{24\pi^3} \int_{q_1^2}^{q_2^2} \bar{\lambda}(q^2, m_{\tilde{\pi}^0}^2; m_{\tilde{\pi}^{\pm}}^2)^{3/2} \left(1 - \frac{3m_l^2}{2q^2} + \frac{m_l^6}{2q^6}\right) dq^2, \tag{97}$$

where $q_1^2 = m_l^2$, $q_2^2 = (\Delta m_{\tilde{\pi}})^2$, and m_l is a lepton mass.

Now, we can estimate decay widths of the charged H-pion and, correspondingly, lifetimes, and track lengths in these channels. To do this, we use (96), (97), and $\Delta m_{\tilde{\pi}}$ from (94) and get

$$\begin{aligned} \Gamma(\tilde{\pi}^{\pm} \to \tilde{\pi}^{0}\pi^{\pm}) &= 6 \cdot 10^{-17}\,\mathrm{GeV}, \quad \tau_{\pi} = 1.1 \cdot 10^{-8}\,\mathrm{s}, \quad c\tau_{\pi} \approx 330\,\mathrm{cm}; \\ \Gamma(\tilde{\pi}^{\pm} \to \tilde{\pi}^{0} l^{\pm} \nu_l) &= 3 \cdot 10^{-15}\,\mathrm{GeV}, \quad \tau_l = 2.2 \cdot 10^{-10}\,\mathrm{s}, \quad c\tau_l \approx 6.6\,\mathrm{cm}. \end{aligned} \tag{98}$$

Then, at TeV scale, characteristic manifestations of H-pions can be observed in the Drell–Yan type reactions due to the following fingerprints:

1. large $E_{T,mis}$ reaction due to production of stable $\tilde{\pi}^0$ and neutrino from $\tilde{\pi}^{\pm}$ and/or $W^{\pm}$ decays, or two leptons from charged H-pion and W decays (this is reaction of associated production, W, $\tilde{\pi}^{\pm}$, $\tilde{\pi}^0$ final state of the process);
2. large $E_{T,mis}$ due to creation of two stable $\tilde{\pi}^0$ and neutrino from decay of charged H-pions, $\tilde{\pi}^{\pm}$, one lepton from $\tilde{\pi}^{\pm}$ and two quark jets from $W^{\pm}$ decay (the same final state with particles W, $\tilde{\pi}^{\pm}$, $\tilde{\pi}^0$);
3. large $E_{T,mis}$ due to two stable neutral H-pions and neutrino from charged H-pion decay, two leptons (virtual Z, and $\tilde{\pi}^{+}$, $\tilde{\pi}^{-}$ in the final state);
4. large $E_{T,mis}$ due to two final neutral H-pions and neutrino from $\tilde{\pi}^{\pm}$, one lepton which originated from virtual W, $\tilde{\pi}^{+}$, $\tilde{\pi}^0$ final states.

Besides, H-pion signals can be seen due to two tagged jets in vector-boson-fusion channel in addition to main characteristics of the stable hyperpion—$E_{mis} \sim m_{\tilde{\pi}}$ and accompanying leptons.

Obviously, targeted search for such signals is possible only when we know, at least approximately, the range of hyperpion mass values. These estimates can be obtained by calculating the relic content of hidden mass in the Universe and comparing it with recent astrophysical data. Within the framework of the model, such calculations were made (see below). The possible values of the H-pion mass are in the range from 600–700 GeV to 1200–1400 GeV, while the naturally $\tilde{\sigma}$-meson is quite heavy—we recall that its mass is directly related to the masses of H-pions in the case of small H-$\tilde{\sigma}$ mixing. Cross sections of the reactions above (with large missed energy and momentum) are too small to be detected without special and careful analysis of specific events with predicted signature. The number of these events is also evidently small, and the signal can be hardly extracted from the background because there is a lot of events with decaying W-bosons and, correspondingly, neutrino or quark jets.

So, another interesting process to probe into the model of this type is the production and decay of a scalar H-meson $\tilde{\sigma}$ at the LHC; this production is possible at the tree level, however, the cross section is strongly damped due to small mixing. A small value of the mixing angle, θ_s, suppresses the cross section by an extra multiplier $\sin^2\theta_s$ in comparison with the standard Higgs boson production.

However, at one-loop level it is possible to get single and double H-sigma in the processes of vector-vector fusion. Namely, in $V^*V'^* \to \tilde{\sigma}, 2\tilde{\sigma}$ and/or in the decay through hyperquark triangle loop Δ, i.e., $V^* \to \Delta \to V'\tilde{\sigma}, 2\tilde{\sigma}$. Here, V^* and V' are intermediate or final vector bosons.

Now, a heavy H-sigma can decay via loops of hyperquarks or hyperpions $\tilde{\sigma} \to V_1V_2$, where $V_{1,2} = \gamma, Z, W$. Besides, the main decay modes of H-sigma are $\tilde{\sigma} \to \tilde{\pi}^0\tilde{\pi}^0,\ \tilde{\pi}^+\tilde{\pi}^-$; these are described by tree-level diagrams that predict large decay width for $m_{\tilde{\sigma}} \geqslant 2m_{\tilde{\pi}}$. As we will see below (from the DM relic abundance analysis), at some values of $\tilde{\pi}$ and $\tilde{\sigma}$ masses these channels are opened. In the small mixing limit, the width is

$$\Gamma(\tilde{\sigma} \to \tilde{\pi}\tilde{\pi}) = \frac{3u^2\lambda_{11}^2}{8\pi m_{\tilde{\sigma}}}\left(1 - \frac{4m_{\tilde{\pi}}^2}{m_{\tilde{\sigma}}^2}\right), \qquad (99)$$

and it depends strongly on the parameter λ_{11}.

An initial analysis of the model parameters was carried out in [68], using the value λ_{11} (it is denoted there as λ_{HC}) and u, from (99) we get: $\Gamma(\tilde{\sigma} \to \tilde{\pi}\tilde{\pi}) \gtrsim 10\,\text{GeV}$ when $m_{\tilde{\sigma}} \gtrsim 2m_{\tilde{\pi}}$.

The smallness of H–$\tilde{\sigma}$ mixing, as it is dictated by conformal approximation, results in the multiplier $\sin^2\theta_s$ for all tree-level squared amplitudes for decay widths. Then, for $\tilde{\sigma}$ decay widths we have

$$\begin{aligned}
\Gamma(\tilde{\sigma} \to f\bar{f}) =& \frac{g_W^2\sin^2\theta_s}{32\pi} m_{\tilde{\sigma}} \frac{m_f^2}{M_W^2}(1 - 4\frac{m_f^2}{m_{\tilde{\sigma}}^2})^{3/2}, \\
\Gamma(\tilde{\sigma} \to ZZ) =& \frac{g_W^2\sin^2\theta_s}{16\pi c_W^2}\frac{M_Z^2}{m_{\tilde{\sigma}}}(1 - 4\frac{m_Z^2}{m_{\tilde{\sigma}}^2})^{1/2}[1 + \frac{(m_{\tilde{\sigma}}^2 - 2M_Z^2)^2}{8M_Z^4}], \\
\Gamma(\tilde{\sigma} \to W^+W^-) =& \frac{g_W^2\sin^2\theta_s}{8\pi}\frac{M_W^2}{m_{\tilde{\sigma}}}(1 - 4\frac{m_W^2}{m_{\tilde{\sigma}}^2})^{1/2}[1 + \frac{(m_{\tilde{\sigma}}^2 - 2M_W^2)^2}{8M_W^4}].
\end{aligned} \qquad (100)$$

Here, m_f is a mass of standard fermion f and $c_W = \cos\theta_W$.

Recall that the two-photon decay of the Higgs boson is the very main channel in which the deviation of the experimental data from the predictions of the SM was originally found. Analogically, we consider a $\tilde{\sigma} \to \gamma\gamma$ decay which occurs through loops of heavy hyperquarks and H-pions; the width has the following form:

$$\Gamma(\tilde{\sigma} \to \gamma\gamma) = \frac{\alpha^2 m_{\tilde{\sigma}}}{16\pi^3}|F_Q + F_{\tilde{\pi}} + F_{\tilde{a}} + F_W + F_{\text{top}}|^2. \qquad (101)$$

Here, F_Q, $F_{\tilde{\pi}}$, F_W, and F_{top} are contributions from the H-quark, H-pion, W-boson, and top-quark loops; they can be presented as follows:

$$\begin{aligned}
F_Q &= -2\kappa\frac{M_Q}{m_{\tilde{\sigma}}}[1 + (1 - \tau_Q^{-1})f(\tau_Q)], \\
F_{\tilde{\pi}} &= \frac{g_{\tilde{\pi}\tilde{\sigma}}}{m_{\tilde{\sigma}}}[1 - \tau_{\tilde{\pi}}^{-1}f(\tau_{\tilde{\pi}})], \ g_{\tilde{\pi}\tilde{\sigma}} \approx u\lambda_{11}, \\
F_{\tilde{a}} &= \frac{g_{\tilde{a}\tilde{\sigma}}}{m_{\tilde{\sigma}}}[1 - \tau_{\tilde{a}}^{-1}f(\tau_{\tilde{a}})], \ g_{\tilde{a}\tilde{\sigma}} \approx u\lambda_{12}, \\
F_W &= -\frac{g_W\sin\theta_s m_{\tilde{\sigma}}}{8M_W}[2 + 3\tau_W^{-1} + 3\tau_W^{-1}(2 - \tau_W^{-1})f(\tau_W)], \\
F_{\text{top}} &= \frac{4}{3}\frac{g_W\sin\theta_s M_t^2}{m_{\tilde{\sigma}}M_W}[1 + (1 - \tau_t^{-1})f(\tau_t)],
\end{aligned} \qquad (102)$$

and

$$f(\tau) = \arcsin^2 \sqrt{\tau}, \;\; \tau < 1,$$
$$f(\tau) = -\frac{1}{4}\left[\ln\frac{1+\sqrt{1-\tau^{-1}}}{1-\sqrt{1-\tau^{-1}}} - i\pi\right]^2, \;\; \tau > 1. \tag{103}$$

As it is seen, contributions from W- and t-quark loops are induced by non-zero $\tilde{\sigma}$–H mixing. Taking necessary parameters from [68], the width is evaluated as $\Gamma(\tilde{\sigma} \to \gamma\gamma) \approx 5$–$10$ MeV.

Obviously, the process $p\bar{p} \to \tilde{\sigma} \to$ all should be analyzed quantitatively after integration of cross section of quark subprocess with partonic distribution functions. It is reasonable, however, to get an approximate value of the vector boson fusion cross section $VV \to \tilde{\sigma}(s) \to$ all, $V = \gamma, Z, W$.

The useful procedure to calculate the cross section with a suitable accuracy is the method of factorization [99]; this approach is simple and for the cross section estimation it suggests a clear recipe:

$$\sigma(VV \to \tilde{\sigma}(s)) = \frac{16\pi^2\Gamma(\tilde{\sigma}(s) \to VV)}{9\sqrt{s}\,\bar{\lambda}^2(M_V^2, M_V^2; s)}\,\rho_{\tilde{\sigma}}(s), \tag{104}$$

where $\tilde{\sigma}(s)$ is $\tilde{\sigma}$ in the intermediate state having energy $\sqrt{s}$. A partial decay width is denoted as $\Gamma(\tilde{\sigma}(s) \to VV)$. The density of probability, $\rho_{\tilde{\sigma}}(s)$, can be written as

$$\rho_{\tilde{\sigma}}(s) = \frac{1}{\pi}\frac{\sqrt{s}\,\Gamma_{\tilde{\sigma}}(s)}{(s - M_{\tilde{\sigma}}^2)^2 + s\,\Gamma_{\tilde{\sigma}}^2(s)}. \tag{105}$$

Here, $\Gamma_{\tilde{\sigma}}(s)$ is the total width of virtual $\tilde{\sigma}$-meson having $M_{\tilde{\sigma}} = \sqrt{s}$. At this energy we get exclusive cross section changing the numerator in (105), namely $\Gamma_{\tilde{\sigma}} \to \Gamma(\tilde{\sigma} \to V'V') = \Gamma_{\tilde{\sigma}} \cdot Br(\tilde{\sigma} \to V'V')$; for the cross section now we have

$$\sigma(VV \to \tilde{\sigma} \to V'V') = \frac{16\pi}{9}\frac{Br(\tilde{\sigma} \to VV)Br(\tilde{\sigma} \to V'V')}{m_{\tilde{\sigma}}^2(1 - 4M_V^2/m_{\tilde{\sigma}}^2)}$$
$$\approx \frac{16\pi}{9m_{\tilde{\sigma}}^2} \cdot Br(\tilde{\sigma} \to VV)Br(\tilde{\sigma} \to V'V'). \tag{106}$$

Now, when $M_{\tilde{\sigma}}^2 \gg M_V^2$ the cross section considered is determined by the branchings of H-sigma decay and the value of $M_{\tilde{\sigma}}$. Note, if $2m_{\tilde{\pi}} > M_{\tilde{\sigma}}$ H-sigma dominantly decays through following channels $\tilde{\sigma} \to WW, ZZ$. In this case, we get for $\tilde{\sigma}$ a narrow peak ($\Gamma \lesssim 10$–100 MeV).

As we said earlier, up to the present, there are no signals from the LHC about the existence of a heavy scalar state that mixes with the Higgs boson. Therefore, we are forced to estimate the mass of the H-sigma relying on astrophysical data on the DM concentration. Namely, we can consider H-pions as stable dark matter particles and then take into account the connection of their mass with the mass of $\tilde{\sigma}$-meson in the (almost) conformal limit.

Now, we should use the cross section which is averaged over energy resolution. As a result, the value of the cross section is reduced significantly. More exactly, for $2m_{\tilde{\pi}} < M_{\tilde{\sigma}}$ the dominant channel is $\tilde{\sigma} \to \tilde{\pi}\tilde{\pi}$ with a wide peak ($\Gamma \sim 10$ GeV). So, $Br(\tilde{\sigma} \to VV)$ is small and consequently the cross section of H-meson prodution is estimated as very small.

Thus, with a sufficiently heavy (with mass (2–3) TeV) second scalar meson, the main fingerprint of its emergence in the reaction is a wide peak induced by the strong decay $\tilde{\sigma} \to 2\tilde{\pi}$. It is accompanied by final states with two photons, leptons and quark jets originating from decays of WW, ZZ, and standard $\pi^{\pm}$. Besides, it occurs with some specific decay mode of $\tilde{\sigma}$ with two final stable $\tilde{\pi}^0$. This channel is specified by a large missed energy; charged final H-pions result in a signature with missed energy plus charged leptons.

As it was shown, existence of global U(1) hyperbaryon symmetry leads to the stability of the lightest neutral H-diquark. In the scenario considered, we suppose that charged H-diquark states decay to the neutral stable one and some other particles. Moreover, we also assume that these charged H-diquarks are sufficiently heavy, so their contributions into processes at (1–2) TeV are negligible [80].

Thus, having two different stable states—neutral H-pion and the lightest H-diquark with conserved H-baryon number—we can study a possibility to construct dark matter from these particles. This scenario with two-component dark matter is an immanent consequence of symmetry of this type of SM extension. Certainly, emergence of a set of pNG states together with heavy hyperhadrons needs careful and detailed analysis. At first stage, the mass splitting between stable components of the DM, not only H-pions, should be considered. Importantly, the model does not contain stable H-baryon participating into electroweak interactions. It means that any constraints for the DM relic concentration are absent for this case.

Note, $\tilde{\pi}^0$ and B^0 have the same tree level masses, so it is important to analyze the mass splitting $\Delta M_{B\tilde{\pi}} = m_{B^0} - m_{\tilde{\pi}^0}$. As it follows from calculations, this parameter depends on electroweak contributions only, all other (strong) diagrams are canceled mutually. Then, we get:

$$\begin{aligned}\Delta M_{B\tilde{\pi}} =& \frac{-g_2^2 m_{\tilde{\pi}}}{16\pi^2}\left[8\beta^2 - 1 - (4\beta^2-1)\ln\frac{m_{\tilde{\pi}}^2}{\mu^2} + 2\frac{M_W^2}{m_{\tilde{\pi}}^2}\left(\ln\frac{M_W^2}{\mu^2} - \beta^2\ln\frac{M_W^2}{m_{\tilde{\pi}}^2}\right)\right.\\ &\left. - 8\frac{M_W}{m_{\tilde{\pi}}}\beta^3\left(\arctan\frac{M_W}{2m_{\tilde{\pi}}\beta} + \arctan\frac{2m_{\tilde{\pi}}^2 - M_W^2}{2m_{\tilde{\pi}}M_W\beta}\right)\right],\end{aligned} \tag{107}$$

where $\beta = \sqrt{1 - \frac{M_W^2}{4m_{\tilde{\pi}}^2}}$. An important point is that $\Delta M_{B\tilde{\pi}}$ dependence on a renormalization point results from the coupling of the pNG states with H-quark currents of different structure at close but not the same energy scales. Thus, the dependence of the characteristics of the DM on the renormalization parameter is necessarily considered when analyzing the features of the DM model.

We remind that it is assumed that not-pNG H-hadrons (possible vector H-mesons, etc.) manifest itself at much more larger energies. It results from the smallness of the scale of explicit SU(4) symmetry breaking comparing with the scale of dynamical symmetry breaking. This hierarchy of scales copies the QCD construction.

Now, since the effects of hyperparticles at the collider are small, and an interesting scenario of a two-component DM arises, let us consider in more detail the possibility of describing the dark matter candidates in the framework of this model [100,101]. At the same time, we should note the importance of previous studies of the DM scenarios based on vector-like technicolor in the papers ref. [65,68,69,71,72,102–104]. Several quite optimistic versions of the DM description (including technineutrons, B-baryons, etc.) were considered, which, however, did not have a continuation, since they relied on a number of not quite reasonable assumptions—in particular, that B-baryons form the triplet.

When we turn to the hypercolor model, it will be necessary not only to calculate the total annihilation cross section for dark matter states, but to analyze the entire kinetics of freezing out of DM particles. The reason is that the mass splittings in the H-pions multiplet and between masses of two components are small (in the last case this parameter can be suggested as small). Then, the coupled system of five Boltzmann kinetic equations should be solved. Namely, two states of the neutral H-baryon, B^0, $\bar{B}^0$, neutral H-pion and also two charged H-pions should be considered. Such cumbersome kinetics are a consequence of proximity of masses of all particles participating in the process of formation of residual DM relic concentration. So, the co-annihilation processes [105] can contribute significantly to the cross section of annihilation. It had been shown also in previous vector-like scenario [72].

Now, for each component of the DM and co-annihilating particles, $i, j = \tilde{\pi}^+, \tilde{\pi}^-, \tilde{\pi}^0; \mu, \nu = B, \bar{B}$, we have the basic Boltzmann Equations (108) and (109) (neglecting reactions of type $iX \leftrightarrow jX$):

$$
\begin{aligned}
\frac{da^3 n_i}{a^3 dt} = & -\sum_j <\sigma v>_{ij} \left(n_i n_j - n_i^{eq} n_j^{eq}\right) - \sum_j \Gamma_{ij} \left(n_i - n_i^{eq}\right) - \\
& \sum_{j,\mu,\nu} <\sigma v>_{ij\to\mu\nu} \left(n_i n_j - \frac{n_i^{eq} n_j^{eq}}{n_\mu^{eq} n_\nu^{eq}} n_\mu n_\nu\right) + \\
& \sum_{j,\mu,\nu} <\sigma v>_{\mu\nu\to ij} \left(n_\mu n_\nu - \frac{n_\mu^{eq} n_\nu^{eq}}{n_i^{eq} n_j^{eq}} n_i n_j\right).
\end{aligned} \tag{108}
$$

We also get

$$
\begin{aligned}
\frac{da^3 n_\mu}{a^3 dt} = & -\sum_\nu <\sigma v>_{\mu\nu} \left(n_\mu n_\nu - n_\mu^{eq} n_\nu^{eq}\right) + \\
& \sum_{\nu,i,j} <\sigma v>_{ij\to\mu\nu} \left(n_i n_j - \frac{n_i^{eq} n_j^{eq}}{n_\mu^{eq} n_\nu^{eq}} n_\mu n_\nu\right) - \\
& \sum_{\nu,i,j} <\sigma v>_{\mu\nu\to ij} \left(n_\mu n_\nu - \frac{n_\mu^{eq} n_\nu^{eq}}{n_i^{eq} n_j^{eq}} n_i n_j\right),
\end{aligned} \tag{109}
$$

where

$$
\begin{aligned}
<\sigma v>_{ij} = & <\sigma v> (ij \to XX), \\
<\sigma v>_{ij\to\mu\nu} = & <\sigma v> (ij \to \mu\nu), \\
\Gamma_{ij} = & \Gamma(i \to jXX).
\end{aligned} \tag{110}
$$

Because of decays of charged H-pions, the main parameters in this calculation are total densities of $\tilde{\pi}$, B^0 and $\bar{B}^0$, namely, $n_{\tilde{\pi}} = \sum_i n_i$ and $n_B = \sum_\mu n_\mu$. Using an equilibrium density, n_{eq}, for describing co-annihilation, we estimate $n_i/n \approx n_i^{eq}/n^{eq}$. Then, the system of equations can be rewritten as

$$
\begin{aligned}
\frac{da^3 n_\pi}{a^3 dt} = & <\bar{\sigma v}>_{\tilde{\pi}} \left(n_{\tilde{\pi}}^2 - \left(n_{\tilde{\pi}}^{eq}\right)^2\right) - <\sigma v>_{\tilde{\pi}\tilde{\pi}} \left(n_{\tilde{\pi}}^2 - \frac{\left(n_{\tilde{\pi}}^{eq}\right)^2}{\left(n_B^{eq}\right)^2} n_B^2\right) + \\
& <\sigma v>_{BB} \left(n_B^2 - \frac{\left(n_B^{eq}\right)^2}{\left(n_{\tilde{\pi}}^{eq}\right)^2} n_{\tilde{\pi}}^2\right),
\end{aligned} \tag{111}
$$

$$
\begin{aligned}
\frac{da^3 n_B}{a^3 dt} = & <\bar{\sigma v}>_B \left(n_B^2 - \left(n_B^{eq}\right)^2\right) + <\sigma v>_{\tilde{\pi}\tilde{\pi}} \left(n_{\tilde{\pi}}^2 - \frac{\left(n_{\tilde{\pi}}^{eq}\right)^2}{\left(n_B^{eq}\right)^2} n_B^2\right) - \\
& <\sigma v>_{BB} \left(n_B^2 - \frac{\left(n_B^{eq}\right)^2}{\left(n_{\tilde{\pi}}^{eq}\right)^2} n_{\tilde{\pi}}^2\right),
\end{aligned} \tag{112}
$$

where

$$
\begin{aligned}
<\bar{\sigma v}>_{\tilde{\pi}} = & \frac{1}{9} \sum_{i,j} <\sigma v>_{ij}, \quad <\bar{\sigma v}>_B = \frac{1}{4} \sum_{\mu,\nu} <\sigma v>_{\mu\nu}, \\
<\sigma v>_{\tilde{\pi}\tilde{\pi}} = & \frac{1}{9}(<\sigma v> (\tilde{\pi}^0 \tilde{\pi}^0 \to B\bar{B}) + 2 <\sigma v> (\tilde{\pi}^+ \tilde{\pi}^- \to B\bar{B})), \\
<\sigma v>_{BB} = & \frac{1}{2}(<\sigma v> (B\bar{B} \to \tilde{\pi}^0 \tilde{\pi}^0) + <\sigma v> (B\bar{B} \to \tilde{\pi}^- \tilde{\pi}^+)).
\end{aligned} \tag{113}
$$

Now, it is reasonable to consider the ratio $m_{\tilde{\pi}}/M_B \approx 1$ using a suitable value of the renormalization parameter in the mass splitting between $m_{\tilde{\pi}^0}$ and M_{B^0}, more exactly, $\Delta M_{B^0\tilde{\pi}^0}/m_{\tilde{\pi}^0} \lesssim 0.02$. Then, the system of kinetic equations simplifies further. Having $n_B^{eq}/n_{\tilde{\pi}}^{eq} = 2/3$, we come to the following form of equations:

$$\frac{da^3 n_{\tilde{\pi}}}{a^3 dt} =< \bar{\sigma v} >_{\tilde{\pi}} \left(n_{\tilde{\pi}}^2 - \left(n_{\tilde{\pi}}^{eq}\right)^2\right) - < \sigma v >_{\tilde{\pi}\tilde{\pi}} \left(n_{\tilde{\pi}}^2 - \frac{9}{4} n_B^2\right) + \\ < \sigma v >_{BB} \left(n_B^2 - \frac{4}{9} n_{\tilde{\pi}}^2\right), \tag{114}$$

$$\frac{da^3 n_B}{a^3 dt} =< \bar{\sigma v} >_B \left(n_B^2 - \left(n_B^{eq}\right)^2\right) + < \sigma v >_{\tilde{\pi}\tilde{\pi}} \left(n_{\tilde{\pi}}^2 - \frac{9}{4} n_B^2\right) - \\ < \sigma v >_{BB} \left(n_B^2 - \frac{4}{9} n_{\tilde{\pi}}^2\right). \tag{115}$$

When masses of the DM components are close to each other, it is necessary to take into account the temperature dependence [105] in the cross sections $< \sigma v >_{\tilde{\pi}\tilde{\pi}}$ and $< \sigma v >_{BB}$ of the processes:

$$< \sigma v >_{BB} \approx < (a + b v^2) v > = \frac{2}{\sqrt{\pi x}} \left(a + \frac{8b}{x}\right), \tag{116}$$

where $x = m_{\tilde{\pi}}/T$, v is the relative velocity of final particles.

There are commonly used notations, which are convenient for solve the system, $Y = n/s$ and $x = m_{\tilde{\pi}}/T$, where s is the density of entropy. Then, neglecting small terms $\Delta M_{\tilde{\pi}}/M_B$ we have:

$$\frac{dY_{\pi}}{dx} = g(x,T) \left[\lambda_{\tilde{\pi}}((Y_{\tilde{\pi}}^{eq})^2 - Y_{\tilde{\pi}}^2) - \lambda_{\tilde{\pi}\tilde{\pi}} \left(Y_{\tilde{\pi}}^2 - \frac{9}{4} Y_B^2\right) + \lambda_{BB} \left(Y_B^2 - \frac{4}{9} Y_{\tilde{\pi}}^2\right)\right], \tag{117}$$

$$\frac{dY_B}{dx} = g(x,T) \left[\lambda_B((Y_B^{eq})^2 - Y_B^2) + \lambda_{\tilde{\pi}\tilde{\pi}} \left(Y_{\tilde{\pi}}^2 - \frac{9}{4} Y_B^2\right) - \lambda_{BB} \left(Y_B^2 - \frac{4}{9} Y_{\tilde{\pi}}^2\right)\right]. \tag{118}$$

The energy density is determined by a set of relativistic degrees of freedom, this function can be written in a convenient form as

$$g(x,T) = \frac{\sqrt{g(T)}}{x^2} \left\{1 + \frac{1}{3} \frac{d(\log g(T))}{d(\log T)}\right\} \\ \simeq \frac{115}{2} + \frac{75}{2} \tanh\left[2.2\left(\log_{10} T + 0.5\right)\right] + 10 \tanh\left[3\left(\log_{10} T - 1.65\right)\right]. \tag{119}$$

Here, we use an approximated value of this parameter, which works in the numerical solution with a good accuracy [101] and better than known approximation $g(T) \approx 100$; also, there are standard notations from [106]: $\lambda_i = 2.76 \times 10^{35} m_{\tilde{\pi}} < \sigma v >_i$, $Y_{\tilde{\pi}}^{eq} = 0.145(3/g(T)) x^{3/2} e^{-x}$, $Y_B^{eq} = 0.145(2/g(T)) x^{3/2} e^{-x}$,

The DM relic density, Ωh^2, is expressed in terms of relic abundance and critical mass density, ρ and ρ_{crit}:

$$\Omega h^2 = \frac{\rho}{\rho_{crit}} h^2 = \frac{m s_0 Y_0}{\rho_{crit}} h^2 \simeq 0.3 \times 10^9 \frac{m}{\text{GeV}} Y_0. \tag{120}$$

Present time values are denoted by the subscript "0".

After the replacement $W = \log Y$ [106], the system of kinetic equations is solved numerically. As it is shown in detail in [101], there is a set of regions in a plane of H-pion and H-sigma masses, where it is possible to get the value of the DM relic density in a correspondence with the modern astrophysical data. More exactly, the H-pion fraction is described by the following intervals: $0.1047 \leq \Omega h_{HP}^2 + \Omega h_{HB}^2 \leq 0.1228$ and $\Omega h_{HP}^2/(\Omega h_{HP}^2 + \Omega h_{HB}^2) \leq 0.25$). There are also some slightly different areas having all parameters nearly the same. However, H-pions make up just over a quarter of dark matter density, more exactly, $0.1047 \leq \Omega h^2_{HP} + \Omega h^2_{HB} \leq 0.1228$ and $0.25 \leq \Omega h^2_{HP}/(\Omega h^2_{HP} + \Omega h^2_{HB}) \leq 0.4$. Certainly, there are regions of parameters which are forbidden by restrictions by XENON collaboration [13,107,108].

It is important that there are no regions of parameters where the H-pion component dominates in the dark matter density. The reason is that this hyperpion component interacts with vector bosons, Z, W, at the tree level and, consequently, annihilates into ordinary particles much faster than stable B^0-baryons. The latter particles do not interact with standard vector bosons directly but only at loop level through H-quark and H-pion loops. It is a specific feature of SU(4) vector-like model having two stable pNG states.

At this stage of analysis (without an account of loop contributions from $B^0 - B^-$ annihilation), there are three allowable regions of parameters (masses):

Area 1: here $M_{\tilde{\sigma}} > 2m_{\tilde{\pi}^0}$ and $u \geq M_{\tilde{\sigma}}$; at small mixing, $s_\theta \ll 1$, and large mass of H-pions we get a reasonable value of the relic density and a significant H-pion fraction;

Area 2: here again $M_{\tilde{\sigma}} > 2m_{\tilde{\pi}^0}$ and $u \geq M_{\tilde{\sigma}}$ but $m_{\tilde{\pi}} \approx$ 300–600 GeV; H-pion fraction is small here, approximately, (10–15)%;

Area 3: $M_{\tilde{\sigma}} < 2m_{\tilde{\pi}}$—this region is possible for all values of parameters, but decay $\tilde{\sigma} \to \tilde{\pi}\tilde{\pi}$ is prohibited and two-photon signal from reaction $pp \to \tilde{\sigma} \to \gamma\gamma X$ would have to be visible at the LHC. Simultaneously, H-pion fraction can be sufficiently large, up to 40% for large $m_{\tilde{\pi}^0} \sim 1$ TeV and small angle of mixing.

Thus, from kinetics of two components of hidden mass it follows that the mass of these particles can vary in the interval (600–1000) GeV in agreement with recent data on the DM relic abundance. Having these values, it is possible to consider some manifestations of the hidden mass structure in the model.

Particularly, the inelastic interactions of high-energy cosmic rays with the DM particles can be interesting for studying the hidden mass distribution using signals of energetic leptons (neutrino) or photons which are produced in this scattering process [101].

Cosmic ray electrons can interact with the H-pion component via a weak boson in the process $e\tilde{\pi}^0 \to \nu_e\tilde{\pi}^-$, then charged $\tilde{\pi}^-$ will decay. In the narrow-width approximation we get for the cross section: $\sigma(e\tilde{\pi}^0 \to \nu_e\tilde{\pi}^0 l\nu_l') \approx \sigma((e\tilde{\pi}^0 \to \nu_e\tilde{\pi}^-) \cdot Br(\tilde{\pi}^- \to \tilde{\pi}^0 l\nu_l')$, branchings of charged hyperpion decay channels are: $Br(\tilde{\pi}^- \to \tilde{\pi}^0 e\nu_e') \approx 0.01$ and also $Br(\tilde{\pi}^- \to \tilde{\pi}^0\pi^-) \approx 0.99$.

Considering final charged hyperpion $\tilde{\pi}^-$ near its mass shell, standard light charged pion produces neutrino $e\nu_e$ and $\mu\nu_\mu$ with following probabilities: $\approx 1.2 \times 10^{-6}$ and ≈ 0.999, correspondingly.

Then, in this reaction, an energetic cosmic electron produces electronic neutrino due to vertex $We\nu_e$, and soft secondary $e'\nu_e'$ or $\mu\nu_\mu$ arise from charged H-pion decays. Now, there are final states with $Br(\tilde{\pi}^0\nu_e\mu'\nu_\mu') \approx 0.99$ and $Br(\tilde{\pi}^0\nu_e e'\nu_e') \approx 10^{-2}$. Obviously, we use here some simple estimations, they can be justified in the framework of the factorization approach [99]. Characteristic values of H-pion mass which can be used for the analysis are, for example, $m_{\tilde{\pi}^0} = 800$ GeV and 1200 GeV.

As it results from calculations, at initial electron energies in the interval $E_e =$ (100–1000) GeV the cross section of the process decreases from $O(10)$ nb up to $O(0.1)$ nb having maximum at small angles between electron and the neutrino emitted, i.e., inelastic neutrino production occurs in the forward direction (for more detail see figures in Ref. [101]). In this approximation, the energy of the neutrino produced is proportional to the energy of the incident electron and depends on the mass of the dark matter particle very weakly. The neutrino flux is calculated by integrating of spectrum, dN/dE_ν, this flux depends on H-pion mass very weakly. In the interval (50–350) GeV it decreases most steeply, and then, down to energies $\sim$1 TeV the fall is smoother.

Certainly, integrating the spectrum dN/dE_ν, we can estimate the number of neutrino landing on the surface of IceTop [109,110] which is approximately one squared kilometer.

Even taking into account some coefficients to amplify the DM density near the galaxy center for the symmetric Einasto profile, we have found the number of such neutrino events per year as very small, $N_\nu = (6\text{–}7)$, in comparison with the corresponding number of events for neutrino with energies in the multi-TeV region. Note, the Einasto profile modified in such manner reproduces well the hidden mass density value near Galaxy center in concordance with other DM profiles [111]. In any case, such small number of neutrino events at IceCube does not allow to study this interaction of cosmic rays with the DM effectively. Any other DM profile gives practically the same estimation of number of events for neutrino with these energies. Indeed, cross section of νN interaction for small neutrino energies $\sim(10^2\text{–}10^3)$ GeV is much lower than for neutrino energies $\sim(10^1\text{–}10^5)$ TeV. Consequently, all parameters of the signal detected, particularly, deposition of energy, intensity of Cherenkov emission, are noticeably worse. Because of the absence of good statistics of neutrino events, it is practically impossible to measure the neutrino spectrum of the predicted form. Probability of neutrino detection can be estimated in the concept of an effective area of the detector [110,112–115]). In our case this probability is small, $P = 10^{-10}\text{–}10^{-8}$ [101], so we need some additional factor that can increase the flux of neutrino substantially.

In principle, some factors amplifying these weak signals of cosmic electron scattering off the DM can be provided by inhomogeneities in the hidden mass distribution, i.e., so called clumps [116–121]. The scattering of cosmic rays off clusters of very high density [122] can result in amplifying neutrino flux substantially [123,124].

Though the scattering process suggested can be seen due to specific form of neutrino flux, the expected number of events is too small to be measured in experiments at modern neutrino observatories. The weakeness of the signal is also resulted from effective bremsstrahlung of electrons and the smallness of electron fraction in cosmic rays, $\sim$1%. Therefore, they are not so good probe for the DM structure; only if there are sharply non-homogeneous spatial distribution of hidden mass, the signal of production of energetic neutrino by cosmic electrons can be detected. It is an important reason to study inelastic scattering of cosmic protons, because they are more energetic and have a much larger flux.

Besides, an important information of the nature and profile of hidden mass should be manifested in a specific form of the DM annihilation gamma spectrum from clumps [125,126]. This signal can be significantly amplified due to increasing of density of hidden mass inside clumps, corresponding cross section depends on the squared density in contrary with the energy spectrum of final particles (neutrino, for example) which is resulted from scattering. In the last case, cross section is proportional to a first degree of the DM density.

Indeed, in the vector-like model with the two lowest in mass neutral stable states, one from these components does not participate in the scattering reaction with leptons, so the flux of final particles is diminished. There are, however, annihilation channels of both components into charged secondaries which emit photons. Some important contributions into this process describe so called virtual internal bremsshtrahlung (VIB). This part of photon spectrum containing information on the DM structure may be about 30%. Consequently, a feature of the DM structure in the model (particularly, existence of two components with different tree-level interactions) can result in some characteristic form of the annihilation diffuse spectra.

Introducing a parameter which determines H-pion fraction in the DM density,

$$\kappa = \frac{\Omega h^2_{HP}}{\Omega h^2_{HP} + \Omega h^2_{HB}}, \tag{121}$$

full annihilation spectrum is written as

$$\frac{d(\sigma v)}{dE_\gamma} E_\gamma^2 = \kappa^2 \frac{d(\sigma v_{\tilde{\pi}^0 \tilde{\pi}^0})}{dE_\gamma} E_\gamma^2 + (1-\kappa)^2 \frac{d(\sigma v_{B^0 \bar{B}^0})}{dE_\gamma} E_\gamma^2. \tag{122}$$

Contributions of the DM components to the total cross section of production of diffuse photons differ because of distinction in tree-level interaction with weak bosons. Annihilation of hyperpions into charged states (in particular, W-bosons) gives the most intensive part of diffuse photon flux. However, B^0-baryons can provide a significant fraction of this flux in some regions of the model parameters, namely, if the DM particles have mass $\approx$600 GeV. It should be noted also that σ-meson mass affects the cross section value changing it noticeably: from $-10\,\%$ to $+50\,\%$, approximately. This effect is seen better for the DM component mass $\approx$800 GeV when contribution from B^0 is not so prominent. Obviously, contributions to the gamma flux intensity from annihilation of different DM components contribute to the gamma flux in correspondence with the model content and structure. Thus, there appears a sign of the existence of two Dark matter components, observed in the form of a specific humped curve of the photon spectrum, due to virtual internal bremsstrahlung subprocesses. This effect should be considered in detail, because it is necessary to have much more astrophysics data together with an accurate analysis of all possible contributions to the spectrum for various regions of parameters. Certainly, because of high density of interacting particles, reactions of annihilation into photons in the DM clumps can be seen much better, and it also should be studied.

There is a set of possible observing consequences of the DM particle origin, structure and interactions produced by vector-like extension of the SM. Some of them have been analyzed quantitatively, while the analysis of others is still in progress. As there are plenty of additional heavy degrees of freedom in this model, they induce new effects that should be considered to predict observable and measurable phenomena. Moreover, the numerical estimations of model parameters and analysis of the effects above are based on some assumptions about spectrum of hyperhadrons. Particularly, it is suggested that charged di-hyperquark states, $B^{\pm}$, have masses which are much larger than neutral-state masses. It allows us to eliminate a lot of possible subprocesses with these particles and simplify substantially the system of kinetic equations for the DM components. This approach is quite reasonable.

Extension of the vector-like model symmetry, from SU(4) to SU(6), unambiguously results in a much larger number of additional H-hadrons which spawn a great quantity of new processes and effects. As noted above, there is an invariance of the model physical Lagrangian with respect to some additional symmetries, as a result of which we obtain a number of stable states and it is necessary to study their possible manifestations. Consideration of a new variant of the vector-like model of H-quarks is at the very beginning, therefore now we can define only some possible scenarios.

In the scenario with $Y = Y_S = 0$, two stable neutral states, H-pion, B^0 and also the lightest charged $B^{\pm}$ occur, as it is dictated by hyper-G-parity. In this case, we again have the opportunity to construct hidden mass from several components, as it was done in the previous version of SU(4) symmetry. However, a quantitative analysis of the mass difference for B-diquarks is necessary in order to assess the importance of the co-annihilation process for them. Assuming this mass splitting to be small, one can predict that the characteristics of a two-component DM in this scenario will not differ much from the previous version. Namely, we expect the masses of all dark matter components to be in the interval (0.8–1.2) TeV providing corresponding DM density.

Very interesting consequences follow from an occurrence of the stable charged state. First, the charged H-hadrons interact electromagnetically with cosmological plasma, so the hidden mass can be split from the plasma much later in comparison with the purely neutral DM. Second, there should be tree-level annihilation of these DM states into photons with an observable flux of specific form. Certainly, these conclusions make sense if relative concentration of the charged component is not small. Known data on the gamma spectrum from cosmic telescopes should help to establish necessary restrictions for the scenario parameters.

Moreover, the stable charged H-hadron can be seen in the collider experiments at corresponding energies. These heavy particles in the final states and neutral stable particles should be observed in the characteristic events with large missed energy. The cross sections of reactions and energies of detectable secondaries (hadronic jets and/or leptons) depend on the mass splitting between neutral and charged states.

The charged stable H-hadron should also be prominent in the scattering off the nuclei in underground experiments, we can expect that the corresponding cross section will be larger then in the stable neutral component scattering due to exchanges via vector bosons, not only through intermediate scalar mesons. However, known restrictions for measurable cross sections which follow from experiments at the underground setup will predict then more heavy stable particles in this model.

Note also that the next possible scenario with $Y_Q = 0$, $Y_S = \pm 1/2$ is less interesting for describing the DM properties because of the absence of stable states. (There is, possibly, a very special case with the one H-baryon state stable, the case should be considered separately, this work is in progress.).

Considering the physical Lagrangian for SU(6) vector-like extension, we find an important feature of the model: in this case there arise interactions of K-doublets and B-states with standard vector bosons (see the section above). These interactions can both amplify channels of new particles production at the collider and increase cross section of the DM annihilation and co-annihilation. Then, possible value of the DM components mass should also be larger to provide a suitable hidden mass density. In any case, these scenarios should be carefully analyzed before we can formulate a set of predictions for collider and astrophysical measurements. As it is seen, the vector-like extensions of the SM allow us to suggest some interesting scenarios with new stable heavy objects—H-hadrons—which can manifest itself both in events with large missed energy at the LHC and in astrophysical signals such as spectrum of photons and/or leptons from various sources in the Galaxy.

4. Dark Atom Physics and Cosmology

The approach of dark atoms, proposed and developed in [127–129], had followed the idea by Sh.L. Glashow [52] on dark matter species as electromagnetically bound systems of new stable charged particles. The potential danger for this approach is the possibility of overproduction of anomalous hydrogen, being a bound state of a heavy +1 charged particle with ordinary electrons. Hence +1 charged particles should be unstable to avoid such an overproduction. Moreover, primordial heavy stable -1 charged particles, being in deficit relative to primordial helium are all captured by primordial helium nuclei, forming a +1 chareged ion, as soon as helium is produced in the course of Big Bang nucleosynthesis [130]. Therefore charged dark atom constituents should have even charge, which is a double charge in the simplest case.

The abundance of particles with charge +2, bound with ordinary electrons, should be suppressed to satisfy the experimental upper limits on the anomalous helium. They should be either produced in deficit relative to the corresponding -2 charged particles [127,128,131], or there should be some special mechanism, suppressing the abundance of anomalous helium in the terrestrial matter [129].

These constraints distinguish negative even charged particles as possible constituents of dark atoms. Particles with charge -2 are captured by primordial helium and form O-helium dark atom. Particles X with even negative charge form X-nuclearites: the -4 charged capture two helium nuclei and form X-berillium (XBe), with the charge -6 capture three helium nuclei to form X-carbon (XC) and with the charge -8–X-oxygen (XO), in which four helium nuclei are bound with -8 charged particle. The existing examples of O or X particles exhibit their leptonic or lepton-like nature, and the properties and effects of the coresponding dark atoms are determined by their nuclear-interacting helium shells. It naturally puts OHe and X-nulearites in the list of hadronic dark matter candidates.

4.1. Dark Atoms Structure, Effects and Probes

General analysis of the bound states of massive negatively charged particle with nuclei was proposed in [132–134]. It assumed a simplified description of nuclei as homogeneously charged spheres and that the charged particle doesn't possess strong interaction. The structure of the corresponding bound state depends on the value of parameter $a = ZZ_o \alpha A m_p R$, where Z,

$$R \sim 1.2 A^{1/3}/(200\,\mathrm{MeV}) \tag{123}$$

and A are, respectively, charge, radius and atomic number of the nucleus. In the Equation (123) Z_o is the charge of particle, α is the fine structure constant and m_p stands for the proton mass. For $0 < a < 1$ the bound state looks like Bohr atom with negatively charged particle in the core and nucleus moving along the Bohr orbit. At $2 < a < \infty$ the bound states look like Thomson atoms, in which the body of nucleus oscillates around the heavy negatively charged particle (see e.g., [14]).

In the case of OHe $Z = 2$, $Z_o = 2$ and $a = ZZ_o\alpha Am_pR \leq 1$, which proves its Bohr-atom-like structure [14,127,128]. For point-like charge distribution in helium nucleus the OHe binding energy is given by

$$E_b = \frac{1}{2}Z^2Z_o^2\alpha^2Am_p \tag{124}$$

and the radius of Bohr orbit in this "atom" [14,127,128] is

$$r_o = \frac{1}{Z_oZ_{He}\alpha 4m_p} = 2\times 10^{-13}\,\mathrm{cm}, \tag{125}$$

being of the order of and even a bit smaller than the size of He nucleus. Therefore non-point-like charge distribution in He leads to a significant correction to the OHe binding energy.

For large nuclei or large particle charge, the system looks like Thomson atom with the particle inside the nuclear droplet. The binding energy can be estimated in this case with the use of harmonic oscillator approximation [14,132–134]

$$E_b = \frac{3}{2}(\frac{ZZ_o\alpha}{R} - \frac{1}{R}(\frac{ZZ_o\alpha}{Am_pR})^{1/2}). \tag{126}$$

In the approximation $R_{He} \approx r_o$ one can easily find from the Equation (126) that binding energy of He with X-particle with charge Z_o is given by

$$E_{He} = 2.4\,\mathrm{MeV}(1 - \frac{1}{Z_o^{1/2}})Z_o. \tag{127}$$

It gives $E_{He} = 4.8\,\mathrm{MeV}$ for X-berillium, 8.6 MeV for X-carbon and 12.8 MeV for X-oxygen.

X-nuclearites look similar to O-nuclearites—neutral bound states of heavy nuclei and multiple O^{--} particles, compensating nuclear charge [135]. However, X-nuclearites consist of a single multiple charged lepton-like particle bound with the corresponding number of helium nuclei and hence their structure needs special study.

4.2. Models of Stable Multiple Charged Particles

4.2.1. Double Cherged Stable Particles of Fourth Generation

The existence of the fourth sequential generation can follow from heterotic string phenomenology. Its quarks and leptons can possess a new conserved charge [14,127]. Conservation of this charge can provide stability of the lightest quark of the 4th generation [14,127]. If it is the U-quark, sphaleron transitions in the early Universe can establish excess of $\bar{U}$ antiquarks at the observed baryon asymmetry. Then $(\bar{U}\bar{U}\bar{U})$ with the charge -2 can be formed. It binds with ^4He in atom-like state of O-helium [127]. Origin of X-particles with larger charges seem highly unprobable in this model.

As we discussed above in Section 2 the experimental data puts constraints on possible deviation of 125 GeV Higgs boson from the predictions of the Standard model. It excludes the full strength coupling of this boson to fourth generation quarks and leptons. The suppression of these couplings implies some other nature of the mass of fourth generation e.g., due to another heavier Higgs boson.

4.2.2. Stable Charged Techniparticles in Walking Technicolor

In the lack of positive result of SUSY searches at the LHC the possibilities of non-supersymmetric solutions for the problems of the Standard model become of special interest. The minimal walking technicolor model (WTC) [14,104,136–140] proposes the composite nature of Higgs boson. In this approach divergence of Higgs boson mass is cut by the scale of technicolor confinement. This scale also determines the scale of the electroweak symmetry breaking. Possible extensions of the minimal WTC model to improve the correspondence of this approach to the recent LHC data are discussed in [141].

WTC involves two techniquarks,U and D. They transform under the adjoint representation of a SU(2) technicolor gauge group. A neutral techniquark–antiquark state is associated with the Higgs boson. Six bosons UU, UD, DD, and their antiparticles are technibaryons. If the technibaryon number TB is conserved, the lightest technibaryon should be stable.

Electric charges of UU, UD and DD are not fixed. They are given in general by $q+1$, q, and $q-1$, respectively. Here q is an arbitrary real number [14,128]. Compensation of anomalies requires in addition technileptons ν' and ζ that are technicolor singlets with charges $(1-3q)/2$ and $(-1-3q)/2$, respectively. Conservation of technilepton number L' provides stability of the lightest technilepton.

Owing to their nontrivial SU(2) electroweak charges techniparticles participate in sphaleron transitions in the early universe. Sphalerons support equilibrium relationship between TB, baryon number B, of lepton number L, and L'. When the rate of sphaleron transitions becomes smaller than the rate of expansion the excess of stable techniparticles is frozen out. It was shown in [128,131] that there is a balance between the excess of negatively charged particles over the corresponding positively charegd particles and the observed baryon asymmetry of the Universe. These negatively charged massive particles are bound in neutral atoms with primordial helium immediately after Big Bang nucleosynthesis.

In the case of $q = 1$ three possibilities were found for a dark atom scenario based on WTC [14,128,131]. If TB is conserved there can be excess of stable antitechnibaryons $\bar{U}\bar{U}$ with charge -2. If technilepton number L' is conserved, the excess of stable technilepton ζ with charge -2 is possible. In both cases, stable -2 charged particles can capture primordial $^4He^{++}$ nuclei and form O-helium atoms, dominating in the observed dark matter. If both TB and L' are conserved a two-component techni-O-helium dark matter scenario is possible.

Finally, the excessive technibaryons and technileptons can have opposite sign. Then two types of dark atoms, $(^4He^{++}\zeta^{--})$ and $(\zeta^{--}(UU)^{++})$, are possible. The former is nuclear interating O-helium, while the latter is weakly interacting and severely constrained by direct dark matter searches. Hence, WIMP-like $(\zeta^{--}(UU)^{++})$ is subdominant in this two-component scenario, while O-helium is the dominant component of dark matter.

In all the three cases it was shown that there are parameters of the model at which the techniparticle asymmetries have proper sign and value, explaining the O-helium dark matter density [128,131].

The case of multiple $-2n$ charged particles remains still unexplored for $n > 1$. We have marked bold possible multiple charged candidates for stable charged constituents of X-nuclearites in Table 3. The analysis of possible structures of corresponding X-nuclearites and their cosmological evolution and possible impact are now under way.

Table 3. List of possible integer charged techniparticles. Candidates for even charged constituents of dark atoms are marked bold.

q	$UU(q+1)$	$UD(q)$	$DD(q-1)$	$\nu'(\frac{1-3q}{2})$	$\zeta(\frac{-1-3q}{2})$
1	**2**	1	0	−1	**−2**
3	**4**	3	**2**	**−4**	−5
5	**6**	5	**4**	−7	**−8**
7	**8**	7	**6**	**−10**	−11

4.3. Effects of Hadronic Dark Matter

4.3.1. Cosmology of Hadronic Dark Matter

The considered BSM models make only a small step beyond the physics of the Standard Model and do not contain the physical basis for inflation and baryosynthesis that may provide some specific features of the cosmological scenario and mechanisms of generation of primordial density fluctuations, in particular. Therefore we assume in our cosmological scenario a standard picture of inflation and baryosynthesis with the adiabatic spectrum of density fluctuations, generated at the inflational stage. After the spectrum of fluctuations is generated, it causes density fluctuations within the cosmological horizon and their evolution depends on the matter/radiation content, equation of state and possible mechanisms of damping. The succession of steps to formation of our hadronic and hadron-like states in the early Universe needs special detailed study, but qualitatively it is similar to the evolution of tera-particles studied in [130].

One can divide possible forms of dark matter in hadronic and hadron-like models on two possible types. The case of new stable hadrons or composite dark matter like O-helium corresponds to SIMPs, while candidates without or with strongly suppressed QCD interaction are closer to WIMPs. In the latter case, the cosmological scenario should follow the main features of the Standard ΛCDM model with possible specifics related with the multicomponent WIMP-like candidates.

In the former case, SIMP interactions with plasma support thermal equilibrium with radiation at the radiation dominance (RD) stage. The radiation pressure acts on the plasma and then is transferred to the O-helium gas. It converts O-helium density fluctuations in acoustic waves, preventing their growth [14].

At temperature $T < T_{od} \approx 1 S_3^{2/3}$ keV SIMPs decouple from plasma and radiation, since [127,128]

$$n_B \langle \sigma v \rangle \, (m_p/m_o) t < 1. \tag{128}$$

Here m_o is the mass of the SIMP particle and we denote $S_3 = m_o/(1\,\mathrm{TeV})$. In the Equation (128) $v = \sqrt{2T/m_p}$ is the baryon thermal velocity. with the use of the analogy with OHe case we took according to [14,127,128]

$$\sigma \approx \sigma_o \sim \pi r_o^2 \approx 10^{-25}\,\mathrm{cm}^2, \tag{129}$$

where r_o is given by Equation (125). Then, SIMP gas decouples from plasma and plays the role of dark matter in formation of the large scale structure (LSS). At $t \sim 10^{12}$ s corresponding to $T \leq T_{RM} \approx 1$ eV, SIMPs start to dominate in the Universe, triggering the LSS formation. The details of the corresponding dark matter scenario are determined by the nature of SIMPs and need special study. Qualitatively, conversion in sound waves leads to suppression on the corresponding scales and the spectrum acquires the features of warmer than cold dark matter scenario [14,127,128]. Decoupled from baryonic matter SIMP gas doesn't follow formation of baryonic objects, forming dark matter halos of galaxies.

In spite of strong (hadronic) cross section SIMP gas is collisionless on the scale of galaxies, since its collision timescale is much larger than the age of the Universe. The baryonic matter is transparent for SIMPs at large scales. Indeed, $n\sigma R = 8 \times 10^{-5} \ll 1$ in a galaxy with mass $M = 10^{10} M_\odot$ and radius $R = 10^{23}$ cm. Here $n = M/4\pi R^3$ and $\sigma = 2 \times 10^{-25}\,\mathrm{cm}^2$ is taken as the geometrical cross section for SIMP collisions with baryons. Therefore, SIMPs should not follow baryonic matter in formation of the baryonic objects. SIMPs can be captured only by sufficiently dense matter proto-object clouds and objects, like planets and stars (see [135]).

4.3.2. Probes for Hadronic Dark Matter

In the charge symmetric case, SIMP collisions can lead to indirect effects of their annihilation, like in the case of WIMP annihilation first considered in [142], and contribute by its products to gamma background and cosmic rays. Effects of annihilation are not possible for asymmetric dark matter, but its inelastic collisions can produce cosmic particles and radiation. For example, OHe excitations in such collisions can result in pair production in the course of de-excitation and the estimated emission in positron annihilation line can explain the excess, observed by INTEGRAL in the galactic bulge [143]. The realistic estimation of the density of dark matter in the center of galaxy makes such explanation possible for O^{--} mass near 1.25 TeV [144].

In the two-component dark atom model, based on the walking technicolor, a subdominant WIMP-like component $UU\zeta$ is present, with metstable technibaryon UU, having charge +2. Decays of this technibaryon to the same sign (positive) lepton pairs can explain excess of high energy cosmic positrons observed by PAMELA and AMS02 [145]. However, any source of positrons inevitably is also the source of gamma radiation. Therefore the observed level of gamma background puts upper limit on the mass of UU, not exceeding 1 TeV [144].

These upper limits on the mass of stable double charged particles challenges their search at the LHC (see [144,146] for review and references).

Owing to their hadronic interaction, SIMP particles are captured by the Earth and slowed down in the terrestrial matter (see e.g., [14]). After thermalization they drift towards the center of the Earth with velocity

$$V = \frac{g}{n_A \sigma_{tr} v} \approx 200\, S_3 A_{med}^{-1/6} \frac{1\,\mathrm{g/\,cm^3}}{\rho}\ \mathrm{cm/\,s} \tag{130}$$

Here $A_{med} \sim 30$ is the average atomic weight in terrestrial surface matter. In the Equation (130)

$$n_A = \frac{\rho}{A_{med} m_p} = 6 \times 10^{23} / A_{med} \frac{\rho}{1\,\mathrm{g/\,cm^3}}$$

is the number density of terrestrial atomic nuclei and the transport geometrical cross section of collisions on matter nuclei with radius R, given by Equation (123), is given by

$$\sigma_{tr} = \pi R^2 \frac{A_{med} m_p}{m_o}.$$

We denote by m_o the mass of the SIMP particle, $S_3 = m_o/(1\,\mathrm{TeV})$, $v = \sqrt{2T/A_{med} m_p}$ as the thermal velocity of matter nuclei and $g = 980\ \mathrm{cm/\,s^2}$.

At a depth L below the Earth's surface, the drift timescale is $t_{dr} \sim L/V$. Here V is the drift velocity given by Equation (130). The incoming flux changes due to the orbital motion of the Earth. It should lead to the corresponding change in the equilibrium underground concentration of SIMPs. At the depth $L \sim 10^5$ cm the timescale of this change of SIMP concentration is given by

$$t_{dr} \approx 5 \times 10^2 A_{med}^{1/6} \frac{\rho}{1\,\mathrm{g/\,cm^3}} S_3^{-1}\ \mathrm{s}.$$

Thermalized due their elastic collisions with matter, SIMPs are too slow to cause in the underground detectors any significant effect of nuclear recoil, on which the strategy of direct WIMP searches is based. However, a specific type of inelastic processes, combined with annual modulation of SIMP concentration can explain positive results of DAMA/NaI and DAMA/LIBRA experiments [147–151] in their apparent contradiction with negative results of other experiments [152–156].

In the case of OHe such explanation was based on the existence of its 3 keV bound state with sodium nuclei [7,14,157]. Annual modulations in transitions to this state can explain positive results of the DAMA experiments. The rate of OHe radiative capture by a nucleus in a medium with temperature T is determined by electric dipole transition and given by [14,157]

$$\sigma v = \frac{f \pi \alpha}{m_p^2} \frac{3}{\sqrt{2}} (\frac{Z}{A})^2 \frac{T}{\sqrt{A m_p E}}. \tag{131}$$

Here A and Z are atomic number and charge of nucleus, E is the energy level. The factor $f = 1.410^{-3}$ accounts for violation of isospin symmetry in this electric dipole transition since He nucleus in *O*He is scalar and isoscalar. Since the rate of the *O*He radiative capture is proportional to the temperature (or to the product of mass of nucleus and square of relative velocity in the non-equilibrium case) the effect of such capture shpould be suppressed in cryogenic detectors. On the other hand, the existence of a low energy bound state was found in [157] only for intermediate mass nuclei and excluded for heavy nuclei, like xenon. It can explain the absence of the signal in such experiments as XENON100 [155] or LUX [156]. The confirmation of these results should follow from complete and self-consistent quantum mechanical description of *O*He interaction with nuclei, which still remains an open problem for the dark atom scenario.

4.4. Open Problems of Hadronic Dark Matter

In spite of uncertainty in the description of the interaction with matter of hypercolor motivated dark matter candidates, they are most probably similar to WIMPs and thus can hardly resolve the puzzles of direct dark matter searches.

One should note that in the simplest case hadronic dark matter can also hardly provide solution for these puzzles. Slowed down in the terrestrial matter SIMP ellastic collisions cannot cause significant nuclear recoil in the underground detectors, while inelastic nuclear processes should lead to energy release in the MeV range and cannot provide explanation for the signal detected by DAMA experiments in a few keV range. This puzzle may be resolved by some specifics of structure of SIMPs and their interaction with nuclei.

Such specifics was proposed in the dark atom model and the solution of a low energy bound state of the *O*He-nucleus system was found. This solution was based on the existence of a dipole repulsive barrier that arises due to *O*He polarization by the nuclear attraction of the approaching nucleus and provides a shallow potential well, in which the low energy level is possible for intermediate mass nuclei. However, the main open problem of the dark atom scenario is the lack of the correct quantum mechanical treatment of this feature of *O*He nuclear interaction [158]. In the essence, this difficulty lies in the necessity to take into account simultaneous effect of nuclear attraction and Coulomb repulsion in the absence of the usual simplifying conditions of the atomic physics (smallness of the ratio of the core and the shell as well as the possibility of perturbative treatment of the electromagnetic interaction of the electronic shell). In any case, strongly interacting dark matter cannot cross the matter from the opposite side of the Earth and it should inevitably lead to diurinal modulation of the *O*He concentration in the underground detector and the corresponding events. The role of this modulation needs special study for the conditions of the DAMA/NaI and DAMA/LIBRA experiments.

The lack of correct quantum mechanical treatment of *O*He nuclear physics also leaves open the question on the dominance of elastic collisions with the matter, on which *O*He scenario is based and on the possible role of inelastic processes in this scenario. Indeed, screening the electric charge of the α-particle, *O*He interactions after Big Bang nucleosynthesis can play a catalyzing role in production of primordial heavy elements, or influence stellar nucleosynthesis. Therefore the lack of developed *O*He nuclear physics prevents detailed analysis of possible role of *O*He in nucleosynthesis, stellar evolution and other astrophysical processes, as well as elaboration of the complete dark atom scenario.

The attractive feature of the dark atom model is the possibility to explain the excess of positron annihilation line emission, observed by INTEGRAL, and the excess of high energy fraction of cosmic positrons, detected by PAMELA and AMS02. These explanations are possible for double charged particles with the mass below 1.3 TeV, challenging the probe of the existence of its even-charged constituents in the direct searches at the LHC [146], in which effects of two-photon annihilation of bound multiple charged particles should be also taken into account [159].

5. Composite Dark Matter in the Context of Cosmo–Particle Physics

Observational cosmology offers strong evidence in favor of a new physics that challenges its discovery and thorough investigation. Cosmo–particle physics [1–3,6] elaborates methods to explore new forms of matter and their physical properties. Physics of dark matter plays an important role in this process and we make here a small step in the exploration of possible forms of this new physics.

We consider some scenarios of SM extensions with new strongly interacting heavy particles which are suggested as DM candidates (new stable hadrons, hyperhadrons and nuclear interacting dark atoms). It makes a minor step beyond the physics of the Standard model and doesn't provide mechanisms for inflation and baryosynthesis. But even such a modest step provides many new interesting physical and astrophysical phenomena.

Within the framework of the hadronic scenario, we analyzed the simplest extensions of the SM quark sector with new heavy quark in fundamental representation of color $SU_C(3)$ group. It was shown that the prediction of new heavy hadrons, which consist of new and standard quarks, does not contradict to cosmological constraints.

An appearance of new heavy quarks resulted from some additional vacuum symmetry breaking at a scale which is much larger than EW scale. Namely, in the mirror model (it is one of possible scenarios with new heavy fermions), there should be an extra Higgs doublet with v.e.v. $\sim(10^6–10^{10})$ GeV to provide masses for new fermions [22]. Analogously, an origin of heavy singlet quark as a consequence of the chain of transitions like $E(6) \to SO(10) \to SU(5)$ also stems from an additional Higgs doublet and corresponding symmetry breaking at a high scale. We suppose that both of these variants of heavy fermion emergence take place at the end of the inflation stage (or after it) due to some first Higgs transition. Then, massive non-stable states decay contributing to the quark-gluon plasma, while stable massive quarks go on interacting with photons. Later, after the second (electroweak) Higgs transition, remaining "light" quarks become massive, and (at the hadronic epoch) hadrons can be formed as bound states of the quarks. New "heavy-light" mesons also produced after the EW symmetry breaking. At the same time, we can just appeal to some special unknown dynamics which should break the symmetry at some higher scale. It may cause nontrivial features of cosmological scenario, which deserve further analysis.

There also arises a question: can the heavy dark matter occurrence substantially modify the distribution of the energy density produced by the acoustic waves at the inflation stage? As we suppose, the most intensive inflation sound waves had passed through the plasma before the first Higgs transition, an emergence of massive particles after the passing of the waves (when the inflation is finished) would only insufficiently change the density in already-formed areas with high energy density. In other words, this primary DM does not qualitatively change the whole pattern of the density distribution imprinting in CMB.

Here we show that the schemes with extensions of quark sector are in agreement with the precision electro-weak constraints on new physics effects. Using an effective Lagrangian of low-energy hadron interaction, we get the asymptotics of potential of interaction. In principle, both lighter mesons $M = (Q\bar{q})$ and fermions $B = (Qqq)$ can be stable.However, fermionic states can burn out in collisions with nucleons and new heavy mesons.

The main argument in favor of choosing mesons (two-quark pseudoscalars) as the DM candidates is that they have a repulsive potential for interactions with nucleons. This is due to the absence of one-pion exchanges, since the corresponding vertex is forbidden by parity conservation. Such a vertex for new fermions is not forbidden, and, at large distances, the potential of their interaction can be attractive. As a result, the new fermions can form bound states with nucleons, but not mesons (at low energies). In other words, at long distances between new heavy hadrons and also between them and nucleons a repulsive forces arise, and a potential barrier prevents the formation of bound states of new and standard hadrons at low energies. From the experimental limits on anomalous hydrogen and helium we can conclude that M^0 and $\bar{M}^0$ abundances are not equal, and this asymmetry is opposite to ordinary baryon asymmetry. An alternative scenario (symmetrical abundance of new quarks) can be permissible when new hadrons are superheavy (with mass $M > 10$ TeV).

As it was shown here, at the galaxy scale even hadronic DM behaves as collision-less gas. Therefore, both bosons and fermions should be distributed in the galactic halo. The difference between them can appear only after the DM capture by stars, when the difference in statistics can be important for a DM core inside the star. Indeed, type of the DM should influence on the formation and evolution of stars. Certainly, for scalars (with the repulsive interaction) and fermions (with an attractive potential at low energies), these processes are different. Discussion of these interesting problems is, however, beyond the scope of this paper.

Taking into account experimental data on masses of standard heavy-light mesons we evaluated the mass-splitting of charged M^- and neutral M^0 components and calculated width of charged meson which occurs large, so the charge component is long-lived, $\tau \gg 1\,s$. Starting from the relic concentration of dark matter and the expression for annihilation cross-section, the mass of the dark matter candidate was determined. The estimation of mass without SGS enhancement gives the value of mass near 20 TeV and an account of this effect increases this value up to 10^2 TeV. These estimations are in agreement with the evaluations of mass in the scenarios with baryonic DM, which are considered in literature. Thus, in the LHC experiments, superheavy new hadrons cannot be produced and directly detected in the nearest future. Moreover, it is difficult to observe superheavy hadrons when searching and studying an anomalous hydrogen and helium. However, as it was noted early, charged hadron M^- having a large lifetime can be directly detected in the process of M^0N scattering off energetic nucleons. In the calculation of annihilation cross-section we take into account some peculiarities of SGS effect. We note also that annihilation cross-section was considered here at the level of sub-processes. It means, to analyze features of the hadronic DM in more detail we need to clarify the annihilation mechanism.

We can conclude that the extensions of the SM with additional heavy fermions and vector-like interactions are perspective both from the theoretical point of view (they demonstrate an interesting structure of dynamical symmetry group and a wide spectrum of states) and an area for the checking of model predictions in collider experiments and in astrophysics.

We have also considered some particular representatives in the class of SM extensions with an additional hypercolor gauge group that confines a set of new vector-like fermions, H-quarks. If the hypercolor group is chosen as symplectic one, the global symmetry of the model is SU($2n_F$) (n_F is a number of H-quark flavors), which is larger than the chiral group and predicts consequently a spectrum of H-hadrons that contains not only heavy analogues of QCD hadrons, but also new states such as heavy diquarks (H-baryons). We have considered the cases of two and three H-flavors.

The analysis of the oblique parameters showed that in order to comply with the restrictions of precision data of the SM, it is necessary to fulfil certain conditions imposed on this type of extensions. For example, there are lower bounds on the masses of new particles arising in the framework of the model, and the mixing parameter of scalars, Higgs boson and H-sigma. The study of the symmetry properties led to the conclusion that two stable neutral objects exist in the SU(4) scenario (a more general SU(6) version with partially composite Higgs is being studied). Thus, the model predicts a two-component dark matter structure. The complexity of the DM has been discussed repeatedly [7,14,131,160–164], starting from the early works [165–167] but in the case of the vector-like model, the emergence of two components is a consequence of the symmetry of the model, and not an artificial assumption to explain some of the observed features in the measured spectra of cosmic photon or lepton fluxes. Such inclusion of various DM components in the unique theoretical framework can be a natural consequence of the realistic extensions of the SM symmetry, as it was shown in the gauge models of broken quark-lepton family symmetry [168,169].

Hoping for the detection of specific signals from new H-hadrons at the collider (when extending the energy interval and improving statistics), it is necessary also to deepen an analysis of the DM particle effects in astrophysics. It is especially important to study the channels of interaction of the neutral component of the H-diquark with ordinary fermions—these reactions define significantly of the observed features of leptonic and photonic spectra measured by space telescopes.

The started analysis of the DM scenarios in the framework of the SU(6) extension promises to be very interesting since the dark matter in this case may seem to contain another H-hadron component. Of course, in hyper-color models, the hyper-interaction itself is not studied yet, but its scale is noticeably higher than the achieved energies at the collider, and the corresponding loop contributions of hyperquarks, hypergluons and heavy H-hadrons are too small to noticeably affect the spectra observed in astrophysics. It seems that an analysis of high-energy cosmic rays scattering off the DM, the consideration of the annihilation of DM particles in various clump models, have good prospects. Analysis of photon and lepton signals will allow not only to assess the validity and possibilities of the proposed options for the SM extension, but also to understand how the DM multicomponent structure arising in them manifests itself in the observed data and gives an information on the DM distribution in the galaxy.

The dark atom scenario is based on the minimal extension of the SM content, involving only hypothetical stable even charged particles and reducing most of observable DM effects to the properties of the helium shell of OHe and its nuclear interactions. Such effects can provide nontrivial solutions for the puzzles of direct and indirect dark matter searches and it looks like this model can be made fully predictible on the basis of the known physics. However, the nontrivial features of OHe interaction with nuclei still leave an open problem a self-consistent quantitative analysis of the corresponding scenario.

To conclude, we have discussed a set of nontrivial dark matter candidates that follow from the BSM models, involving QCD color, hypercolor and technicolor physics. These predictions provide interesting combinations of collider, non-collider, astrophysical and cosmological signatures that can lead to thorough investigation of these models of new physics by the methods of cosmo–particle physics. We cannot expect that our models can give answers to all the problems of the physical basis of the Universe, but the observational fact that we live in the Universe full of unknowm forms of matter and energy stimulates our efforts to approach the mystery of their puzzling nature.

Author Contributions: The authors contributed equally to this work.

Funding: The work was supported by a grant of the Russian Science Foundation (Project No-18-12-00213).

Acknowledgments: Authors are sincerely grateful to M. Bezuglov for fruitful cooperation in scientific research and discussion of the results.

Conflicts of Interest: The authors declare no conflict of interest.

Appendix A. Algebras su(6) and sp(6)

The generators Λ_α, $\alpha = 1, \ldots, 35$ of SU(6) satisfy the following relations:

$$\Lambda_\alpha^\dagger = \Lambda_\alpha, \qquad \mathrm{Tr}\,\Lambda_\alpha = 0, \qquad \mathrm{Tr}\,\Lambda_\alpha \Lambda_\beta = \frac{1}{2}\delta_{\alpha\beta}. \tag{A1}$$

It is convenient for us to separate two subsets of the generators. The first one forms a sub-algebra $sp(6) \subset su(6)$—it includes matrices $\Sigma_{\dot\alpha}$, $\dot\alpha = 1, \ldots, 21$ satisfying a relation

$$\Sigma_{\dot\alpha}^T M_0 + M_0 \Sigma_{\dot\alpha} = 0, \tag{A2}$$

where the antisymmetric matrix M_0 is chosen as in Equation (56).

We choose these generators as follows:

$$\Sigma_a = \frac{1}{2\sqrt{2}}\begin{pmatrix} 0 & \tau_a & 0 \\ \tau_a & 0 & 0 \\ 0 & 0 & 0 \end{pmatrix}, \quad \Sigma_{3+a} = \frac{i}{2\sqrt{2}}\begin{pmatrix} 0 & \tau_a & 0 \\ -\tau_a & 0 & 0 \\ 0 & 0 & 0 \end{pmatrix}, \quad \Sigma_{6+a} = \frac{1}{2}\begin{pmatrix} 0 & 0 & 0 \\ 0 & 0 & 0 \\ 0 & 0 & \tau_a \end{pmatrix},$$

$$\Sigma_{9+a} = \frac{1}{2\sqrt{2}}\begin{pmatrix} \tau_a & 0 & 0 \\ 0 & \tau_a & 0 \\ 0 & 0 & 0 \end{pmatrix}, \quad \Sigma_{12+a} = \frac{1}{4}\begin{pmatrix} 0 & 0 & \tau_a \\ 0 & 0 & \tau_a \\ \tau_a & \tau_a & 0 \end{pmatrix}, \quad \Sigma_{15+a} = \frac{i}{4}\begin{pmatrix} 0 & 0 & \tau_a \\ 0 & 0 & -\tau_a \\ -\tau_a & \tau_a & 0 \end{pmatrix},$$

$$\Sigma_{19} = \frac{1}{2\sqrt{2}}\begin{pmatrix} 1 & 0 & 0 \\ 0 & -1 & 0 \\ 0 & 0 & 0 \end{pmatrix}, \quad \Sigma_{20} = \frac{1}{4}\begin{pmatrix} 0 & 0 & 1 \\ 0 & 0 & -1 \\ 1 & -1 & 0 \end{pmatrix}, \quad \Sigma_{21} = \frac{i}{4}\begin{pmatrix} 0 & 0 & 1 \\ 0 & 0 & 1 \\ -1 & -1 & 0 \end{pmatrix},$$

where τ_a, $a = 1, 2, 3$ are the Pauli matrices.

The remainder of the generators, β_α, $\alpha = 1, \ldots, 14$, belong to the coset SU(6)/Sp(6) and satisfy a relation

$$\beta_\alpha^T M_0 = M_0 \beta_\alpha. \tag{A3}$$

These can be defined as follows:

$$\beta_1 = \frac{1}{2\sqrt{2}}\begin{pmatrix} 0 & 1 & 0 \\ 1 & 0 & 0 \\ 0 & 0 & 0 \end{pmatrix}, \quad \beta_2 = \frac{i}{2\sqrt{2}}\begin{pmatrix} 0 & -1 & 0 \\ 1 & 0 & 0 \\ 0 & 0 & 0 \end{pmatrix}, \quad \beta_{2+a} = \frac{1}{2\sqrt{2}}\begin{pmatrix} \tau_a & 0 & 0 \\ 0 & -\tau_a & 0 \\ 0 & 0 & 0 \end{pmatrix},$$

$$\beta_6 = \frac{1}{2\sqrt{6}}\begin{pmatrix} 1 & 0 & 0 \\ 0 & 1 & 0 \\ 0 & 0 & -2 \end{pmatrix}, \quad \beta_{6+a} = \frac{1}{4}\begin{pmatrix} 0 & 0 & \tau_a \\ 0 & 0 & -\tau_a \\ \tau_a & -\tau_a & 0 \end{pmatrix}, \quad \beta_{10} = \frac{1}{4}\begin{pmatrix} 0 & 0 & 1 \\ 0 & 0 & 1 \\ 1 & 1 & 0 \end{pmatrix},$$

$$\beta_{10+a} = \frac{i}{4}\begin{pmatrix} 0 & 0 & \tau_a \\ 0 & 0 & \tau_a \\ -\tau_a & -\tau_a & 0 \end{pmatrix}, \quad \beta_{14} = \frac{i}{4}\begin{pmatrix} 0 & 0 & 1 \\ 0 & 0 & -1 \\ -1 & 1 & 0 \end{pmatrix}.$$

References

1. Sakharov, A.D. Cosmoparticle physics—A multidisciplinary science. *Vestnik AN SSSR* **1989**, *4*, 39–40.
2. Khlopov, M.Y. Fundamental cross-disciplinary studies of microworld and Universe. *Vestn. Russ. Acad. Sci.* **2001**, *71*, 1133-1137.
3. Khlopov, M.Y. *Cosmoparticle Physics*; World Scientific: New York, NY, USA; London, UK; Hong Kong, China; Singapore, 1999.
4. Khlopov, M.Y. Beyond the standard model in particle physics and cosmology: Convergence or divergence? In *Neutrinos and Explosive Events in the Universe*; Springer: Berlin, Germany, 2005; pp. 73–82.
5. Khlopov, M.Y. Cosmoparticle physics: Cross-disciplinary study of physics beyond the standard model. *Bled Workshops Phys.* **2006**, *7*, 51–62.
6. Khlopov, M.Y. *Fundamentals of Cosmoparticle Physics;* CISP-Springer: Cambridge, UK, 2012.
7. Khlopov, M.Y. Fundamental Particle Structure in the Cosmological Dark Matter. *Int. J. Mod. Phys. A* **2013**, *28*, 1330042. [CrossRef]
8. Bertone, G. *Particle Dark Matter: Observations, Models and Searches*; Cambridge Univiversity Press: Cambridge, UK, 2010.
9. Frenk, C.S.; White, S.D.M. Dark matter and cosmic structure. *Ann. Phys.* **2012**, *524*, 507–534. [CrossRef]
10. Gelmini, G.B. Search for Dark Matter. *Int. J. Mod. Phys. A* **2008**, *23*, 4273–4288. [CrossRef]
11. Aprile, E.; Profumo, S. Focus on dark matter and particle physics. *New J. Phys.* **2009**, *11*, 105002. [CrossRef]
12. Feng, J.L. Dark Matter Candidates from Particle Physics and Methods of Detection. *Ann. Rev. Astron. Astrophys.* **2010**, *48*, 495–545. [CrossRef]

13. Aprile, E.; Aprile, E.; Aalbers, J.; Agostini, F.; Alfonsi, M.; Amaro, F.D.; Anthony, M.; Arneodo, F.; Barrow, P.; Baudis, L.; Bauermeister, B.; et al. First Dark Matter Search Results from the XENON1 Experiment. *Phys. Rev. Lett.* **2017**, *119*, 181301. [CrossRef] [PubMed]
14. Khlopov, M. Cosmological Reflection of Particle Symmetry. *Symmetry* **2016**, *8*, 81. [CrossRef]
15. Maltoni, M.; Novikov, V.A.; Okun, L.B.; Rozanov, A.N.; Vysotsky, M.I. Extra quark-lepton generation and precision meazurements. *Phys. Lett. B* **2000**, *476*, 107–115. [CrossRef]
16. Belotsky, K.M.; Fargion, D.; Khlopov, M.Y.; Konoplich, R.V.; Ryskin, M.G.; Shibaev, K.I. Heavy hadrons of 4th family hidden in our Universe and close to detection. *Gravit. Cosmol.* **2005**, *11*, 3–15.
17. Khlopov, M.Y. Physics of dark matter in the light of dark atoms. *Mod. Phys. Lett. A.* **2011**, *26*, 2823–2839. [CrossRef]
18. Khlopov, M.Y. Introduction to the special issue on indirect dark matter searches. *Int. J. Mod. Phys. A* **2014**, *29*, 1443002. [CrossRef]
19. Cudell, J.R.; Khlopov, M. Dark atoms with nuclear shell: A status review. *Int. J. Mod. Phys. D* **2015**, *24*, 1545007. [CrossRef]
20. Buchkremer, M.; Schmidt, A. Long-lived Heavy Quarks: A Review. *Adv. High Energy Phys.* **2013**, *2013*, 690254. [CrossRef]
21. Luca, V.; Mitridate, A.; Redi, M.; Smirnov, J.; Strumia, A. Colored Dark Matter. *Phys. Rev. D* **2018**, *97*, 115024. [CrossRef]
22. Bazhutov, Y.N.; Vereshkov, G.M.; Kuksa, V.I. Experimental and Theoretical Premises of New Stable Hadron Existence. *Int. J. Mod. Phys. A* **2017**, *2*, 1759188. [CrossRef]
23. Pati, J.C.; Salam, A. Lepton number as the fourth colour. *Phys. Rev. D* **1974**, *10*, 275–289. [CrossRef]
24. Barger, V.; Deshpande, N.G.; Phillips, R.J.; Whisnant, K. Extra fermions in E_6 superstring models. *Phys. Rev. D* **1986**, *33*, 1902–1924. [CrossRef]
25. Angelopoulos, V.D.; Ellis, J.; Kowalski, H.; Nanopoulos, D.V.; Tracas, N.D.; Zwirner, F. Search for new quarks suggested by superstring. *Nucl. Phys. B* **1987**, *292*, 59–92. [CrossRef]
26. Langacker, P.; London, D. Mixing between ordinary and exotic fermions. *Phys. Rev. D* **1988**, *38*, 886–906. [CrossRef]
27. Beylin, V.A.; Vereshkov, G.M.; Kuksa, V.I. Mixing of singlet quark with standard ones and the properties of new mesons. *Phys. Atom. Nucl.* **1992**, *55*, 2186–2192.
28. Rattazzi, R. Phenomenological implications of a heavy isosinglet up-type quark. *Nucl. Phys. B* **1990**, *335*, 301–310. [CrossRef]
29. Beylin, V.; Kuksa, V. Dark Matter in the Standard Model Extension with Singlet Quark. *Adv. High Energy Phys.* **2018**, *2018*, 8670954. [CrossRef]
30. Khlopov, M.Y.; Shibaev, R.M. Probes for 4th Generation Constituents of Dark Atoms in Higgs Boson Studies at the LHC. *Adv. High Energy Phys.* **2014**, *2014*, 406458. [CrossRef]
31. Ilyin, V.A.; Maltoni, M.; Novikov, V.A.; Okun, L.B.; Rozanov, A.N.; Vysotsky, M.I. On the search for 50 GeV neutrinos. *Phys. Lett. B* **2001**, *503*, 126–132. [CrossRef]
32. Novikov, V.A.; Okun, L.B.; Rozanov, A.N.; Vysotsky, M.I. Extra generations and discrepancies of electroweak precision data. *Phys. Lett. B* **2002**, *529*, 111–116. [CrossRef]
33. Eberhardt, O.; Herbert, G.; Lacker, H.; Lenz, A.; Menzel, A.; Nierste, U.; Wiebusch, M. Impact of a Higgs Boson at a Mass of 126 GeV on the Standard Model with Three and Four Fermion Generations. *Phys. Rev. Lett.* **2012**, *109*, 241802. [CrossRef] [PubMed]
34. Botella, F.J.; Branco, G.C.; Nebot, M. The Hunt for New Physics in the Flavor Sector with up vector-like quarks. *J. High Energy Phys.* **2012**, *2012*, 40. [CrossRef]
35. Kumar Alok, A.; Banerjee, S.; Kumar, D.; Sankar, S.U.; London, D. New-physics signals of a model with vector singlet up-type quark. *Phys. Rev. D* **2015**, *92*, 013002. [CrossRef]
36. Llorente, J.; Nachman, B. Limits on new coloured fermions using precision data from Large Hadron Collider. *Nucl. Phys. B* **2018**, *936*, 106–117. [CrossRef]
37. Peskin, M.E.; Takeuchi, T. Estimations of oblique electroweak corrections. *Phys. Rev. D* **1992**, *46*, 381–409. [CrossRef]
38. Burgess, C.P.; Godfrey, S.; Konig, H.; London, D.; Maksymyk, I. A global Fit to Extended Oblique Parameters. *Phys. Lett. B* **1994**, *326*, 276–281. [CrossRef]

39. Tanabashi, M.; Hagiwara, K.; Hikasa, K.; Nakamura, K.; Sumino, Y.; Takahashi, F.; Tanaka, J.; Agashe, K.; Aielli, G.; Amsler, C. The Review of Particle Physics. *Phys. Rev. D* **2018**, *98*. 03001. [CrossRef]
40. Dover, C.B.; Gaisser, T.K.; Steigman, G. Cosmological constraints on new stable hadrons. *Phys. Rev. Lett.* **1979**, *42*, 1117–1120. [CrossRef]
41. Wolfram, S. Abundances of stable particles produced in the early universe. *Phys. Lett. B* **1979**, *82*, 65–68. [CrossRef]
42. Starkman, G.D.; Gould, A.; Esmailzadeh, R.; Dimopoulos, S. Opening the window on strongly interacting dark matter. *Phys. Rev. D* **1990**, *41*, 3594–3603. [CrossRef]
43. Javorsek, D.; Elmore, D.; Fischbach, E.; Granger, D.; Miller, T.; Oliver, D.; Teplitz, V. New experimental limits on strongly interacting massive particles at the TeV scale. *Phys. Rev. Lett.* **2001**, *87*, 231804. [CrossRef] [PubMed]
44. Mitra, S. Uranus's anomalously low excess heat constrains strongly interacting dark matter. *Phys. Rev. D* **2004**, *70*, 103517. [CrossRef]
45. Mack, G.D.; Beacom, J.F.; Bertone, G. Towards closing the window on strongly interacting dark matter: Far-reaching constraints from Earth's heat flow. *Phys. Rev. D* **2007**, *76*, 043523. [CrossRef]
46. Wandelt, B.D.; Dave, R.; Farrar, G.R.; McGuire, P.C.; Spergel, D.N.; Steinhardt, P.J. Self-Interacting Dark Matter. In Proceedings of the Fourth International Symposium Sources and Detection of Dark Matter and Dark Energy in the Universe, Marina del Rey, CA, USA, 23–25 February 2000; pp. 263–274.
47. McGuire, P.C.; Steinhardt, P.J. Cracking Open the Window for Strongly Interacting Massive Particles as the Halo Dark Matter. In Proceedings of the 27th International Cosmic Ray Conference, Hamburg, Germany, 7–15 August 2001; p. 1566.
48. Zaharijas, G.; Farrar, G.R. A window in the dark matter exclusion limits. *Phys. Rev. D* **2005**, *72*, 083502. [CrossRef]
49. McCammon, D.; Almy, R.; Deiker, S.; Morgenthaler, J.; Kelley, R.L.; Marshall, F.J.; Moseley, S.H.; Stahle, C.K; Szymkowiak, A.E. A sounding rocket payload for X-ray astronomy employing high-resolution microcalorimeters. *Nucl. Instrum. Methods A* **1996**, *370*, 266–268. [CrossRef]
50. McCammon, D.; Almy, R.; Apodaca, E; Tiest, W.B.; Cui, W.; Deiker, S.; Galeazzi, M.; Juda, M.; Lesser, A.; Miharaet, T. A high spectral resolution observation of the soft X-ray diffuse background with thermal detectors. *Astrophys. J.* **2002**, *576*, 188–203. [CrossRef]
51. Belotsky, K.; Khlopov, M.; Shibaev, K. Stable quarks of the 4th family? In *The Physics of Quarks: New Research*; Nova Science Publishers: New Your, NY, USA, 2008; p. 28.
52. Glashow, S.G. A Sinister Extension of the Standard Model to $SU(3) \times SU(2) \times SU(2) \times U(1)$. In Proceedings of the XI Workshop on Neutrino Telescopes, Venice, Italy, 22–25 February 2005; p. 9.
53. Smith, P.F.; Bennet, J.R.J.; Homer, G.J.; Lewin, J.D.; Walford, H.E.; Smith, W.A. A search for anomalous hydrogen in enriched D_2O, using a time-of-flight spectrometer. *Nucl. Phys. B* **1982**, *206*, 333–348. [CrossRef]
54. Muller, P.; Wang, L.B.; Holt, R.J.; Lu, Z.T.; O'Connor, T.P.; Schiffer, J.P. Search for Anomalously Heavy Isotopes of Helium in the Earth's Atmosphere. *Phys. Rev. Lett.* **2004**, *92*, 022501. [CrossRef]
55. Vereshkov, G.M.; Kuksa, V.I. $U(1) \times SU(3)$-gauge model of baryon-meson interactions. *Phys. Atom. Nucl.* **1991**, *54*, 1700–1704.
56. Sommerfeld, A. *Atombau und Spektrallinien*; F. Vieweg und Sohn: Brunswick, Germany, 1921.
57. Gamow, G. Zur Quantentheorie des Atomkernes. *Z. Phys.* **1928** *51*, 204–212. [CrossRef]
58. Sakharov, A.D. Interaction of an electron and positron in pair production. *Sov. Phys. Usp.* **1991**, *34*, 375–377. [CrossRef]
59. Belotsky, K.M.; Khlopov, M.Y.; Shibaev, K.I. Sakharov's enhancement in the effect of 4th generation neutrino. *Gravit. Cosmol.* **2000**, *6*, 140–150.
60. Arbuzov, A.B.; Kopylova, T.V. On relativization of the Sommerfeld-Gamow-Sakharov factor. *J. High Energy Phys.* **2012**, *2012*, 9. [CrossRef]
61. Feng, J.L.; Kaplinghat, M.; Yu, H.B. Sommerfeld Enhancement for Thermal Relic Dark Matter. *Phys. Rev. D* **2010**, *82*, 083526. [CrossRef]
62. Khoze, V.V.; Plascencia, A.D.; Sakurai, K. Simplified models of dark matter with a long-lived co-annihilation partner. *J. High Energy Phys.* **2017**, *2017*, 41. [CrossRef]
63. Asadi, P.; Baumgart, M.; Fitzpatric, P.J. Capture and decay of EW WIMPonium. *J. Cosmol. Astropart. Phys.* **2017**, *2017*, 5, [CrossRef]

64. Huo, R.; Matsumoto, S.; Tsai, J.L.S.; Yanagida, T.T. A scenario of heavy but visible baryonic dark matter. *J. High Energy Phys.* **2016**, *2016*, 162. [CrossRef]
65. Cirelli, M.; Strumia, A.; Tamburini, M. Cosmology and Astrophysics of Minimal Dark Matter. *Nucl. Phys. B* **2007**, *787* 152–175. [CrossRef]
66. Blum, K.; Sato, R.; Slatyer, T.R. Self-consistent calculation of the Sommerfeld enhancement. *J. Cosmol. Astropart. Phys.* **2016**, *2016*, 21. [CrossRef]
67. Kilic, C.; Okui, T.; Sundrum, R. vector-like Confinement at the LHC. *J. High Energy Phys.* **2010**, *2010*, 18. [CrossRef]
68. Pasechnik, R.; Beylin, V.; Kuksa, V.; Vereshkov, G. Chiral-symmetric technicolor with standard model Higgs boson. *Phys. Rev. D* **2013**, *88*, 075009. [CrossRef]
69. Pasechnik, R.; Beylin, V.; Kuksa, V.; Vereshkov, G. Vector-like technineutron Dark Matter: Is a QCD-type Technicolor ruled out by XENON100? *Eur. Phys. J. C* **2014**, *74*, 2728. [CrossRef]
70. Lebiedowicz, P.; Pasechnik, R.; Szczurek, A. Search for technipions in exclusive production of diphotons with large invariant masses at the LHC. *Nucl. Phys. B* **2014**, *881*, 288–308. [CrossRef]
71. Pasechnik, R.; Beylin, V.; Kuksa, V.; Vereshkov, G. Composite scalar Dark Matter from vector-like $SU(2)$ confinement. *Int. J. Mod. Phys. A* **2016**, *31*, 1650036. [CrossRef]
72. Beylin, V.; Bezuglov, M.; Kuksa, V. Analysis of scalar Dark Matter in a minimal vector-like Standard Model extension. *Int. J. Mod. Phys. A* **2017**, *32*, 1750042. [CrossRef]
73. Antipin, O.; Redi, M. The half-composite two Higgs doublet model and the relaxion. *J. High Energy Phys.* **2015**, *2015*, 1–16. [CrossRef]
74. Agugliaro, A.; Antipin, O.; Becciolini, D.; De Curtis, S.; Redi, M. UV-complete composite Higgs models. *Phys. Rev. D* **2017**, *95*, 035019. [CrossRef]
75. Barducci, D.; De Curtis, S.; Redi, M.; Tesi, A. An almost elementary Higgs: Theory and practice. *J. High Energy Phys.* **2018**, *2018*, 17. [CrossRef]
76. Antipin, O.; Redi, M.; Strumia, A. Dynamical generation of the weak and Dark Matter scales from strong interactions. *J. High Energy Phys.* **2015**, *2015*, 157. [CrossRef]
77. Mitridate, A.; Redi, M.; Smirnov, J.; Strumia, A. Dark matter as a weakly coupled dark baryon. *J. High Energy Phys.* **2017**, *2017*, 210. [CrossRef]
78. Appelquist, T.; Brower, R.C.; Buchoff, M.I.; Fleming, G.T.; Jin, X.Y.; Kiskis, J.; Kribs, G.D.; Neil, E.T.; Osborn, J.C.; Rebbi, C.; et al. Stealth dark matter: Dark scalar baryons through the Higgs portal. *Phys. Rev. D* **2015**, *92*, 075030. [CrossRef]
79. Cai, C.; Cacciapaglia, G.; Zhang, H.H. Vacuum alignment in a composite 2HDM. *J. High Energy Phys.* **2019**, *2019*, 130. [CrossRef]
80. Beylin, V.; Bezuglov, M.; Kuksa, V.; Volchanskiy, N. An analysis of a minimal vector-like extension of the Standard Model. *Adv. High Energy Phys.* **2017**, *2017*, 1765340. [CrossRef]
81. Low, I.; Skiba, W.; Tucker-Smith, D. Little Higgses from an antisymmetric condensate. *Phys. Rev. D* **2002**, *66*, 072001. [CrossRef]
82. Csaki, C.; Hubisz, J.; Kribs, G.D.; Meade, P.; Terning, J. Variations of little Higgs models and their electroweak constraints. *Phys. Rev. D* **2003**, *68*, 035009. [CrossRef]
83. Gregoire, T.; Tucker-Smith, D.; Wacker, J.G. What precision electroweak physics says about the SU(6)/Sp(6) little Higgs. *Phys. Rev. D* **2004**, *69*, 115008. [CrossRef]
84. Han, Z.; Skiba, W. Little Higgs models and electroweak measurements. *Phys. Rev. D* **2005**, *72*, 035005. [CrossRef]
85. Brown, T.; Frugiuele, C.; Gregoire, T. UV friendly T-parity in the SU(6)/Sp(6) little Higgs model. *J. High Energy Phys.* **2011**, *2011*, 108. [CrossRef]
86. Gopalakrishna, S.; Mukherjee, T.S.; Sadhukhan, S. Status and Prospects of the Two-Higgs-Doublet SU(6)/Sp(6) little-Higgs Model and the Alignment Limit. *Phys. Rev. D* **2016**, *94*, 015034. [CrossRef]
87. Pauli, W. On the conservation of the lepton charge. *Nuovo Cimento* **1957**, *6*, 204–215. [CrossRef]
88. Gürsey, F. Relation of charge independence and baryon conservation to Pauli's transformation. *Nuovo Cimento* **1958**, *7*, 411–415. [CrossRef]
89. Vysotskii, M.I.; Kogan, Y.I.; Shifman, M.A. Spontaneous breakdown of chiral symmetry for real fermions and the $N = 2$ supersymmetric Yang–Mills theory. *Sov. J. Nucl. Phys.* **1985**, *42*, 318–324.

90. Verbaarschot, J. The Supersymmetric Method in Random Matrix Theory and Applications to QCD. In *Proceedings of the 35th Latin American School of Physics on Supersymmetries in Physics and its Applications (ELAF 2004)*; Bijker, R., Castaños, O., Fernández, D., Morales-Técotl, H., Urrutia, L., Villarreal, C., Eds.; American Institute of Physics: Melville, NY, USA, 2004; Volume 744, pp. 277–362. [CrossRef]
91. Campbell, D.K. Partially conserved axial-vector current and model chiral field theories in nuclear physics. *Phys. Rev. C* **1979**, *19*, 1965–1970, [CrossRef]
92. Dmitrašinović, V.; Myhrer, F. Pion-nucleon scattering and the nucleon Σ term in an extended linear Σ model. *Phys. Rev. C* **2000**, *61*, 025205. [CrossRef]
93. Delorme, J.; Chanfray, G.; Ericson, M. Chiral Lagrangians and quark condensate in nuclei. *Nucl. Phys. A* **1996**, *603*, 239–256. [CrossRef]
94. Gasiorowicz, S.; Geffen, D.A. Effective Lagrangians and Field Algebras with Chiral Symmetry. *Rev. Mod. Phys.* **1969**, *41*, 531–573. [CrossRef]
95. Parganlija, D.; Kovács, P.; Wolf, G.; Giacosa, F.; Rischke, D.H. Meson vacuum phenomenology in a three-flavor linear sigma model with (axial-)vector mesons. *Phys. Rev. D* **2013**, *87*, 014011. [CrossRef]
96. Bai, Y.; Hill, R.J. Weakly interacting stable hidden sector pions. *Phys. Rev. D* **2010**, *82*, 111701. [CrossRef]
97. Antipin, O.; Redi, M.; Strumia, A.; Vigiani, E. Accidental Composite Dark Matter. *J. High Energy Phys.* **2015**, *2015*, 39. [CrossRef]
98. Ryttov, T.A.; Sannino, F. Ultra Minimal Technicolor and its DarkMatter TIMP. *Phys. Rev. D* **2006**, *78*, 115010. [CrossRef]
99. Kuksa, V.; Volchanskiy, N. Factorization in the model of unstable particles with continuous masses. *Eur. J. Phys.* **2013**, *11*, 182–194. [CrossRef]
100. Beylin, V.; Bezuglov, M.; Kuksa, V. Two-component Dark Matter in the vector-like hypercolor extension of the Standard Model. *EPJ Web Conf.* **2019**, *201*, 09001. [CrossRef]
101. Beylin, V.; Bezuglov, M.; Kuksa, V.; Tretyakov, E.; Yagozinskaya, A. On the scattering of a high-energy cosmic ray electrons off the dark matter. *Int. J. Mod. Phys. A* **2019**, *34*, 1950040. [CrossRef]
102. Foadi, R.; Frandsen, M.T.; Sannino, F. Technicolor dark matter. *Phys. Rev. D* **2009**, *80*, 037702. [CrossRef]
103. Frandsen, M.T.; Sannino, F. Isotriplet technicolor interacting massive particle as dark matter. *Phys. Rev. D* **2010**, *81*, 097704. [CrossRef]
104. Gudnason, S.B.; Kouvaris, C.; Sannino, F. Dark matter from new technicolor theories. *Phys. Rev. D* **2006**, *74*, 095008. [CrossRef]
105. Griest, K.; Seckel, D. Three exceptions in the calculation of relic abundances. *Phys. Lett. B* **1991**, *43*, 3191. [CrossRef]
106. Steigman, G.; Dasgupta, B.; Beacom, J.F. Precise relic WIMP abundance and its impact on searches for dark matter annihilation. *Phys. Rev. D* **2012**, *86*, 023506. [CrossRef]
107. Akerib, D.S.; Araújo, H.M.; Bai, X.; Bailey, A.J.; Balajthy, J.; Beltrame, P.; Bernard, E.P.; Bernstein, A.; Biesiadzinski, T.P.; Boulton, E.M.; et al. Results on the Spin-Dependent Scattering of Weakly Interacting Massive Particles on Nucleons from the Run 3 Data of the LUX Experiment. *Phys. Rev. Lett.* **2016**, *116*, 161302. [CrossRef]
108. Aalbers, J.; Agostini, F.; Alfonsi, M.; Amaro, F.D.; Amsler, C.; Aprile, E.; Arazi, L.; Arneodo, F.; Barrow, P.; Baudis, L.; et al. DARWIN: Towards the ultimate dark matter detector. *J. Cosmol. Astropart. Phys.* **2016**, *2016*, 17, [CrossRef]
109. Ishihara, A.; IceCube Collab. Extremely high energy neutrinos in six years ofIceCube data. *J. Phys. Conf. Ser.* **2016**, *718*, 062027, [CrossRef]
110. Aartsen, M.G.; IceCube Collab. Evidence for High-Energy Extraterrestrial Neutrinos at the IceCube Detector. *Science* **2013**, *342*, 1242856, [CrossRef]
111. Cirelli, M. Dark Matter Indirect searches: Phenomenological and theoretical aspects. *J. Phys. Conf. Ser.* **2013**, *447*, 012006. [CrossRef]
112. Chiarusi, T.; Spurio, M. High-Energy Astrophysics with Neutrino Telescopes. *Eur. Phys. J. C* **2010**, *65*, 649–701, [CrossRef]
113. Karle, A. (IceCube Collab). IceCube. In Proceedings of the 31st International Cosmic Rays Conference ICRC2009, Lodz, Pland, 7–15 July 2009.

114. Niederhausen, H.; Xu, Y. (IceCube Collab). High Energy Astrophysical Neutrino FluxMeasurement Using Neutrino-induced Cascades Observed in 4 Years of IceCube Data. In Proceedings of the 35th International Cosmic Rays Conference ICRC2017, Busan, Korea, 10–20 July 2017; p. 968.
115. Ahlers, M.; Helbing, K.; Perez de los Heros, C. Probing Particle Physics with IceCube. *Eur. Phys. J. C* **2018**, *78*, 924, [CrossRef]
116. Berezinsky, V.; Dokuchaev, V.; Eroshenko, Y. Small-scale clumps in the galactic halo and dark matter annihilation. *Phys. Rev. D* **2003**, *68*, 103003, [CrossRef]
117. Berezinsky, V.; Dokuchaev, V.; Eroshenko, Y. Remnants of dark matter clumps. *Phys. Rev. D* **2008**, *77*, 083519. [CrossRef]
118. Berezinsky, V.; Dokuchaev, V.; Eroshenko, Y. Formation and internal structure of superdense dark matter clumps and ultracompact minihaloes. *J. Cosmol. Astropart. Phys.* **2013**, *2013*, 59, [CrossRef]
119. Berezinsky, V.; Dokuchaev, V.; Eroshenko, Y. Small-scale clumps of dark matter. *Phys. Uspekhi* **2014**, *57*, 1–36, [CrossRef]
120. Erkal, D.; Koposov, S.; Belokurov, V. A sharper view of Pal 5's tails: Discovery of stream perturbations with a novel non-parametric technique. *Mon. Not. R. Astron. Soc.* **2017**, *470*, 60–84, [CrossRef]
121. Buckley, M.R.; Di Franzo, A. Collapsed Dark Matter Structures. *Phys. Rev. Lett.* **2018**, *120*, 051102, [CrossRef]
122. Colb, E.W.; Tkachev, I.I. Large-amplitude isothermal fluctuations and high-density dark-matter clumps. *Phys. Rev. D* **1994**, *50*, 769–773, [CrossRef]
123. Tasitsiomi, A.; Olinto, A.V. Detectability of neutralino clumps via atmospheric Cherenkov telescopes. *Phys. Rev. D* **2002**, *66*, 083006. [CrossRef]
124. Aloisio, R.; Blasi, P.; Olinto, A.V. Gamma-Ray Constraints on Neutralino Dark Matter Clumps in the Galactic Halo. *Astrophys. J.* **2004**, *601*, 47–53, [CrossRef]
125. Belotsky, K.; Kirillov, A.; Khlopov, M. Gamma-ray evidences of the dark matter clumps. *Gravit. Cosmol.* **2014**, *20*, 47–54, [CrossRef]
126. Belotsky, K.; Kirillov, A.; Khlopov, M. Gamma-ray effects of dark forces in dark matter clumps. *Adv. High Energy Phys.* **2014**, *2014*, 651247, [CrossRef]
127. Khlopov, M.Y. Composite dark matter from 4th generation. *JETP Lett.* **2006**, *83*, 1–4. [CrossRef]
128. Khlopov, M.Y.; Kouvaris, C. Strong interactive massive particles from a strong coupled theory. *Phys. Rev. D* **2008**, *77*, 065002. [CrossRef]
129. Fargion, D.; Khlopov, M.; Stephan, C. Cold dark matter by heavy double charged leptons? *Class. Quantum Grav.* **2006**, *23*, 7305.
130. Fargion, D.; Khlopov, M. Tera-leptons shadows over sinister Universe. *Gravit. Cosmol.* **2013**, *19*, 219–231. [CrossRef]
131. Khlopov, M.Y.; Kouvaris, C. Composite dark matter from a model with composite Higgs boson. *Phys. Rev. D* **2008**, *78*, 065040. [CrossRef]
132. Cahn, R.N.; Glashow, S. Chemical Signatures for Superheavy Elementary Particles. *Science* **1981**, *213*, 607–611. [CrossRef]
133. Pospelov, M. Particle Physics Catalysis of Thermal Big Bang Nucleosynthesis. *Phys. Rev. Lett.* **2007**, *98*, 231301. [CrossRef]
134. Kohri, K.; Takayama, F. Big bang nucleosynthesis with long-lived charged massive particles. *Phys. Rev. D* **2007**, *76*, 063507. [CrossRef]
135. Gani, V.A.; Khlopov, M.Y.; Voskresensky, D.N. Double charged heavy constituents of dark atoms and superheavy nuclear objects. *Phys. Rev. D* **2019**, *99*, 015024. [CrossRef]
136. Sannino, F.; Tuominen, K. Orientifold theory dynamics and symmetry breaking. *Phys. Rev. D* **2005**, *71*, 051901. [CrossRef]
137. Hong, D.K.; Hsu, S.D.H.; Sannino, F. Composite higgs from higher representations. *Phys. Lett. B* **2004**, *597*, 89–93. [CrossRef]
138. Dietrich, D.D.; Sannino, F.; Tuominen, K. Light composite higgs from higher representations versus electroweak precision measurements: Predictions for LHC. *Phys. Rev. D* **2005**, *72*, 055001. [CrossRef]
139. Dietrich, D.D.; Sannino, F.; Tuominen, K. Light composite higgs and precision electroweak measurements on the Z resonance: An update. *Phys. Rev. D* **2006**, *73*, 037701. [CrossRef]
140. Gudnason, S.B.; Kouvaris, C.; Sannino, F. Towards working technicolor: Effective theories and dark matter. *Phys. Rev. D* **2006**, *73*, 115003. [CrossRef]

141. Goertz, F. Composite Higgs theory. *Proc. Sci.* **2018**. [CrossRef]
142. Zeldovich, Y.B,; Klypin, A.A.; Khlopov, M.Y.; Chechetkin, V.M. Astrophysical bounds on the mass of heavy stable neutral leptons. *Sov. J. Nucl. Phys.* **1980**, *31*, 664–669.
143. Cudell, J.-R.; Khlopov, M.Y.; Wallemacq, Q. Dark atoms and the positron-annihilation-line excess in the galactic bulge. *Adv. High Energy Phys.* **2014**, *2014*, 869425. [CrossRef]
144. Khlopov, M. Direct and Indirect Probes for Composite Dark Matter. *Front. Phys.* **2019**, *7*, 4. [CrossRef]
145. Belotsky, K.; Khlopov, M.; Kouvaris, C.; Laletin, M. Decaying Dark Atom constituents and cosmic positron excess. *Adv. High Energy Phys.* **2014**, *2014*, 214258. [CrossRef]
146. Bulekov, O.V.; Khlopov, M.Y.; Romaniouk, A.S.; Smirnov, Y.S. Search for Double Charged Particles as Direct Test for Dark Atom Constituents. *Bled Workshops Phys.* **2017**, *18*, 11–24.
147. Bernabei, R.; Belli, P.; Cerulli, R.; Montecchia, F.; Amato, M.; Ignesti, G.; Incicchitti, A.; Prosperi, D.; Dai, C.; He, H.; et al. Search for WIMP annual modulation signature: Results from DAMA/NaI-3 and DAMA/NaI-4 and the global combined analysis. *Phys. Lett. B* **2000**, *480*, 23–31. [CrossRef]
148. Bernabei, R.; Belli, P.; Cappella, F.; Cerulli, R.; Montecchia, F.; Nozzoli, F.; Incicchitti, A.; Prosperi, D.; Dai, C.; Kuang, H.; et al. Dark matter search. *Riv. Nuovo Cim.* **2003**, *26*, 1–73. [CrossRef]
149. Bernabei, R.; Belli, P.; Cappella, F.; Cerulli, R.; Dai, C.; d'Angelo, A.; He, H.; Incicchitti, A.; Kuang, H.; Ma, J.; et al. First results from DAMA/LIBRA and the combined results with DAMA/NaI. *Eur. Phys. J. C* **2008** , *56*, 333–355. [CrossRef]
150. Bernabei, R.; Belli, P.; d'Angelo, S.; Di Marco, A.; Montecchia, F.; Cappella, F.; d'Angelo, A.; Incicchitti, A.; Caracciolo, V.; Castellano, S.; et al. Dark Matter investigation by DAMA at Gran Sasso. *Int. J. Mod. Phys. A* **2013**, *28*, 1330022. [CrossRef]
151. Bernabei, R.; Sheng, X.D.; Ma, X.H.; Cerulli, R.; Marco, A.D.; Dai, C.J.; Merlo, V.; Montecchia, F.; Incicchitti, A.; Belli, P.; et al. New Model independent Results from the First Six Full Annual Cycles of DAMA/LIBRA-Phase2. *Bled Workshops Phys.* **2018**, *19*, 27–57.
152. Abrams, D.; Akerib, D.; Barnes, P. Exclusion limits on the WIMP nucleon cross-section from the cryogenic dark matter search. *Phys. Rev. D* **2002**, *66*, 122003. [CrossRef]
153. Akerib, D.; Armel-Funkhouser, M.; Attisha, M.; Young, B. Exclusion limits on the WIMP-nucleon cross section from the first run of the cryogenic dark matter search in the soudan underground laboratory. *Phys. Rev. D* **2005**, *72*, 052009. [CrossRef]
154. Ahmed, Z.; Akerib, D.; Arrenberg, S. Search for weakly interacting massive particles with the first five-tower data from the cryogenic dark matter search at the soudan underground laboratory. *Phys. Rev. Lett.* **2009**, *102*, 011301. [CrossRef]
155. Aprile, E.; Arisaka, K.; Arneodo, F. First dark matter results from the XENON100 experiment. *Phys. Rev. Lett.* **2010**, *105*, 131302. [CrossRef]
156. Akerib, D.S.; Alsum, S.; Araújo, H.M.; Bai, X.; Balajthy, J.; Beltrame, P.; Bernard, E.P.; Bernstein, A.; Biesiadzinski, T.P.; Boulton, E.M.; et al. Search for annual and diurnal rate modulations in the LUX experiment. *Phys. Rev. D* **2018**, *98*, 062005. [CrossRef]
157. Khlopov, M.; Mayorov, A.; Soldatov, E. Dark Atoms of the Universe: Towards OHe nuclear physics. *Bled Workshops Phys.* **2010**, *11*, 73–88.
158. Cudell, J.-R.; Khlopov, M.Y.; Wallemacq, Q. The nuclear physics of OHe. *Bled Workshops Phys.* **2012**, *13*, 10.
159. Jäger, S.; Kvedarait, S.; Perez, G.; Savoray, I. Bounds and Prospects for Stable Multiply Charged Particles at the LHC. *arXiv* **2018**, arXiv:1812.03182.
160. Profumo, S.; Sigurdson, K.; Ubaldi, L. Can we discover multi-component WIMP dark matter? *J. Cosmol. Astropart. Phys.* **2009**, *2009*, 16, [CrossRef]
161. Aoki, M.; Duerr, M.; Kubo, J.; Takano, H. Multi-Component Dark Matter Systems and Their Observation Prospects. *Phys. Rev. D* **2012**, *86*, 076015, [CrossRef]
162. Esch, S.; Klasen, M.; Yaguna, C.E. A minimal model for two-component dark matter. *J. High Energy Phys.* **2014**, *2014*, 108, [CrossRef]
163. Bhattacharya, S.; Drozd, A.; Grzadkowski, B.; Wudka, J. Two-Component Dark Matter. *J. High Energy Phys.* **2013**, *2013*, 158, [CrossRef]
164. Ahmed, A.; Duch, M.; Grzadkowski, B.; Iglikli, M. Multi-component dark matter: The vector and fermion case. *Eur. Phys. J. C* **2018**, *78*, 905. [CrossRef]

165. Doroshkevich, A.G.; Khlopov, M.Y. On the physical nature of dark matter of the Universe. *Sov. J. Nucl. Phys.* **1984**, *39*, 551–553.
166. Doroshkevich, A.G.; Khlopov, M.Y. Formation of structure in the universe with unstable neutrinos. *Mon. Not. R. Astron. Soc.* **1984**, *211*, 279–282. [CrossRef]
167. Khlopov, M. Physical arguments, favouring multicomponent dark matter. In *Dark Matter in Cosmology, Clocks and Tests Of Fundamental Laws*; Guiderdoni, B., Greene, G., Hinds, D., Van Tran Thanh, J., Eds.; Frontieres: Gif-sur-Yvette, France, 1995; pp. 133–138.
168. Berezhiani, Z.; Khlopov, M.Y. Theory of broken gauge symmetry of families. *Sov. J. Nucl. Phys.* **1990**, *51*, 739–746.
169. Sakharov, A.S.; Khlopov, M.Y. Horizontal unification as the phenomenology of the theory of "everything". *Phys. Atom. Nucl.* **1994**, *57*, 651–658.

symmetry

MDPI

Article

New Holographic Dark Energy Model in Brans-Dicke Theory

M. Sharif [1], Syed Asif Ali Shah [1] and Kazuharu Bamba [2,*]

1 Department of Mathematics, University of the Punjab, Quaid-e-Azam Campus, Lahore 54590, Pakistan; msharif.math@pu.edu.pk (M.S.); asifalishah695@gmail.com (S.A.A.S.)

2 Division of Human Support System, Faculty of Symbiotic Systems Science, Fukushima University, Fukushima 960-1296, Japan

* Correspondence: bamba@sss.fukushima-u.ac.jp; Tel.:+81-24-503-3263

Received: 17 April 2018; Accepted: 4 May 2018; Published: 11 May 2018

Abstract: We study the cosmic evolution of the Bianchi type I universe by using new holographic dark energy model in the context of the Brans-Dicke theory for both non-interacting and interacting cases between dark energy and dark matter. We evaluate the equation of state for dark energy ω_D and draw the $\omega_D - \dot{\omega}_D$ plane, where the dot denotes the time derivative. It is found that a stage in which the cosmic expansion is accelerating can be realized in both cases. In addition, we investigate the stability of the model by analyzing the sound speed. As a result, it is demonstrated that for both cases, the behavior of the sound speed becomes unstable. Furthermore, with the Om-diagnostic tool, it is shown that the quintessence region of the universe can exist.

Keywords: brans-dicke theory; dark energy model; cosmological parameters

PACS: 04.50.kd; 95.36.+x

1. Introduction

In 1961, Brans and Dicke [1] proposed an alternative to General Relativity theory, by absorbing Mach's principle (which states that inertial forces experienced by a body in non-uniform motion are determined by the distribution of matter in the universe) into gravity, named as Brans-Dicke (BD) theory. In this theory, the dynamic of gravity is represented by a scalar field, while the metric tensor solely incorporates the spacetime structure. As a consequence, gravity couples with a time-dependent scalar field, $\psi(t)$, corresponding to the inverse of Newton's gravitational constant, $G(t)$, through a coupling parameter ω.

On the other hand, the holographic dark energy (HDE) approach appears to play a fundamental role in cosmic evolution (for recent reviews on the issue of dark energy and the theories of modified gravity, see, for instance, [2–8]). The holographic principle states that the number of degrees of freedom in a bounded system should be finite and associated to its boundary area. According to this principle, there is a theoretical relation between infrared and ultraviolet cutoffs. This model was originally proposed by Li [9] who used the basic concept of holographic principle in the background of Quantum Gravity. He concluded that, in a system having an ultraviolet cutoff and size L , the total energy should not be more than that of a black hole with the same size, leading to $L^3\rho_D \leq LM_p^2$ (where $M_p = (8\pi G_{eff})^{\frac{-1}{2}}$ and ρ_D represent the reduced Planck mass and the energy density of HDE, respectively). He also investigated three choices of L which are assumed to give an infrared cutoff. First, he assumed $L = H^{-1}$ introducing the most natural choice for the infrared cutoff in the formalism of HDE, but this choice could not illustrate the accelerated expansion of the universe in General Relativity [10] and BD gravity [11]. As a Second cutoff, the particle horizon radius was selected, which also failed to explain the current cosmic behavior. A future event horizon was the third choice, which eventually managed to yield the desired results. However, in this case, extra care is needed, since future singularities may lead to an undesirable phenomenology of the Universe (see, e.g., [12,13]).

More recently, Granda and Oliveros [14] considered a HDE model in which the energy density depends on the Hubble parameter and its derivative, being referred to as the new holographic dark energy (NHDE) model. Oliveros and Acero [15] studied this NHDE approach in the vicinity of the FRW metric with a non-linear interaction between DE and DM, and discussed the portrait of the equation of state (EoS) parameter ω_D, on the $\omega_D - \dot{\omega}_D$ plane, as well as its behavior with the aid of the Om-diagnostic tool. Fayaz et al. [16] investigated HDE model in a Bianchi Type I with (BI) universe within the context of the generalized teleparallel theory and found phantom/quintessence regions of the universe. Sadri and Vakili [17] studied the NHDE approach in BD theory using a FRW universe model with logarithmic scalar field, and analyzed the EoS/deceleration parameter, statefinder and Om-diagnostic tool for both non-interacting/interacting case. Jahromi et al. [18] studied the generalized entropy formalism and used a NHDE model to illustrate the evolution of the universe through its cosmological parameters.

The effects of anisotropy in the universe can be studied in the framework of an anisotropic BI model. Reddy et al. [19] analyzed homogeneous and axially symmetric BI models in BD theory and found that the deceleration parameter is negative, leading to an accelerated expansion of the universe. Setare [20] studied the HDE model with non-flat FRW metric in BD cosmology and found that the EoS parameter demonstrates a phantom-like region and crosses the phantom divide line. Kumar and Singh [21] used exact solutions describing BI cosmological models to study the cosmic evolution in a scalar-tensor theory. Setare and Vanegas [22] investigated an interacting HDE model and discussed cosmological implications. Sharif and Kausar [23,24] examined the dynamical behavior of a Bianchi universe with anisotropic fluid in $f(R)$ gravity. Sharif and Waheed [25] studied the evolution of a BI model in BD theory, using isotropic, anisotropic, as well as magnetized anisotropic fluid, and found that the latter may attain isotropy to the universe. Milan and Singh [26] discussed an HDE model with infrared cutoff as a future event horizon, as well as a logarithmic form of BD scalar field for the FRW universe in BD theory. Felegary et al. [27] studied the dynamics of an interacting HDE model in BD cosmology as regards the future event horizon cutoff, as well as its Hubble-horizon counterpart, and discussed the coincidence problem.

In this paper, we consider the NHDE approach for a BI universe and study the associated cosmic evolution in the background of BD theory. It should be noted that the present HDE model under consideration is corresponding to a kind of particular case of the investigations in [28] and its generalized considerations [29]. The outline of the paper is as follows. In Section 2, we study the NHDE model for non-interacting as well as in the interacting case and investigate the associated cosmological parameters. We also analyze the stability of the NHDE model through the corresponding sound speed. Section 3 deals with the Om-diagnostic tool, to study the cosmic evolution. Finally, we summarize our results in Section 4.

2. NHDE Model and BD Theory

The action for the BD theory is [1]

$$S = \int d^4x\sqrt{-g}\left(\frac{\omega}{\psi}g^{ij}\partial_i\psi\partial_j\psi - \psi R + L_m\right), \tag{1}$$

where R and L_m represent the Ricci scalar and matter Lagrangian density, respectively. The field equations for BD theory are

$$G_{ij} = \frac{1}{\psi}(T^{(m)}_{ij} + T^{(\psi)}_{ij}), \tag{2}$$

$$\Box\psi = \frac{T^{(m)}}{2\omega+3}, \tag{3}$$

where

$$T^{(m)}_{ij} = (\rho+p)u_i u_j - p g_{ij},$$

$$T^{(\psi)}_{ij} = \psi_{,i;j} - g_{ij}\Box\psi + \frac{\omega}{\psi}(\psi_{,i}\psi_{,j} - \frac{1}{2}g_{ij}\psi^{,\alpha}\psi_{,\alpha}), \quad T^{(m)} = g^{ij}T^{(m)}_{ij}.$$

Here, G_{ij} indicates the Einstein tensor, $\Box$ is the d'Alembertian operator while $T^{(m)}_{ij}$ and $T^{(\psi)}_{ij}$ are the energy-momentum tensors for matter distribution and scalar field, respectively. Equations (2) and (3) represent the field equations for BD theory and equation of evolution for the scalar field, respectively. We consider homogeneous and anisotropic locally rotationally symmetric BI universe model as

$$ds^2 = dt^2 - A^2(t)dx^2 - B^2(t)(dy^2 + dz^2), \tag{4}$$

where A and B indicate the scale factors in spatial directions.

The corresponding field equations for BI model are

$$\frac{2\dot{A}\dot{B}}{AB} + \frac{\dot{B}^2}{B^2} + \left(\frac{\dot{A}}{A} + 2\frac{\dot{B}}{B}\right)\frac{\dot{\psi}}{\psi} - \frac{\omega}{2}\frac{\dot{\psi}^2}{\psi^2} = \frac{\rho_m + \rho_D}{\psi}, \tag{5}$$

$$\frac{2\ddot{B}}{B} + \frac{\dot{B}^2}{B^2} + \frac{\ddot{\psi}}{\psi} + \frac{2\dot{B}}{B}\frac{\dot{\psi}}{\psi} + \frac{\omega}{2}\frac{\dot{\psi}^2}{\psi^2} = -\frac{p_D}{\psi}, \tag{6}$$

$$\frac{\ddot{B}}{B} + \frac{\dot{A}\dot{B}}{AB} + \frac{\ddot{A}}{A} + \left(\frac{\dot{A}}{A} + \frac{\dot{B}}{B}\right)\frac{\dot{\psi}}{\psi} + \frac{\ddot{\psi}}{\psi} + \frac{\omega}{2}\frac{\dot{\psi}^2}{\psi^2} = -\frac{p_D}{\psi}, \tag{7}$$

where dot represents derivative with respect to t. In the above equations, ρ_D and ρ_m indicate DE and DM energy densities, respectively, while p_D is the pressure of DE. For the scalar field ψ, the wave equation in (3) takes the form

$$\ddot{\psi} + \left(\frac{\dot{A}}{B} + 2\frac{\dot{B}}{B}\right)\dot{\psi} - \frac{\rho_m + \rho_D - 3p_D}{2\omega+3} = 0. \tag{8}$$

For the sake of simplicity, we take $A = B^m, m \neq 1$, consequently, Equations (5), (7) and (8) turn out to be

$$(2m+1)\frac{\dot{B}^2}{B^2} + (m+2)\frac{\dot{B}}{B}\frac{\dot{\psi}}{\psi} - \frac{\omega}{2}\frac{\dot{\psi}^2}{\psi^2} = \frac{\rho_m + \rho_D}{\psi}, \tag{9}$$

$$(m+1)\frac{\ddot{B}}{B} + m^2\frac{\dot{B}^2}{B^2} + (m+1)\frac{\dot{B}}{B}\frac{\dot{\psi}}{\psi} + \frac{\dot{B}}{B}\frac{\dot{\psi}}{\psi} + \frac{\ddot{\psi}}{\psi} + \frac{\omega}{2}\frac{\dot{\psi}^2}{\psi^2} = -\frac{p_D}{\psi}, \tag{10}$$

$$\ddot{\psi} + (m+2)\frac{\dot{B}}{B}\dot{\psi} - \frac{\rho_m + \rho_D - 3p_D}{2\omega+3} = 0. \tag{11}$$

Due to non-linear field equations, we suppose power-law model for the scalar field as $\psi(t) = \psi_0 B^{\alpha}$, $\alpha > 0$ and ψ_0 are constants. Subtracting Equation (10) from (6) and using power-law relation, we obtain a differential equation for the scale factor B as

$$\frac{\ddot{B}}{B} + (m+\alpha+1)\frac{\dot{B}^2}{B^2} = 0,$$

whose integration leads to

$$B(t) = (m+\alpha+2)^{\frac{1}{m+\alpha+2}}(c_1 t + c_2)^{\frac{1}{m+\alpha+2}}, \tag{12}$$

where c_1 and c_2 are integration constants. Consequently, we have

$$A(t) = ((m+\alpha+2)(c_1 t + c_2))^{\frac{m}{m+\alpha+2}}.$$

For our line element, the mean Hubble parameter is given as

$$H = \frac{1}{3}\left(\frac{\dot{A}}{A} + 2\frac{\dot{B}}{B}\right).$$

Using $A = B^m$ and Equation (12), the above equation yields

$$H = \frac{c_1(m+2)}{(m+\alpha+2)(c_1 t + c_2)}. \tag{13}$$

In the following, we discuss non-interacting and interacting cases of NHDE and investigate cosmological parameters graphically.

2.1. Non-Interacting Case

In this section, we discuss the case when DE and DM do not interact, the corresponding conservation equations are

$$\dot{\rho}_D + 3(1+\omega_D)\rho_D H = 0, \tag{14}$$

$$\dot{\rho}_m + 3\rho_m H = 0, \tag{15}$$

where $\omega_D = \frac{p_D}{\rho_D}$ is the EoS parameter through which we analyze different universe eras. The energy density of HDE model is defined as

$$\rho_D = 3n^2 M_p^2 L^{-2},$$

where n is a dimensionless constant. The energy density of NHDE model is given by

$$\rho_D = 3n^2 M_p^2 L^{-2}\left(1 - \frac{\epsilon L}{3r_c}\right), \tag{16}$$

here $r_c = \frac{M_p^2}{2M_5^3}$ is the crossover length scale while $\epsilon = \pm 1$ denotes self-accelerated and normal branches of solution. If $L \ll 3r_c$, the above energy density reduces to the energy density of HDE model. The fractional energy densities in their usual form are given as

$$\Omega_m = \frac{\rho_m}{\rho_{cr}} = \frac{4\omega\rho_m}{3\psi^2 H^2}, \tag{17}$$

$$\Omega_D = \frac{\rho_D}{\rho_{cr}} = c^2\left(1 - \frac{\epsilon}{3Hr_c}\right), \tag{18}$$

where $r_c = \frac{1}{2H\sqrt{\Omega_{rc}}}$. Taking time derivative of Equation (16), we obtain

$$\dot{\rho}_D = \rho_D\left(2\frac{\dot{\psi}}{\psi} + 2\frac{\dot{H}}{H}\right) + \frac{c^2\epsilon\psi^2\sqrt{\Omega_{rc}}\dot{H}H}{2\omega}. \tag{19}$$

Using Equation (13), we have

$$\frac{\dot{H}}{H^2} = \frac{-3(m+\alpha+2)}{m+2}. \tag{20}$$

Differentiating Equation (18) with respect to t, we find

$$\dot{\Omega}_D = \frac{2}{3}\epsilon c^2 \sqrt{\Omega_{rc}} \frac{\dot{H}}{H}.$$

Using Equations (14) and (19), the EoS parameter turns out to be

$$\omega_D = -1 - \frac{2\alpha}{m+2} + \left(\frac{\Omega_D + c^2}{\Omega_D}\right)\left(\frac{m+\alpha+2}{m+2}\right). \tag{21}$$

We fix the value of fractional density of DE as $\Omega_D = 0.73$ [30] while other parameters are fixed as $c = 0.8$, $m = -1.55$ and $\alpha > 0$. Using these values in the above equation, we see that $\omega_D < 0$ which corresponds to accelerated behavior of the universe. Caldwell and Linder [31] investigated that the quintessence model of DE can be separated into two distinct regions, i.e., thawing and freezing regions through $\omega_D - \dot{\omega}_D$ plane. The thawing region is characterized when $\dot{\omega}_D > 0, \omega_D < 0$ while the freezing region is determined for $\dot{\omega}_D < 0, \omega_D < 0$. Taking the time derivative of Equation (21), it follows that

$$\dot{\omega}_D = -\frac{2}{3}\epsilon\sqrt{\Omega_{rc}} \frac{c^4 c_1 (m+\alpha+2)}{\Omega_D^2 (c_1 t + c_2)(m+2)}. \tag{22}$$

In this scenario, we plot $\omega_D - \dot{\omega}_D$ plane for two values of integration constant c_2 (Figure 1). The left plot indicates that positive value of c_2 leads to $\dot{\omega}_D > 0, \omega_D < 0$ which corresponds to thawing region. The right graph is plotted for $c_2 = -10$ showing that negative value of c_2 yields freezing region for NHDE model.

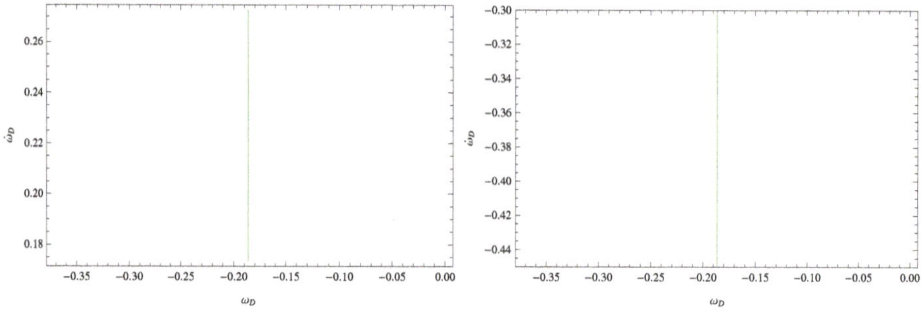

Figure 1. Plot of $\omega_D - \dot{\omega}_D$ plane with $c_2 = 5$ (**left**), $c_2 = -10$ (**right**), $c_1 = 1$, $c = 0.8$, $m = -1.55$ and $\alpha = 3.5$ for non-interacting case.

Now, we analyze stability of the NHDE model using squared speed of sound given as

$$v_s^2 = \frac{\dot{p}_D}{\dot{\rho}_D} = \omega_D + \dot{\omega}_D \frac{\rho_D}{\dot{\rho}_D}. \tag{23}$$

The model is unstable for $v_s^2(t) < 0$ while $v_s^2(t) > 0$ leads to stability. Using Equations (19), (21) and (22) in (23), it follows that

$$\begin{aligned} v_s^2 &= -1 - \tfrac{2\alpha}{m+2} + \left(\tfrac{\Omega_D + c^2}{\Omega_D}\right)\left(\tfrac{m+\alpha+2}{m+2}\right) \\ &- 2\epsilon\sqrt{\Omega_{rc}} \frac{c^4 c_1 (m+\alpha+2)((m+\alpha+2)(c_1 t + c_2))^{\frac{2\alpha}{m+\alpha+2}} (1 - \frac{2}{3}\epsilon\sqrt{\Omega_{rc}})}{\Omega_D^2 (c_1 t + c_2)(m+2)} \\ &\times \Bigg[3((m+\alpha+2)(c_1 t + c_2))^{\frac{2\alpha}{m+\alpha+2}} (1 - \tfrac{2}{3}\epsilon\sqrt{\Omega_{rc}}) \big(\tfrac{2\alpha c_1}{(m+\alpha+2)(c_1 t + c_2)} \\ &- \tfrac{2c_1}{c_1 t + c_2}\big) - 2c_1 \epsilon (m+\alpha+2)^{\frac{2}{m+\alpha+2}-3} (c_1 t + c_2)^{\frac{2}{m+\alpha+4}-3} \Bigg]^{-1}. \end{aligned} \tag{24}$$

The graph of $v_s^2(t)$ versus t is shown in Figure 2, where the unit of time t is taken as second. The change in free parameters does not affect the behavior of $v_s^2(t)$, so we show only one plot here. It is found that $v_s^2(t) < 0$, representing that our model is unstable.

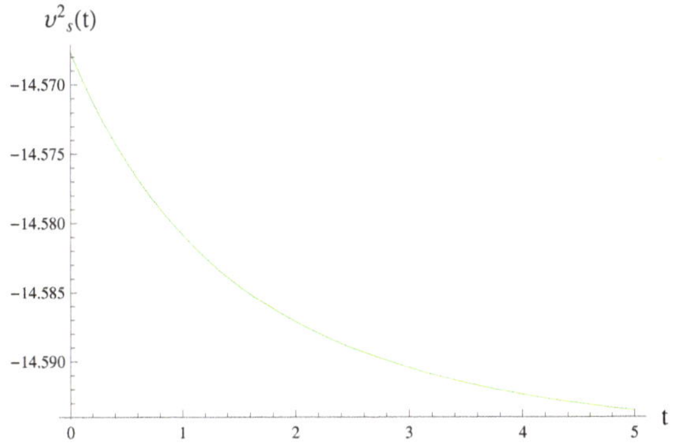

Figure 2. Plot of $v_s^2(t)$ versus t (the unit of which is second) with $c_1 = 1$, $c_2 = 5$, $c = 0.8$, $m = -1.55$ and $\alpha = 3.5$ for non-interacting case.

2.2. Interacting Case

Here we study the case when both dark components, i.e., DM and DE, interact with each other. In this case, the continuity equations are given by

$$\dot{\rho}_D + 3(1+\omega_D)\rho_D H = -\Gamma, \tag{25}$$

$$\dot{\rho}_m + 3\rho_m H = \Gamma, \tag{26}$$

where $\Gamma = 3b^2 H\rho_D$ is a particular interacting term with the interacting parameter b^2. Using Equations (20) and (25), the EoS parameter is given by

$$\omega_D = -1 - b^2 - \frac{2\alpha}{m+2} + \left(\frac{\Omega_D + c^2}{\Omega_D}\right)\left(\frac{m+\alpha+2}{m+2}\right). \tag{27}$$

In order to observe the behavior of the EoS parameter, we fix the constants Ω_D, m, c, α as for the previous case while the interacting parameter will be varied. We observe that ω_D exhibits similar behavior as in non-interacting case, i.e., it demonstrates accelerated behavior of the universe. Taking derivative of the above equation with respect to t, we have

$$\dot{\omega}_D = -\frac{2}{3}\epsilon\sqrt{\Omega_{rc}}\frac{c^4 c_1(m+\alpha+2)}{\Omega_D^2(c_1 t + c_2)(m+2)}. \tag{28}$$

Figure 3 shows the graph of $\omega_D - \dot{\omega}_D$ plane for two values of the interacting parameter. It is found that $\omega_D - \dot{\omega}_D$ plane corresponds to thawing and freezing regions for $c_2 = 5$ and $c_2 = -10$, respectively. Using Equations (19), (27) and (28) in (23), the sound speed parameter takes the form

$$
\begin{aligned}
v_s^2 &= -1-b^2-\tfrac{2\alpha}{m+2}+\left(\tfrac{\Omega_D+c^2}{\Omega_D}\right)\left(\tfrac{m+\alpha+2}{m+2}\right) \\
&- 2\epsilon\sqrt{\Omega_{rc}}\frac{c^4c_1(m+\alpha+2)((m+\alpha+2)(c_1t+c_2))^{\frac{2\alpha}{m+\alpha+2}}(1-\frac{2}{3}\epsilon\sqrt{\Omega_{rc}})}{\Omega_D^2(c_1t+c_2)(m+2)} \\
&\times \Bigg[3((m+\alpha+2)(c_1t+c_2))^{\frac{2\alpha}{m+\alpha+2}}(1-\tfrac{2}{3}\epsilon\sqrt{\Omega_{rc}})(\tfrac{2\alpha c_1}{(m+\alpha+2)(c_1t+c_2)} \\
&- \tfrac{2c_1}{c_1t+c_2})-2c_1\epsilon(m+\alpha+2)^{\frac{2}{m+\alpha+2}-3}(c_1t+c_2)^{\frac{2}{m+\alpha+4}-3}\Bigg]^{-1}.
\end{aligned} \tag{29}
$$

The graph of $v_s^2(t)$ versus t is plotted in Figure 4 which shows that $v_s^2(t) < 0$ demonstrating that our model is not stable. Here, the unit of time t is taken as second.

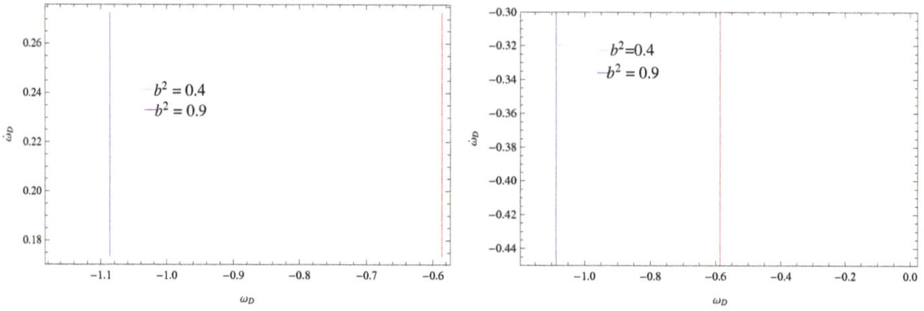

Figure 3. Plot of $\omega_D - \dot{\omega}_D$ plane with $c_2 = 5$ **(left)**, $c_2 = -10$ **(right)**, $c_1 = 1, c = 0.8,\ m = -1.55$ and $\alpha = 3.5$ for interacting case.

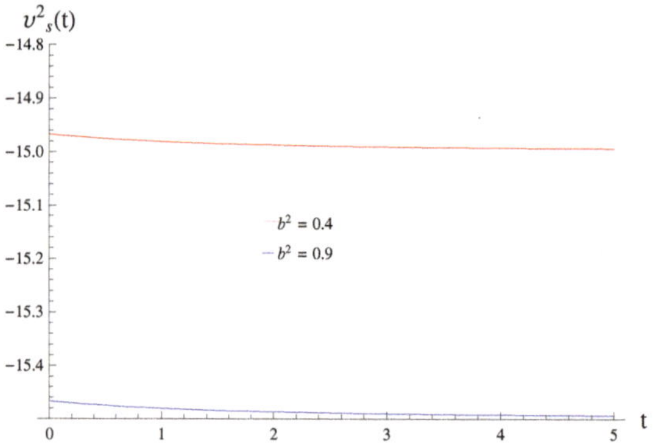

Figure 4. Plot of $v_s^2(t)$ versus t (the unit of which is second) with $c_1 = 1, c_2 = 5, c = 0.8,\ m = -1.55$ and $\alpha = 3.5$ for interacting case.

3. Om-Diagnostic

Here, we study different stages of the universe through the Om-diagnostic tool [32]. This helps to observe the behavior of the DE model and divides it into two sections. The positive values of Om(t) give phantom-like behavior and its negative values correspond to the quintessence region. The Om-diagnostic tool is defined as

$$
Om(t) = \frac{h^2(t)-1}{t^3-1}, \tag{30}
$$

where $h(t) = \frac{H(t)}{H_0}$, H_0 is the Hubble constant. Using Equation (13), the above equation becomes

$$Om(t) = \frac{\frac{(m+2)^2 c_1^2}{9H_0^2((m+\alpha+2)(c_1 t+c_2))^2} - 1}{t^3 - 1}. \tag{31}$$

We see that the Om-diagnostic tool attains negative values in the range $1.02 \le t \le 5$ which shows quintessence behavior of the universe (Figure 5, in which the unit of time t is taken as second).

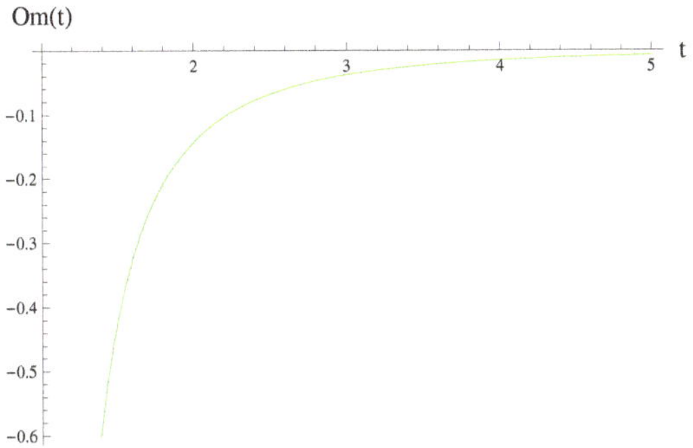

Figure 5. Plot of Om(t) versus t (the unit of which is second) with $c_1 = 1, c_2 = 5, m = -1.55, \alpha = 3.5$ and $H_0 = 68$.

4. Conclusions

In this paper, we investigate the NHDE approach in a BI cosmological model, to discuss the expanding behavior of the universe in the framework of BD theory. For this purpose, the cosmological parameters are evaluated in two scenarios. First, we consider the case where DM and DE do not interact with each other. In this case, the EoS parameter is negative, leading to a universe that experiences accelerated expansion. We have also analyzed its behavior on the $\omega_D - \dot{\omega}_D$ plane, which indicates that our model lies either in the thawing or in the freezing region, corresponding to a positive or a negative value of the associated integration constant, respectively. Furthermore, the stability of the NHDE model is investigated, using the corresponding speed of sound parameter. It is found that this parameter attains negative values, leading to unstable cosmological models.

Second, the interaction between DM and DE is taken into account. In this case, the same parameters as before are formulated yielding that the EoS remains negative for two values of the interaction parameter, b^2, i.e., our universe is in an expanding phase. It is also found that the $\omega_D - \dot{\omega}_D$ plane analysis demonstrates similar behavior for both values of b^2, i.e., our model remains in the thawing and freezing regions. Furthermore, we conclude that such a model does not exhibit stable behavior. Finally, cosmological evolution is discussed also with the aid of the Om-diagnostic tool, which, in our case, exhibits quintessence behavior of the universe. It is quite interesting to mention that our results are consistent with the corresponding isotropic universe model [17].

Author Contributions: M.S. and S.A.A.S. have mainly executed mathematical calculations and consdered the constructions of the paper. K.B. has adjusted and edited the paper and suggested the investitations on the subject of Holographic Dark Energy.

Acknowledgments: The work of KB is partially supported by the JSPS KAKENHI Grant Number JP 25800136 and Competitive Research Funds for Fukushima University Faculty (17RI017).

Conflicts of Interest: The authors declare no conflicts of interest.

References

1. Brans, C.; Dicke, R.H. Mach's principle and a relativistic theory of gravitation. *Phys. Rev.* **1961**, *124*, 925. [CrossRef]
2. Nojiri, S.I.; Odintsov, S.D. Unified cosmic history in modified gravity: From F (R) theory to Lorentz non-invariant models. *Phys. Rep.* **2011**, *505*, 59–144. [CrossRef]
3. Capozziello, S.; De Laurentis, M. Extended theories of gravity. *Phys. Rep.* **2011**, *509*, 167–321. [CrossRef]
4. Nojiri, S.; Odintsov, S.D.; Oikonomou, V.K. Modified gravity theories on a nutshell: Inflation, bounce and late-time evolution. *Phys. Rep.* **2017**, *692*, 1–104. [CrossRef]
5. Capozziello, S.; Faraoni, V. Beyond Einstein gravity: A Survey of gravitational theories for cosmology and astrophysics. *Fundam. Theor. Phys.* **2010**, *170*. [CrossRef]
6. Bamba, K.; Odintsov, S.D. Inflationary cosmology in modified gravity theories. *Symmetry* **2015**, *7*, 220–240. [CrossRef]
7. Bamba, K.; Capozziello, S.; Nojiri, S.I.; Odintsov, S.D. Dark energy cosmology: The equivalent description via different theoretical models and cosmography tests. *Astrophys. Space Sci.* **2012**, *342*, 155–228. [CrossRef]
8. Kleidis, K.; Spyrou, N.K. Dark Energy: The Shadowy Reflection of Dark Matter? *Entropy* **2016**, *18*, 94. [CrossRef]
9. Li, M. A model of holographic dark energy. *Phys. Lett. B* **2004**, *603*, 1–5. [CrossRef]
10. Hsu, SD. Entropy bounds and dark energy. *Phys. Lett. B* **2004**, *594*, 13–16. [CrossRef]
11. Xu, L.; Li, W.; Lu, J. Holographic dark energy in Brans–Dicke theory. *Eur. Phys. J. C* **2009**, *60*, 135–140. [CrossRef]
12. Nojiri, S.I.; Odintsov, S.D.; Tsujikawa, S. Properties of singularities in the (phantom) dark energy universe. *Phys. Rev. D* **2005**, *71*. [CrossRef]
13. Kleidis, K.; Oikonomou, V.K. Effects of finite-time singularities on gravitational waves. *Astrophys. Space Sci.* **2016**, *361*, 326. [CrossRef]
14. Granda, L.N.; Oliveros, A. Infrared cut-off proposal for the holographic density. *Phys. Lett. B* **2008**, *669*, 275–277. [CrossRef]
15. Oliveros, A.; Acero, M.A. New holographic dark energy model with non-linear interaction. *Astrophys. Space Sci.* **2015**, *357*, 12. [CrossRef]
16. Fayaz, V.; Hossienkhani, H.; Pasqua, A.; Amirabadi, M.; Ganji, M. f (T) theories from holographic dark energy models within Bianchi type I universe. *Eur. Phys. J. Plus* **2015**, *130*, 28. [CrossRef]
17. Sadri, E.; Vakili, B. A new holographic dark energy model in Brans-Dicke theory with logarithmic scalar field. *Astrophys. Space Sci.* **2018**, *363*, 13. [CrossRef]
18. Jahromi, A.S.; Moosavi, S.A.; Moradpour, H.; Graça, J.M.; Lobo, I.P.; Salako, I.G.; Jawad, A. Generalized entropy formalism and a new holographic dark energy model. *Phys. Lett. B* **2018**, *780*, 21–24. [CrossRef]
19. Reddy, D.R.; Naidu, R.L.; Rao, V.U. A cosmological model with negative constant deceleration parameter in Brans-Dicke theory. *Int. J. Theor. Phys.* **2007**, *46*, 1443–1448. [CrossRef]
20. Setare, M.R. The holographic dark energy in non-flat Brans–Dicke cosmology. *Phys. Lett. B* **2007**, *644*, 99–103. [CrossRef]
21. Kumar, S.; Singh, C.P. Exact bianchi type-I cosmological models in a scalar-tensor theory. *Int. J. Theor. Phys.* **2008**, *47*, 1722–1730. [CrossRef]
22. Setare, M.R.; Vagenas, E.C. The cosmological dynamics of interacting holographic dark energy model. *Int. J. Mod. Phys. D* **2009**, *18*, 147–157. [CrossRef]
23. Sharif, M.; Kausar, H.R. Anisotropic fluid and Bianchi type III model in f (R) gravity. *Phys. Lett. B* **2011**, *697*, 1–6. [CrossRef]
24. Sharif, M.; Kausar, H.R. Non-vacuum solutions of Bianchi type VI 0 universe in f (R) gravity. *Astrophys. Space Sci.* **2011**, *332*, 463–471. [CrossRef]
25. Sharif, M.; Waheed, S. Anisotropic universe models in Brans–Dicke theory. *Eur. Phys. J. C* **2012**, *72*, 1876. [CrossRef]
26. Srivastava, M.; Singh, C.P. Holographic Dark Energy Model in Brans-Dicke Theory with Future Event Horizon. *arXiv* **2017**, arXiv:1706.06777. [CrossRef]
27. Felegary, F.; Darabi, F.; Setare, M.R. Interacting holographic dark energy model in Brans–Dicke cosmology and coincidence problem. *Int. J. Mod. Phys. D* **2018**, *27*, 1850017. [CrossRef]

28. Nojiri, S.I.; Odintsov, S.D. Unifying phantom inflation with late-time acceleration: Scalar phantom–non-phantom transition model and generalized holographic dark energy. *Gen. Relativ. Gravit.* **2006**, *38*, 1285–1304. [CrossRef]
29. Nojiri, S.I.; Odintsov, S.D. Covariant generalized holographic dark energy and accelerating universe. *Eur. Phys. J. C* **2017**, *77*, 528. [CrossRef]
30. Sharif, M.; Jawad, A. Analysis of pilgrim dark energy models. *Eur. Phys. J. C* **2013**, *73*, doi:10.1140/epjc/s10052-013-2382-1. [CrossRef]
31. Caldwell, R.R.; Linder, E.V. Limits of quintessence. *Phys. Rev. Lett.* **2005**, *95*, doi:10.1103/PhysRevLett.95.141301. [CrossRef] [PubMed]
32. Sahni, V.; Shafieloo, A.; Starobinsky, A.A. Two new diagnostics of dark energy. *Phys. Rev. D* **2008**, *78*, doi:10.1103/PhysRevD.78.103502. [CrossRef]

symmetry

MDPI

Article

Decaying Dark Energy in Light of the Latest Cosmological Dataset

Ivan de Martino

Department of Theoretical Physics and History of Science, University of the Basque Country UPV/EHU, Faculty of Science and Technology, Barrio Sarriena s/n, 48940 Leioa, Spain; ivan.demartino@ehu.eus

Received: 2 August 2018; Accepted: 21 August 2018; Published: 1 September 2018

Abstract: Decaying Dark Energy models modify the background evolution of the most common observables, such as the Hubble function, the luminosity distance and the Cosmic Microwave Background temperature–redshift scaling relation. We use the most recent observationally-determined datasets, including Supernovae Type Ia and Gamma Ray Bursts data, along with $H(z)$ and Cosmic Microwave Background temperature versus z data and the reduced Cosmic Microwave Background parameters, to improve the previous constraints on these models. We perform a Monte Carlo Markov Chain analysis to constrain the parameter space, on the basis of two distinct methods. In view of the first method, the Hubble constant and the matter density are left to vary freely. In this case, our results are compatible with previous analyses associated with decaying Dark Energy models, as well as with the most recent description of the cosmological background. In view of the second method, we set the Hubble constant and the matter density to their best fit values obtained by the *Planck* satellite, reducing the parameter space to two dimensions, and improving the existent constraints on the model's parameters. Our results suggest that the accelerated expansion of the Universe is well described by the cosmological constant, and we argue that forthcoming observations will play a determinant role to constrain/rule out decaying Dark Energy.

Keywords: Dark Energy; statistical analysis; Baryon Acoustic Oscillation (BAO); Supernovae; cosmological model; Hubble constant; Cosmic Microwave Background (CMB) temperature

---

1. Introduction

In the last decades, several observations have pointed out that the Universe is in an ongoing period of accelerated expansion that is driven by the presence of an exotic fluid with negative pressure [1–12]. Its simplest form is a cosmological constant Λ, having an equation of state $w = -1$. More complicated prescriptions lead to the so-called Dark Energy (DE). Although several models have been proposed to explain DE [13–27], the observations have only determined that it accounts for $\sim$ 68% of the total energy-density budget of the Universe, while its fundamental nature is still unknown (see, for instance, the reviews [28,29]). In addition, we should mention that the accelerated expansion of the Universe could be explained by several modifications of the gravitational action. For example, introducing higher order terms of the Ricci curvature in the Hilbert–Einstein Lagrangian, gives rise to an effective matter stress–energy tensor which could drive the current accelerated expansion (see, for example, the reviews [12,30–35]). Another alternative for reproducing the dark energy effects is by introducing non-derivative terms interactions in the action, in addition to the Einstein–Hilbert action term, such that it creates the effect of a massive graviton [36–38].

We are interested in exploring a specific decaying DE model, $\Lambda(z) \propto (1+z)^m$, leading to creation/annihilation of photons and Dark Matter (DM) particles. The model is based on the theoretical framework developed in [39–43], while the thermodynamic features have been developed in [44,45]. Since DE continuously decays into photons and/or DM particles along the cosmic evolution, the relation between the temperature of the Cosmic Microwave Background (CMB) radiation and the redshift is modified.

In the framework of the standard cosmological model, the Universe expands adiabatically and, as consequence of the entropy and photon number conservation, the temperature of the CMB radiation scales linearly with redshift, $\propto (1+z)$. Nevertheless, in those models where conservation laws are violated, the creation or annihilation of photons can lead to distortions in the blackbody spectrum of the CMB and, consequently, to deviations of the standard CMB temperature–redshift scaling relation. Such deviations are usually explored with a phenomenological parameterization, such as $T_{\rm CMB}(z) = T_0(1+z)^{1-\beta}$ proposed in [41], where β is a constant parameter ($\beta = 0$ means adiabatic evolution), and T_0 is the CMB temperature at $z = 0$, which has been strongly constrained with COBE-FIRAS experiment, $T_0 = 2.7260 \pm 0.0013$ K [46]. The parameter β has been constrained using two methodologies: (a) the fine structure lines corresponding to the transition energies of atoms or molecules, present in quasar spectra, and excited by the CMB photon [47]; and (b) the multi-frequency measurements of the Sunyaev-Zel'dovich (SZ) effect [48–50]. Recent results based on data released by the *Planck* satellite and the *South Pole Telescope* (SPT) have led to sub-percent constraints on β which results to be compatible with zero at 1σ level (more details can be found in [11,51–56]).

In this paper, we start with the theoretical results obtained in [44,45]. Such a model has been constrained using luminosity distance measurements from Supernovae Type Ia (SNIa), differential age data, Baryonic Acoustic Oscillation (BAO), the CMB temperature–redshift relation, and the CMB shift parameter. Since the latter depends on the redshift of the last surface scattering, $z_{CMB} \sim 1000$, it represents a very high redshift probe. On the contrary, other datasets were used to probe the Universe at low redshift, $z \lesssim 3.0$. We aim to improve those constraints performing two different analysis: first, we constrain the whole parameter space to study the possibility of the model to alleviate the tension in the Hubble constant (see Section 5.4 in [10] for the latest results on the subject); and, second, we adopt the *Planck* cosmology to improve the constraint on the remaining parameters. Thus, we retain the SNIa, and use the most recent measurements the differential age, BAO, and the CMB temperature–redshift data. In addition, we use luminosity distances data of Gamma Ray Burst (GRB), which allow us to extend the redshift range till $z \sim 8$. Finally, we also use the reduced (compressed) set of parameters from CMB constraints [10].

The paper is organized as follows. In Section 2, we summarize the theoretical framework starting from the general Friedman–Robertson–Walker (FRW) metric, and point out the modification to the cosmological background arising from the violation of the conservation laws. In Section 3, we present the datasets used in the analysis, and the methodology implemented to explore the parameter space. The results are shown and discussed in Section 4 and, finally, in Section 5, we give our conclusions.

2. Theoretical Framework

The starting point is the well-known FRW metric

$$ds^2 = c^2dt^2 - a^2(t)\left[\frac{dr^2}{1-kr^2} + r^2(d\theta^2 + \sin^2\theta d\phi^2)\right], \tag{1}$$

where $a(t)$ is the scale factor and k is the curvature of the space time [57]. In General Relativity (GR), one obtains the following Friedman equations:

$$8\pi G(\rho_{m,tot} + \rho_x) + \Lambda_0 c^2 = 3\left(\frac{\dot{a}}{a}\right)^2 + 3\frac{kc^2}{a^2}, \tag{2}$$

$$\frac{8\pi G}{c^2}(p_{m,tot} + p_x) - \Lambda_0 c^2 = -2\frac{\ddot{a}}{a} - \frac{\dot{a}^2}{a^2} - \frac{kc^2}{a^2}, \tag{3}$$

where the total pressure is $p_{m,tot} = p_\gamma$, the total density is $\rho_{m,tot} = \rho_m + \rho_\gamma$, and ρ_x and p_x are the density and pressure of DE, respectively. Following [44,45], we set both the "bare" cosmological constant Λ_0 and the curvature k equal to 0.

In the standard cosmology, the Bianchi identities hold and the stress–energy momentum $T^{\mu\nu}$ is locally conserved

$$\nabla_\mu T^{\mu\nu} = 0\,. \tag{4}$$

Adopting a perfect fluid, the previous relation can recast as

$$\dot{\rho} + 3(\rho + p)H = 0\,, \tag{5}$$

where $H \equiv \dot{a}/a$ is the definition of the Hubble parameter. Thus, each component is conserved. Nevertheless, due to the photon/matter creation/annihilation happening in the case of decaying DE, the conservation equation is recast in the following relations:

$$\dot{\rho}_m + 3\rho_m H = (1-\epsilon)\, C_x\,, \tag{6}$$

$$\dot{\rho}_\gamma + 3\gamma\rho_\gamma H = \epsilon\, C_x\,, \tag{7}$$

$$\dot{\rho}_x + 3(p_x + \rho_x)H = -C_x\,, \tag{8}$$

where γ is a free parameter determining the equation of state of radiation $p_\gamma = (\gamma - 1)\rho_\gamma$ and, C_x and ϵ account for the decay of DE. C_x describe the physical mechanism leading to the production of particles (see, for instance, the thermogravitational quantum creation theory [40] or the quintessence scalar field cosmology [14]), and ϵ must be small enough in order to have the current density of radiation matching the observational constraints. Assuming $p_x = -\rho_x$, and defining

$$\rho_x = \frac{\Lambda(t)}{8\pi G}\,, \tag{9}$$

the parameter C_x can be obtained from the Equation (8)

$$C_x = -\frac{\dot{\Lambda}(t)}{8\pi G}\,. \tag{10}$$

Following [44,45], one can adopt a power law model

$$\Lambda(t) = B\left(\frac{a(t)}{a(0)}\right)^{-m} = B(1+z)^m\,, \tag{11}$$

then, writing Equation (2) at the present epoch, one can obtain $B = 3H_0^2(1 - \Omega_{m,0})$, where $\Omega_{m,0}$ is the matter density fraction at $z = 0$. It is very straightforward to verify that setting the power law index $m = 0$ leads to the cosmological constant. From Equation (8), it is also possible to write down an effective equation of state for the DE [45]:

$$w_{eff} = \frac{m}{3} - 1\,. \tag{12}$$

Finally, using Equations (2) and (6)–(8), the Hubble parameter can be obtained [43–45]:

$$H(z) \simeq \frac{8\pi G}{3}(\rho_m + \rho_x) = H_0\left[\frac{3(1-\Omega_{m,0})}{3-m}(1+z)^m + \frac{(3\Omega_{m,0} - m)}{3-m}(1+z)^3\right]^{1/2}\,. \tag{13}$$

Let us note that the standard Hubble parameter is recovered by setting $m = 0$ in Equation (13). Having the Hubble parameter allows us to compute the luminosity distance as follows

$$D_L = \frac{(1+z)c}{H_0} \int_0^z \frac{dz'}{E(z')}, \tag{14}$$

where we have defined $E(z) \equiv H(z)/H_0$.

Finally, following the approach originally proposed in [39], combining the Equations (6)–(8), with the equation for the number density conservation

$$\dot{n}_\gamma + 3n_\gamma H = \psi_\gamma , \tag{15}$$

where ψ_γ is the photon source, and the Gibbs Law

$$n_\gamma T_\gamma d\sigma_\gamma = d\rho_\gamma - \frac{\rho_\gamma + p_\gamma}{n_\gamma} dn_\gamma , \tag{16}$$

one obtains, through the use of thermodynamic identities, the following CMB temperature redshift relation (see for more details [43,45]):

$$T_{CMB}(z) = T_0(1+z)^{3(\gamma-1)} \times \left(\frac{(m - 3\Omega_{m,0}) + m(1+z)^{m-3}(\Omega_{m,0} - 1)}{(m-3)\Omega_{m,0}} \right)^{(\gamma-1)} . \tag{17}$$

Again, setting $m = 0$ gives the standard relation $T_{CMB}(z) = T_0(1+z)$. Equations (13), (14) and (17) can be easily implemented to test the decaying DE scenario. To show the effectiveness of these observables in constraining the cosmological model, we depict in Figure 1 their scalings as a function of the redshift for different value of the parameters γ and m, while we set H_0 and Ω_0 to their best fit from *Planck* satellite. In Figure 1a–c, we fix $\gamma = 4/3$ (which represents its standard value) while varying m in the range $[-0.5, 0.5]$ to show its impact on the Hubble constant, the luminosity distance and the CMB temperature. On the contrary, in Figure 1d, we set $m = 0$ (standard value) and vary γ illustrating how much the T_{CMB}-redshift relation is affected. The redshift ranges in the panels are set to the ones of the datasets. Looking at the plots, it is clear that the data will be really sensible to a variation of γ, while m will be more difficult to constrain.

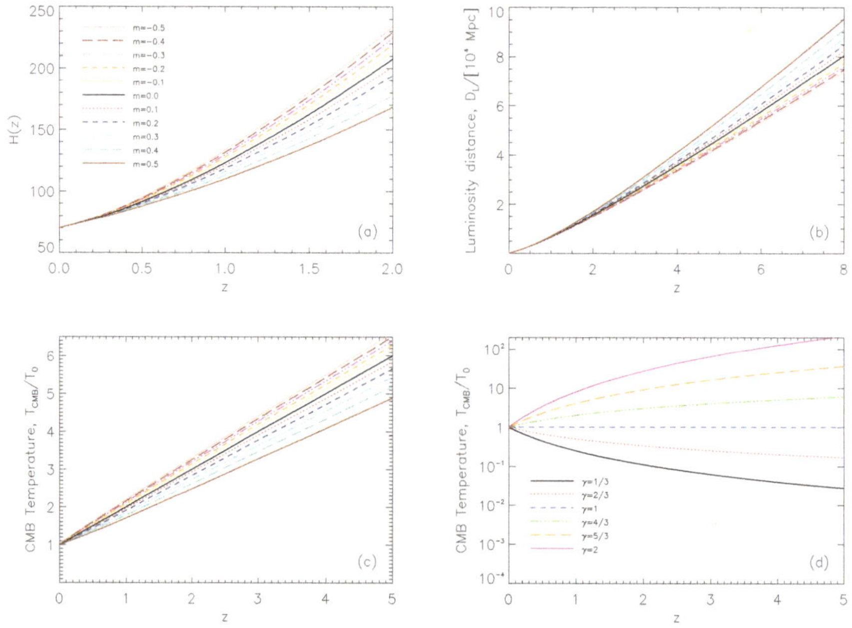

Figure 1. The figure shows as function of redshift the Hubble constant in panel (**a**), the luminosity distance in panel (**b**), and the CMB temperature in panels (**c**) and (**d**). Colors and lines indicate the different values assigned to the parameters m and γ to illustrate their impact on the observables.

3. Methodology and Data

We use measurements of $H(z)$, luminosity distances from SNIa and GRBs, BAO, and the CMB temperature–redshift relation. Then, we predict the theoretical counterparts using Equations (13), (14), and (17), and fit each one to the corresponding dataset computing the likelihood $-2\log\mathcal{L} = \chi^2(\mathbf{p})$, where $\mathbf{p} = [H_0, \Omega_{m,0}, m, \gamma]$ are the parameters of the model. The parameter space is explored using a Monte Carlo Markov Chain (MCMC) based on the Metropolis–Hastings [58,59] sampling algorithm with an adaptive step size to guarantee an optimal acceptance rate between 20% and 50% [60,61], while the convergence is ensured by the Gelman–Rubin criteria [62]. Once the convergence criteria is satisfied, the different chains are merged to compute the marginalized likelihood $\mathcal{L}(\mathbf{p}) = \Pi_k\mathcal{L}(\mathbf{p})$, where k indicates the different datasets, and to constrain the model's parameters. The priors are specified in Table 1.

Table 1. Parameter space explored by the MCMC algorithm.

Parameter	Priors
H_0	$[50.0, 100.0]$
$\Omega_{m,0}$	$[0.0, 1.0]$
m	$[-1, 1]$
γ	$[1.0, 2.0]$

Finally, the expectation value ($\langle p_i \rangle$) of the 1D marginalized likelihood distribution and the corresponding variance are computed as follows [63]

$$\langle p_i \rangle = \int d^{N_s}\mathbf{p}\mathcal{L}(\mathbf{p})p_i, \tag{18}$$

$$\sigma_i^2 = \int d^{N_s}\mathbf{p}\mathcal{L}(\mathbf{p})(p_i - \langle p_i \rangle)^2, \tag{19}$$

where N_s is the dimension of the parameter space.

Finally, the joint likelihood of the independent observables is used to compare decaying DE model with ΛCDM employing the Akaike Information Criteria (AIC) [64]:

$$AIC = -2\log\mathcal{L}_{max} + 2N_p\,, \tag{20}$$

where N_p is the number of parameters. A negative variation of the AIC indicator with respect to the reference model, $\Delta(AIC) = AIC_{dec.DE} - AIC_{\Lambda CDM}$, would indicate the model performs better than ΛCDM.

3.1. Supernovae Type Ia

We use a dataset of 557 Supernovae Type Ia (SNIa) in the redshift range $z = [0, 1.4]$ extracted from the UnionII catalogue (more details can be found in [65]). The observable is the so-called distance modulus μ_{obs}, which is the difference of the apparent and absolute magnitudes. Its theoretical counterpart can be computed starting from the luminosity distance in Equation (14), and it is given by

$$\mu_{th}(z) = 5\log_{10}\hat{D}_L(z) + \mu_0\,, \tag{21}$$

where $\mu_0 = 42.38 - 5\log_{10}h$, with $h \equiv H_0/100$, and $\hat{D}_L(z)$ is given by

$$\hat{D}_L(z) = (1+z)\int_0^z \frac{dz'}{E(z')}. \tag{22}$$

Then, we can define the χ^2 function as

$$-2\log\mathcal{L}_{\rm SN}(\mathbf{p}) = \chi^2_{\rm SN}(\mathbf{p}) = \sum_{i=1}^{557}\left(\frac{\mu_{th}(z_i,\mathbf{p}) - \mu_{obs}(z_i)}{\sigma_\mu(z_i)}\right)^2, \tag{23}$$

where $\sigma_\mu(z)$ is the error on $\mu_{obs}(z)$. Let us note that the parameter μ_0 encodes the dependence by the Hubble constant. Whenever one is not interest in fitting H_0, the marginalized χ^2 function can be defined as [66–69]:

$$\tilde{\chi}^2_{\rm SN}(\mathbf{p}) = \tilde{A} - \frac{\tilde{B}^2}{\tilde{C}}, \tag{24}$$

where

$$\tilde{A} = \sum_{i=1}^{557}\left(\frac{\mu_{th}(z_i,\mathbf{p},\mu_0=0) - \mu_{obs}(z_i)}{\sigma_\mu(z_i)}\right)^2, \tag{25}$$

$$\tilde{B} = \sum_{i=1}^{557}\frac{\mu_{th}(z_i,\mathbf{p},\mu_0=0) - \mu_{obs}(z_i)}{\sigma_\mu^2(z_i)}, \tag{26}$$

$$\tilde{C} = \sum_{i=1}^{557}\frac{1}{\sigma_\mu^2(z_i)}\,. \tag{27}$$

3.2. Differential Ages, $H(z)$

Following [70], we use 30 uncorrelated measurements of expansion rate, $H(z)$, that have been obtained using the differential age method [71–78]. Thus, we define the corresponding χ^2 as

$$-2\log\mathcal{L}_{\mathrm{H}}(\mathbf{p}) = \chi^2_{\mathrm{H}}(\mathbf{p}) = \sum_{i=1}^{30}\left(\frac{H(z_i,\mathbf{p}) - H_{obs,}(z_i)}{\sigma_H(z_i)}\right)^2, \tag{28}$$

where $\sigma_H(z)$ is the error on $H_{obs}(z)$. As stated in Section 3.1, the marginalized χ^2 with respect to H_0 can be also defined using Equation (24), where, for the $H(z)$ dataset, we have

$$\tilde{A} = \sum_{i=1}^{30}\left(\frac{(H(z_i,\mathbf{p},H_0=1) - H_{obs}(z_i)}{\sigma_H(z_i)}\right)^2, \tag{29}$$

$$\tilde{B} = \sum_{i=1}^{30}\frac{H(z_i,\mathbf{p},H_0=1) - H_{obs}(z_i)}{\sigma_H^2(z_i)}, \tag{30}$$

$$\tilde{C} = \sum_{i=1}^{30}\frac{1}{\sigma_H^2(z_i)}. \tag{31}$$

3.3. Baryonic Acoustic Oscillation

It is customary to define the BAO's observable as the following ratio: $\hat{\Xi} \equiv r_d/D_V(z)$, where r_d is the sound horizon at the drag epoch z_d [79]:

$$r_d = \int_{z_d}^{\infty}\frac{c_s(z)}{H(z)}dz, \tag{32}$$

and D_V the spherically averaged distance measure [80]

$$D_V(z) \equiv \left[(1+z)^2 d_A^2(z)\frac{cz}{H(z)}\right]^{1/3}. \tag{33}$$

Following [70], we use data from the 6dFGS [81], the SDSS DR7 [82], the BOSS DR11 [83–85], which are reported in Table I of [70]. Such a dataset is uncorrelated, therefore the likelihood can be straightforwardly computed as

$$-2\log\mathcal{L}_{\mathrm{BAO}}(\mathbf{p}) = \chi^2_{\mathrm{BAO}}(\mathbf{p}) = \sum_{i=1}^{6}\left(\frac{\hat{\Xi}(\mathbf{p},z_i) - \Xi_{obs}(z_i)}{\sigma_{\Xi}(z_i)}\right)^2, \tag{34}$$

where $\sigma_{\Xi}(z)$ is the error on $\Xi(z)$.

3.4. Gamma Ray Burst

We use a dataset of 109 GRB given in [86] which have been already used in other cosmological analysis (see for example [87]). The dataset was compiled using the Amati relation [88–90], and it is formed by 50 GRBs at $z < 1.4$ and 59 GRBs spanning the range of redshift $[0.1, 8.1]$. As it is for SNIa, the observable is the distance modulus, which in the case of GRBs is related to peak energy and the bolometric fluence (for more details, see [86,87]). The theoretical counterpart is computed using Equation (21), and the χ^2 function is defined as follows

$$-2\log\mathcal{L}_{\mathrm{GRB}}(\mathbf{p}) = \chi^2_{\mathrm{GRB}}(\mathbf{p}) = \sum_{i=1}^{109}\left(\frac{\mu_{th}(z_i,\mathbf{p}) - \mu_{obs}(z_i)}{\sigma_\mu(z_i)}\right)^2. \tag{35}$$

3.5. T_{CMB}–Redshift Relation

The last dataset is represented by the measurements of the CMB temperature at different redshifts. We use 12 data points obtained by using multi-frequency measurements of the Sunyaev–Zel'dovich effect produced by 813 galaxy clusters stacked on the CMB maps of the *Planck* satellite [54]. To those data, we add 10 high redshift measurements obtained through the study of quasar absorption line spectra [47]. The full dataset includes 22 data points spanning the redshift range $[0.0, 3.0]$, and they are listed in Table I of [51]. Finally, we predict the theoretical counterpart using Equation (17), and we compute the likelihood as

$$-2\log \mathcal{L}_{\mathrm{T_{CMB}}}(\mathbf{p}) = \chi^2_{\mathrm{T_{CMB}}}(\mathbf{p}) = \sum_{i=1}^{22}\left(\frac{T_{CMB,th}(z_i,\mathbf{p}) - T_{CMB,obs}(z_i)}{\sigma_{T_{CMB}}(z_i)}\right)^2 . \tag{36}$$

3.6. PlanckTT + LowP

The CMB power spectrum is the most powerful tools used to constrain cosmological parameters. However, the calculation of the power spectrum is time consuming, and it is common to use the so-called reduced parameters. It is possible to compress the whole information of the CMB power spectrum into a set of four parameters [91,92]: the CMB shift parameter (R), the angular scale (l_A) of the sound horizon at the redshift of the last scattering surface (z_*), the baryon density, and the scalar spectral index. Here, we rely only on R and l_A which can be compute as follows:

$$R = \sqrt{\Omega_{m,0}} \int_0^{z_*} \frac{dz'}{E(z')}, \tag{37}$$

$$l_A = \frac{\pi D_A(z_*)}{r_s(z_*)}, \tag{38}$$

where r_s is the sound horizon at z_*. In the 2015 data release of *Planck* satellite, the observational values of those parameters are: $[R, l_A] = [1.7488; 301.76] \pm [0.0074; 0.14]$ (for more details see Section 5.1.6 in [93]). Thus, the likelihood can be straightforwardly computed as

$$-2\log \mathcal{L}_{\mathrm{CMB}}(\mathbf{p}) = \chi^2_{\mathrm{CMB}}(\mathbf{p}) = \left(\frac{R_{obs} - R_{th}(\mathbf{p})}{\sigma_R}\right)^2 + \left(\frac{l_{A,obs} - l_{A,th}(\mathbf{p})}{\sigma_{l_A}}\right)^2 . \tag{39}$$

4. Results and Discussions

Following the aforementioned methodology, we carried out two sets of analysis: (A) we fit the whole parameter space composed by the Hubble constant H_0, the matter density parameter $\Omega_{m,0}$, γ and m; and (B) we set $H_0 = 67.37 \pm 0.54$ and $\Omega_{m,0} = 0.3147 \pm 0.0074$ which are the best fit values of joint analysis of the CMB power spectrum and other probes [10], while m and γ stay free to vary. All results are summarized in Table 2, and some comments are deserved.

In Analysis (A), we show that the best fit values of $[H_0, \Omega_{m,0}]$ are consistent with the most common cosmological analysis at low redshift, and $[m, \gamma]$ are compatible with the ones from [44,45] and their standard values at 1σ meaning that DE is well described by a cosmological constant. Interestingly, although our parameter space is larger than previous analysis, we get a comparable precision in m. This fact expresses the constraining powerful of this dataset with respect to the one used in previous analysis. The matter density is always compatible with current constraint from *Planck* 2018 results [10] at $\sim 2\sigma$. Nevertheless, there are two cases in which the central value of $\Omega_{m,0}$ gets closer to the one from *Planck* at $\sim 1\sigma$: (i) when using only $H(z)$ and CMB temperature data; and (ii) when using all datasets. In addition, the central value of the Hubble constant deserves some comments. When we used only $H(z)$ and T_{CMB} datasets, we obtained a lower central value of H_0 that is compatible at 1σ with *Planck* 2018 constraints and at 3σ with recent constraint from SNIa [94,95]. On the contrary, when introducing luminosity distances measurements, the best fits values of H_0 increases showing a tension with *Planck* 2018 results. The agreement of H_0 from the expansion rate data is rather expected since it has been found in other recent analysis [96–98].

Table 2. Maximum likelihood parameters and 1σ uncertainties from the MCMC algorithm and for the following datasets.

Dataset	**$[H_0, \Omega_{m,0}]$ Free**			
	H_0	$\Omega_{m,0}$	m	γ
$H(z)$+T_{CMB}	$66.9^{+2.56}_{-2.34}$	$0.314^{+0.055}_{-0.045}$	$0.07^{+0.16}_{-0.14}$	1.34 ± 0.02
SNIa+$H(z)$+T_{CMB}	$71.02^{+0.85}_{-0.91}$	0.26 ± 0.03	0.01 ± 0.11	1.34 ± 0.02
SNIa+GRB+$H(z)$+T_{CMB}	$71.46^{+0.84}_{-0.85}$	0.25 ± 0.03	$0.03^{+0.10}_{-0.11}$	1.34 ± 0.02
SNIa+GRB+$H(z)$+BAO+T_{CMB}	$70.31^{+0.66}_{-0.62}$	0.30 ± 0.01	0.18 ± 0.06	1.36 ± 0.01
SNIa+GRB+$H(z)$+BAO+T_{CMB}+CMB	69.8 ± 0.6	0.29 ± 0.01	0.01 ± 0.02	1.335 ± 0.005
	$[H_0, \Omega_{m,0}] = [67.37, 0.315]$			
$H(z)$+T_{CMB}			0.08 ± 0.07	1.34 ± 0.01
SNIa+$H(z)$+T_{CMB}			0.05 ± 0.07	1.34 ± 0.01
SNIa+GRB+$H(z)$+T_{CMB}			$0.04^{+0.07}_{-0.08}$	1.34 ± 0.01
SNIa+GRB+$H(z)$+BAO+T_{CMB}			0.05 ± 0.06	1.339 ± 0.009
SNIa+GRB+$H(z)$+BAO+T_{CMB}+CMB			0.01 ± 0.02	1.332 ± 0.005

Interestingly, the central value of m in the analysis including all the background observables is higher and it is compatible with zero only at 3σ. In such case, the power law index is $m = 0.18 \pm 0.06$ which can be recast in term of the equation of state parameter using Equation (12) and obtaining $w_{eff} = -0.94 \pm 0.02$, which is in tension with latest results from Planck satellite ($w = -1.04 \pm 0.1$ [10]). This fact demands a deeper analysis to be done with forthcoming datasets such as LSST, Euclid and WFIRST which will explore the Universe until redshift $z \sim 6$ providing high redshift SNIa and BAO data, and growth factor data with unprecedented precision [99–102]. Finally, in the full analysis including also the CMB constraints, we found a lower value of m which can be translated in $w_{eff} = -0.996 \pm 0.007$, which is perfectly compatible with a cosmological constant. To compare the decaying DE model with ΛCMD, we applied the AIC criteria obtaining $\Delta(AIC) = 1.53$ which slightly favors the standard cosmological model over the decaying DE one.

In the second analysis, H_0 and $\Omega_{m,0}$ are fixed to the *Planck* 2018 best fit values, and the parameters m and γ are fully in agreement with their expected values. Our best constraint of the power law index is $m = 0.01 \pm 0.02$ which means $w_{eff} = -0.996 \pm 0.007$ fully compatible with *Planck* 2018 results, and with a cosmological constant at 1σ. Moreover, to directly compare our results with the ones in [44,45], we carried out another analysis setting $\gamma = 4/3$ and leaving only m as free parameter. The constrained values of m with 1σ error is: $m = 0.004 \pm 0.006$, which represents a factor of ~ 5 improvement in σ_m over previous constraints.

Finally, in Figures 2 and 3, we show the 68% and 95% confidence levels of the whole and reduced parameter space constrained with the full dataset. To avoid overcrowding, in Figure 2, we do not overplot the contours from the several combinations of the datasets listed in Table 2.

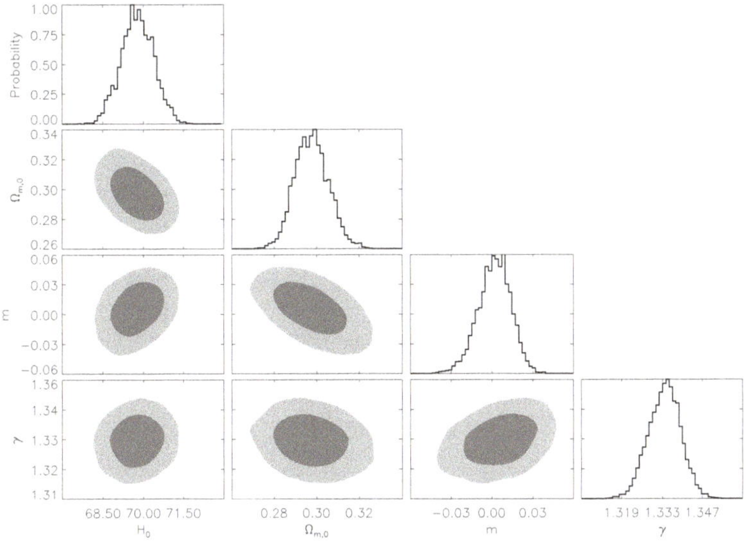

Figure 2. 2D marginalized contours of the model parameters $[H_0, \Omega_0, m, \gamma]$ obtained from the MCMC analysis. The 68% (dark grey) and 95% (light grey) confidence levels are shown for each pair of parameters. In each row, the marginalized likelihood distribution is also shown.

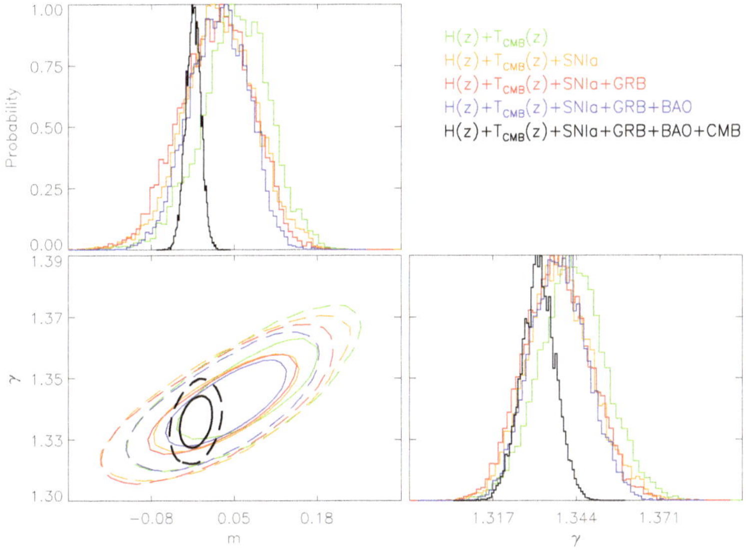

Figure 3. 2D marginalized 68% (solid line) and 95% (dashed line) contours of the model parameters $[m, \gamma]$ obtained from the MCMC analysis.

5. Conclusions

We have studied the decaying DE model introduced in [43–45]. In this model, photons and DM particles can be created or disrupted violating the conservation laws and altering the CMB temperature–redshift scaling relation. The model has been studied using the latest dataset of SNIa, GRB, BAO, $H(z)$, $T_{CMB}(z)$ and $PlanckTT + lowP$ data, which are described in Section 3.

First, we have explored the whole parameter space composed by the Hubble constant, the matter density fraction, and the parameters m and γ introduced in [44]. In this configuration, when using all the background observables, we obtain that the parameter m, which is the power law index of the DE decay law, is compatible with a cosmological constant only at 3σ. Therefore, forthcoming dataset could find a statistically relevant departure from standard cosmology, or alleviate this tension. Nevertheless, it is worth noting that, by adding the CMB constraints, such a tension disappears. Second, we have also studied a reduced parameter space composed by only m and γ, and setting the Hubble constant and the matter density parameter to their best fit values obtained recently by *Planck* satellite [10]. In this case, both parameters are always compatible at 1σ level with standard cosmology. Third, varying only m as in [44,45], we have improved the previous constraints of a factor ~ 5.

Finally, on the one side, we have demonstrated the improved constraining power of current dataset with respect to previous analysis, while, on the other side, we expect that forthcoming higher precision measurements of the CMB temperature at the location of high redshift galaxy clusters and Quasars, high redshift SNIa, improved measurements of BAO and of luminosity distance of GRBs, will be able to confirm or rule out decaying DE models [99–102].

Funding: This research received no external funding.

Acknowledgments: This article is based upon work from COST Action CA1511 Cosmology and Astrophysics Network for Theoretical Advances and Training Actions (CANTATA), supported by COST (European Cooperation in Science and Technology).

Conflicts of Interest: The author declares no conflict of interest.

Abbreviations

The following abbreviations are used in this manuscript:

BAO	Baryon Acoustic Oscillation
DE	Dark Energy
DM	Dark Matter
FRW	Friedman–Robertson–Walker
GR	General Relativity
GRB	Gamma Ray Burst
MCMC	Monte Carlo Markov Chain
SNIa	Supernovae Type Ia
SPT	South Pole Telescope

References

1. Perlmutter, S.; Gabi, S.; Goldhaber, G.; Goobar, A.; Groom, D.E.; Hook, I.M.; Kim, A.G.; Kim, M.Y.; Lee, J.C.; Pain, R.; et al. Measurements of the Cosmological Parameters Omega and Lambda from the First Seven Supernovae at $z \geq 0.35$. *Astrophys. J.* **1997**, *483*, 565.
2. Riess, A.G.; Strolger, L.G.; Tonry, J.; Casertano, S.; Ferguson, H.C.; Mobasher, B.; Challis, P.; Filippenko, A.V.; Jha, S.; Li, W.; et al. Type Ia Supernova Discoveries at $z > 1$ from the Hubble Space Telescope: Evidence for Past Deceleration and Constraints on Dark Energy Evolution. *Astrophys. J.* **2004**, *607*, 665–687.
3. Astier, P.; Guy, J.; Regnault, N.; Pain, R.; Aubourg, E.; Balam, D.; Basa, S.; Carlberg, R.G.; Fabbro, S.; Fouchez, D.; et al. The Supernova Legacy Survey: Measurement of Ω_M, Ω_Λ and w from the first year data set. *Astron. Astrophys.* **2006**, *447*, 31–48.
4. Suzuki, N.; Rubin, D.; Lidman, C.; Aldering, G.; Amanullah, R.; Barbary, K.; Barrientos, L.F.; Botyanszki, J.; Brodwin, M.; Connolly, N.; et al. The Hubble Space Telescope Cluster Supernova Survey. V. Improving the Dark-energy Constraints above $z > 1$ and Building an Early-type-hosted Supernova Sample. *Astrophys. J.* **2012**, *746*, 85.

5. Pope, A.C.; Matsubara, T.; Szalay, A.S.; Blanton, M.R.; Eisenstein, D.J.; Gray, J.; Jain, B.; Bahcall, N.A.; Brinkmann, J.; Budavari, T.; et al. Cosmological Parameters from Eigenmode Analysis of Sloan Digital Sky Survey Galaxy Redshifts. *Astrophys. J.* **2004**, *607*, 655.
6. Percival, W.J.; Baugh, C.M.; Bland-Hawthorn, J.; Bridges, T.; Cannon, R.; Cole, S.; Colless, M.; Collins, C.; Couch, W.; Dalton, G.; et al. The 2dF Galaxy Redshift Survey: The power spectrum and the matter content of the Universe. *Mon. Not. R. Astron. Soc.* **2001**, *327*, 1297–1306.
7. Tegmark, M.; Blanton, M.R.; Strauss, M.A.; Hoyle, F.; Schlegel, D.; Scoccimarro, R.; Vogeley, M.S.; Weinberg, D.H.; Zehavi, I.; Berlind, M.S.; et al. The Three-Dimensional Power Spectrum of Galaxies from the Sloan Digital Sky Survey. *Astrophys. J.* **2004**, *606*, 702.
8. Hinshaw, G.; Larson, D.; Komatsu, E.; Spergel, D.N.; Bennett, C.L.; Dunkley, J.; Nolta, M.R.; Halpern, M.; Hill, R.S.; Odegard, N.; et al. Nine-Year Wilkinson Microwave Anisotropy Probe (WMAP) Observations: Cosmological Parameter Results. *Astrophys. J. Suppl. Ser.* **2013**, *208*, 19.
9. Blake, C.; Kazin, E.A.; Beutler, F.; Davis, T.M.; Parkinson, D.; Brough, S.; Colless, M.; Contreras, C.; Couch, W.; Croom, S.; et al. The WiggleZ Dark Energy Survey: Mapping the distance-redshift relation with baryon acoustic oscillations. *Mon. Not. R. Astron. Soc.* **2011**, *418*, 1707.
10. Planck Collaboration. Planck 2018 results. VI. Cosmological parameters. *arXiv* **2018**, arXiv:1807.06209.
11. de Martino, I.; Génova-Santos, R.; Atrio-Barandela, F.; Ebeling, H.; Kashlinsky, A.; Kocevski, D.; Martins, C.J.A.P. Constraining the Redshift Evolution of the Cosmic Microwave Background Blackbody Temperature with PLANCK Data. *Astrophys. J.* **2015**, *808*, 128.
12. De Martino, I.; Martins, C.J.A.P.; Ebeling, H.; Kocevski, D. Constraining spatial variations of the fine structure constant using clusters of galaxies and Planck data. *Phys. Rev. D* **2016**, *94*, 083008.
13. Peebles, P.J.E.; Ratra, B. Cosmology with a time-variable cosmological 'constant'. *Astrophys. J. Lett.* **1988**, *325*, L17.
14. Ratra, B.; Peebles, P.J.E. Cosmological consequences of a rolling homogeneous scalar field. *Phys. Rev. D* **1988**, *37*, 3406.
15. Sahni, V.; Starobinsky, A.A. The Case for a Positive Cosmological Λ-Term. *Int. J. Mod. Phys.* **2000**, *D9*, 373. doi:10.1142/S0218271800000542
16. Caldwell, R.R. A phantom menace? Cosmological consequences of a dark energy component with super-negative equation of state. *Phys. Lett. B* **2002**, *545*, 23.
17. Padmanabhan, T. Cosmological constant-the weight of the vacuum. *Phys. Rep.* **2003**, *380*, 235.
18. Peebles, P.J.E.; Ratra, B. The cosmological constant and dark energy. *Rev. Mod. Phys.* **2003**, *75*, 559.
19. Demianski, M.; Piedipalumbo, E.; Rubano, C.; Tortora, C. Two viable quintessence models of the Universe: Confrontation of theoretical predictions with observational data. *Astron. Astrophys.* **2005**, *431*, 27.
20. Cardone, V.F.; Tortora, C.; Troisi, A.; Capozziello, S. Beyond the perfect fluid hypothesis for the dark energy equation of state. *Phys. Rev. D* **2006**, **73**, 043508.
21. Capolupo, A. Dark matter and dark energy induced by condensates. *Adv. High Energy Phys.* **2016**, *2016*, doi:10.1155/2016/8089142.
22. Capolupo, A. Quantum vacuum, dark matter, dark energy and spontaneous supersymmetry breaking. *Adv. High Energy Phys.* **2018**, *2018*, doi:10.1155/2018/9840351.
23. Kleidis, K.; Spyrou, N.K. A conventional approach to the dark energy concept. *Astron. Astrophys.* **2011**, *529*, A26.
24. Kleidis, K.; Spyrou, N.K. A conventional form of dark energy. *J. Phys. Conf. Ser.* **2011**, *283*, 012018.
25. Kleidis, K.; Spyrou, N.K. Polytropic dark matter flows illuminate dark energy and accelerated expansion. *Astron. Astrophys.* **2015**, *576*, A23.
26. Kleidis, K.; Spyrou, N.K. Dark energy: The shadowy reflection of dark matter? *Entropy* **2016**, *18*, 94.
27. Kleidis, K.; Spyrou, N.K. Cosmological perturbations in the ΛCDM-like limit of a polytropic dark matter model. *Astron. Astrophys.* **2017**, *606*, A116.
28. Caldwell, R.; Kamionkowski, M. The Physics of Cosmic Acceleration. *Ann. Rev. Nuclear Part. Sci.* **2009**, *59*, 397.
29. Wang, B.; Abdalla, E.; Atrio-Barandela, F.; Pavón, D. Dark Matter and Dark Energy Interactions: Theoretical Challenges, Cosmological Implications and Observational Signatures. *Rep. Prog. Phys.* **2016** , *79*, 9.
30. Capozziello, S.; de Laurentis, M.; Francaviglia, M.; Mercadante, S. From Dark Energy & Dark Matter to Dark Metric. *Found. Phys.* **2009**, *39*, 1161.

31. Nojiri, S.; Odintsov, S.D. Unified cosmic history in modified gravity: From F(R) theory to Lorentz non-invariant models. *Phys. Rep.* **2011**, *505*, 59–144.
32. de Martino, I.; De Laurentis, M.; Capozziello, S. Constraining f(R) gravity by the Large Scale Structure. *Universe* **2015**, *1*, 123.
33. Nojiri, S.; Odintsov, S.D.; Oikonomou, V.K. Modified gravity theories on a nutshell: Inflation, bounce and late-time evolution. *Phys. Rep.* **2017**, *692*, 1–104.
34. Nojiri, S.; Odintsov, S.D. Introduction to modified gravity and gravitational alternative for dark energy. *Int. J. Geom. Meth. Mod. Phys.* **2007**, *4*, 115.
35. Capozziello, S.; De Laurentis, M. Extended Theories of Gravity. *Phys. Rep.* **2011**, *509*, 167.
36. Arraut, I. The graviton Higgs mechanism. *Europhys. Letter* **2015**, *111*, 61001.
37. Arraut, I.; Chelabi, K. Non-linear massive gravity as a gravitational σ-model. *Europhys. Letter* **2016**, *115*, 31001.
38. Arraut, I.; Chelabi, K. Vacuum degeneracy in massive gravity: Multiplicity of fundamental scales. *Mod. Phys. Lett. A* **2017**, *32*, 1750112
39. Lima, J.A.S. Thermodynamics of decaying vacuum cosmologies. *Phys. Rev. D* **1996**, *54*, 2571.
40. Lima, J.A.S.; Alcaniz, J.A.S. Flat Friedmann-Robertson-Walker cosmologies with adiabatic matter creation: kinematic tests. *Astron. Astrophys.* **1999**, *348*, 1.
41. Lima, J.A.S.; Silva, A.I.; Viegas, S.M. Is the radiation temperature-redshift relation of the standard cosmology in accordance with the data? *Mon. Not. R. Astron. Soc.* **2000**, *312*, 747.
42. Puy, D. Thermal balance in decaying Λ cosmologies. *Astron. Astrophys.* **2004**, *422*, 1–9.
43. Ma, Y. Variable cosmological constant model: The reconstruction equations and constraints from current observational data. *Nuclear Phys. B* **2008**, *804*, 262.
44. Jetzer, P.; Puy, D.; Signore, M.; Tortora, C. Limits on decaying dark energy density models from the CMB temperature-redshift relation. *Gen. Relat. Grav.* **2011**, *43*, 1083.
45. Jetzer, P.; Tortora, C. Constraints from the CMB temperature and other common observational data sets on variable dark energy density models. *Phys. Rev. D* **2011**, *84*, 043517.
46. Fixsen, D.J. The Temperature of the Cosmic Microwave Background. *Astrophys. J.* **2009**, *707*, 916.
47. Bahcall, J.N.; Wolf, R.A. Fine-Structure Transitions. *Astrophys. J.* **1968**, *152* , 701.
48. Fabbri, R.; Melchiorri, F.; Natale, V. The Sunyaev-Zel'dovich effect in the millimetric region. *Astrophys. Space Sci.* **1978**, *59*, 223.
49. Rephaeli, Y. On the determination of the degree of cosmological Compton distortions and the temperature of the cosmic blackbody radiation. *Astrophys. J.* **1980**, *241*, 858.
50. Sunyaev, R.A.; Zeldovich, Y.B. The Observations of Relic Radiation as a Test of the Nature of X-Ray Radiation from the Clusters of Galaxies. *Comment Astrophys. Space Phys.* **1972**, *4*, 173.
51. Avgoustidis, A.; Génova-Santos, R.T.; Luzzi, G.; Martins, C.J.A.P. Subpercent constraints on the cosmological temperature evolution. *Phys. Rev. D* **2016**, *93*, 043521.
52. Luzzi, G.; Shimon, M.; Lamagna, L.; Rephaeli, Y.; De Petris, M.; Conte, A.; De Gregori, S.; Battistelli, E. S. Redshift Dependence of the Cosmic Microwave Background Temperature from Sunyaev-Zeldovich Measurements. *Astrophys. J.* **2009**, *705* , 1122.
53. Luzzi, G.; Génova-Santos, R.T.; Martins, C.J.A.P.; De Petris, M.; Lamagna, L. Constraining the evolution of the CMB temperature with SZ measurements from Planck data. *J. Cosmol. Astropart. Phys.* **2015**, *1509*, 011.
54. Hurier, G.; Aghanim, N.; Douspis, M.; Pointecouteau, E. Measurement of the T_{CMB} evolution from the Sunyaev-Zel'dovich effect. *Astron. Astrophys.* **2014**, *561* , A143.
55. Saro, A.; Liu, J.; Mohr, J.J.; Aird, K.A.; Ashby, M.L.N.; Bayliss, M.; Benson, B.A.; Bleem, L.E.; Bocquet, S.; Brodwin, M.; et al. Constraints on the CMB temperature evolution using multiband measurements of the Sunyaev-Zel'dovich effect with the South Pole Telescope. *Mon. Not. R. Astron. Soc.* **2014**, *440*, 2610.
56. De Martino, I.; Atrio-Barandela, F.; da Silva, A.; Ebeling, H.; Kashlinsky, A.; Kocevski, D.; Martins, C.J.A.P. Measuring the Redshift Dependence of the Cosmic Microwave Background Monopole Temperature with Planck Data. *Astrophys. J.* **2012**, *757*, 144.
57. Weinberg, S. *Gravitation and Cosmology: Principles and Applications of the General Theory of Relativity;* Wiley: New York, NY, USA, 1972.
58. Hastings, W.K. Monte Carlo Sampling Methods using Markov Chains and their Applications. *Biometrika* **1970**, *57*, 97.

59. Metropolis, N.; Rosenbluth, A.W.; Rosenbluth, M.N.; Teller, A.H.; Teller, E. Equation of State Calculations by Fast Computing Machines. *J. Chem. Phys.* **1953**, *21*, 1087.
60. Gelman, A.; Roberts, G.O.; Gilks, W.R. Efficient Metropolis jumping rule. *Bayesian Stat.* **1996**, *5*, 599.
61. Roberts, G.O.; Gelman, A.; Gilks, W.R. Weak convergence and optimal scaling of random walk Metropolis algorithms. *Ann. Appl. Probab.* **1997**, *7*, 110.
62. Gelman, A.; Rubin, D.B. Inference from Iterative Simulation Using Multiple Sequences. *Stat. Sci.* **1992**, *7*, 457.
63. Spergel, D.N.; Verde, L.; Peiris, H.V.; Komatsu, E.; Nolta, M.R.; Bennett, C.L.; Halpern, M.; Hinshaw, G.; Jarosik, N.; Kogut, A.; et al. First-Year Wilkinson Microwave Anisotropy Probe (WMAP) Observations: Determination of Cosmological Parameters. *Astrophys. J. Suppl.* **2003**, *148*, 175.
64. Akaike, H. A new look at the statistical model identification. *IEEE Trans. Autom. Control* **1974**, *19*, 716.
65. Amanullah, R.; Lidman, C.; Rubin, D.; Aldering, G.; Astier, P.; Barbary, K.; Burns, M.S.; Conley, A.; Dawson, K.S.; Deustua, S.E.; et al. Spectra and Hubble Space Telescope Light Curves of Six Type Ia Supernovae at $0.511 < z < 1.12$ and the Union2 Compilation. *Astrophys. J.* **2010**, *716*, 712.
66. Di Pietro, E.; Claeskens, J.F. Future supernovae data and quintessence models. *Mon. Not. R. Astron. Soc.* **2003**, *341*, 1299.
67. Nesseris, S.; Perivolaropoulos, L. Comparison of the Legacy and Gold SnIa Dataset Constraints on Dark Energy Models. *Phys. Rev. D* **2005**, *72*, 123519.
68. Perivolaropoulos, L. Constraints on linear negative potentials in quintessence and phantom models from recent supernova data. *Phys. Rev. D* **2005**, *71*, 063503.
69. Wei, H. Constraints on linear negative potentials in quintessence and phantom models from recent supernova data. *Phys. Lett. B* **2010**, *687*, 286.
70. Luković, V.V.; D'Agostino, R.; Vittorio, N. Is there a concordance value for H_0? *Astron. Astrophys.* **2016**, *595*, A109.
71. Jimenez, R.; Loeb, A. Constraining Cosmological Parameters Based on Relative Galaxy Ages. *Astrophys. J.* **2005**, *573*, 37.
72. Simon, J.; Verde, L.; Jimenez, R. Constraints on the redshift dependence of the dark energy potential. *Phys. Rev. D* **2005**, *71*, 123001.
73. Stern, D.; Jimenez, R.; Verde, L.; Stanford, S.A.; Kamionkowski, M. Cosmic Chronometers: Constraining the Equation of State of Dark Energy. II. A Spectroscopic Catalog of Red Galaxies in Galaxy Clusters. *Astrophys. J. Suppl.* **2010**, *188*, 280.
74. Zhang, C.; Zhang, H.; Yuan, S.; Liu, S.; Zhang, T.J.; Sun, Y.C. Four new observational H(z) data from luminous red galaxies in the Sloan Digital Sky Survey data release seven. *Res. Astron. Astrophys.* **2014**, *14*, 1221.
75. Moresco, M. Raising the bar: New constraints on the Hubble parameter with cosmic chronometers at $z \sim 2$. *Mon. Not. R. Astron. Soc.* **2015**, *450*, L16.
76. Moresco, M.; Cimatti, A.; Jimenez, R.; Pozzetti, L.; Zamorani, G.; Bolzonella, M.; Dunlop, J.; Lamareille, F.; Mignoli, M.; Pearce, H.; et al. Improved constraints on the expansion rate of the Universe up to $z \sim 1.1$ from the spectroscopic evolution of cosmic chronometers. *J. Cosmol. Astropart. Phys.* **2012**, *2012*, 1112–1119.
77. Moresco, M.; Pozzetti, L.; Cimatti, A.; Jimenez, R.; Maraston, C.; Verde, L.; Thomas, D.; Citro, A.; Tojeiro, R.; Wilkinson, D. A 6% measurement of the Hubble parameter at $z \sim 0.45$: Direct evidence of the epoch of cosmic re-acceleration. *J. Cosmol. Astropart. Phys.* **2016**, *2016*, 014.
78. Moresco, M.; Verde, L.; Pozzetti, L.; Jimenez, R.; Cimatti, A. New constraints on cosmological parameters and neutrino properties using the expansion rate of the Universe to $z \sim 1.75$. *J. Cosmol. Astropart. Phys.* **2012**, *2012*, 053.
79. Eisenstein, D.J.; Hu, W.; Tegmark, M. Cosmic Complementarity: H_0 and Ω_m from Combining Cosmic Microwave Background Experiments and Redshift Surveys. *Astrophys. J. Lett.* **1998**, *504*, L57.
80. Eisenstein, D.J.; Zehavi, I.; Hogg, D.W.; Scoccimarro, R.; Blanton, M.R.; Nichol, R.C.; Scranton, R.; Seo, H.; Tegmark, M.; Zheng, Z.; et al. Detection of the Baryon Acoustic Peak in the Large-Scale Correlation Function of SDSS Luminous Red Galaxies. *Astrophys. J.* **2005**, *633*, 560.
81. Beutler, F.; Blake, C.; Colless, M.; Jones, D.H.; Staveley-Smith, L.; Campbell, L.; Parker, Q.; Saunders, W.; Watson, F. The 6dF Galaxy Survey: baryon acoustic oscillations and the local Hubble constant. *Mon. Not. R. Astron. Soc.* **2011**, *416*, 3017.
82. Ross, A.J.; Samushia, L.; Howlett, C.; Percival, W.J.; Burden, A.; Manera, M.; The clustering of the SDSS DR7 main Galaxy sample - I. A 4 per cent distance measure at $z = 0.15$. *Mon. Not. R. Astron. Soc.* **2015**, *449*, 835.

83. Anderson, L.; Aubourg, E.; Bailey, S.; Beutler, F.; Bhardwaj, V.; Blanton, M.; Bolton, A.S.; Brinkmann, J.; Brownstein, J.R.; Burden, A.; et al. The clustering of galaxies in the SDSS-III Baryon Oscillation Spectroscopic Survey: baryon acoustic oscillations in the Data Releases 10 and 11 Galaxy samples. *Mon. Not. R. Astron. Soc.* **2014**, *441*, 24.
84. Delubac, T.; Bautista, J.E.; Busca, N.G.; Rich, J.; Kirkby, D.; Bailey, S.; Font-Ribera, A.; Slosar, A.; Lee, K.G.; Pieri, M.M.; et al. Baryon acoustic oscillations in the Lyα forest of BOSS DR11 quasars. *Astron. Astrophys.* **2015**, *574*, A59.
85. Font-Ribera, A.; Kirkby, D.; Busca, N.; Miralda-Escudé, J.; Ross, N.P.; Slosar, A.; Rich, J.; Aubourg, E.; Bailey, S.; Bhardwaj, V.; et al. Quasar-Lyman α forest cross-correlation from BOSS DR11: Baryon Acoustic Oscillations. *J. Cosmol. Astropart. Phys.* **2014**, *5*, 27.
86. Wei, H. Observational constraints on cosmological models with the updated long gamma-ray bursts. *J. Cosmol. Astropart. Phys.* **2010**, *1008*, 020.
87. Haridasu, B.S.; Luković, V.V.; D'Agostino, R.; Vittorio, N. Strong evidence for an accelerating Universe. *Astron. Astrophys.* **2017**, *600*, L1.
88. Amati, L.; Frontera, F.; Guidorzi, C. Extremely energetic Fermi gamma-ray bursts obey spectral energy correlations. *Astron. Astrophys.* **2009**, *508*, 173.
89. Amati, L.; Frontera, F.; Tavani, M.; in't Zand, J.J.M.; Antonelli, A.; Costa, E.; Feroci, M.; Guidorzi, C.; Heise, J.; Masetti, N.; et al. Intrinsic spectra and energetics of BeppoSAX Gamma-Ray Bursts with known redshifts. *Astron. Astrophys.* **2002**, *390*, 81.
90. Amati, L.; Guidorzi, C.; Frontera, F.; Della Valle, M.; Finelli, F.; Landi, R.; Montanari, E. Measuring the cosmological parameters with the $E_{p,i} - E_{iso}$ correlation of gamma-ray bursts. *Mon. Not. R. Astron. Soc.* **2008**, *391*, 577.
91. Kosowsky, A.; Milosavljevic, M.; Jimenez, R. Efficient cosmological parameter estimation from microwave background anisotropies. *Phys. Rev. D* **2002**, *66*, 063007.
92. Wang, Y.; Mukherjee, P. Observational Constraints on Dark Energy and Cosmic Curvature. *Phys. Rev. D* **2007**, *76*, 103533.
93. Planck Collaboration. Planck 2015 results. XIV. Dark energy and modified gravity. *Astron. Astrophys.* **2016**, *594*, A14.
94. Riess, A.G.; Macri, L.M.; Hoffmann, S.L. A 2.4% Determination of the Local Value of the Hubble Constant. *Astrophys. J.* **2016**, *826*, 56.
95. Riess, A.G.; Casertano, S.; Yuan, W.; Macri, L.; Anderson, J.; MacKenty, J.W.; Bowers, J.B.; Clubb, K.I.; Filippenko, A. V.; Jones, D.O.; et al. New Parallaxes of Galactic Cepheids from Spatially Scanning the Hubble Space Telescope: Implications for the Hubble Constant. *arXiv* **2018**, arXiv:1801.01120.
96. Yu, H.; Ratra, B.; Wang, F.Y. Hubble Parameter and Baryon Acoustic Oscillation Measurement Constraints on the Hubble Constant, the Deviation from the Spatially Flat ΛCDM Model, the Deceleration-Acceleration Transition Redshift, and Spatial Curvature. *Astrophys. J.* **2018**, *856*, 3.
97. Gómez-Valent, A.; Amendola, L. H_0 from cosmic chronometers and Type Ia supernovae, with Gaussian Processes and the novel Weighted Polynomial Regression method. *J. Cosmol. Astropart. Phys.* **2018**, *2018*, 051.
98. Wang, D.; Zhang, W. Machine Learning Cosmic Expansion History. *arXiv* **2018**, arXiv:1712.09208.
99. LSST Science Collaborations and LSST Project. *LSST Science Book*, 2nd ed; LSST Science Collaborations and LSST Project: Tucson, AZ, USA, 2009; arXiv:0912.0201.
100. Laureijs, R.; Amiaux, J.; Arduini, S.; Auguères, J.L.; Brinchmann, J.; Cole, R.; Cropper, M.; Dabin, C.; Duvet, L.; Ealet, A.; et al. Euclid Definition Study Report ESA/SRE(2011)12. *arXiv* **2011**, arXiv:1110.3193.
101. Spergel, D.; Gehrels, N.; Baltay, C.; Bennett, D.; Breckinridge, J.; Donahue, M.; Dressler, A.; Gaudi, B.S.; Greene, T.; Guyon, O.; et al. Wide-Field InfrarRed Survey Telescope-Astrophysics Focused Telescope Assets WFIRST-AFTA 2015 Report. *arXiv* **2015**, arXiv:1503.03757.
102. Kashlinsky, A.; Arendt, R.G.; Atrio-Barandela, F.; Helgason, K. Lyman-tomography of cosmic infrared background fluctuations with Euclid: Probing emissions and baryonic acoustic oscillations at $z > 10$. *Astrophys. J. Lett.* **2015**, *813*, L12.

symmetry

MDPI

Article

Dark Energy and Dark Matter Interaction: Kernels of Volterra Type and Coincidence Problem

Alexander B. Balakin * and Alexei S. Ilin

Department of General Relativity and Gravitation, Institute of Physics, Kazan Federal University, Kremlevskaya str. 18, Kazan 420008, Russia; alexeyilinjukeu@gmail.com

* Correspondence: Alexander.Balakin@kpfu.ru; Tel.: +7-917-933-4450

Received: 20 August 2018; Accepted: 15 September 2018; Published: 18 September 2018

Abstract: We study a new exactly solvable model of coupling of the Dark Energy and Dark Matter, in the framework of which the kernel of non-gravitational interaction is presented by the integral Volterra-type operator well-known in the classical theory of fading memory. Exact solutions of this isotropic homogeneous cosmological model were classified with respect to the sign of the discriminant of the cubic characteristic polynomial associated with the key equation of the model. Energy-density scalars of the Dark Energy and Dark Matter, the Hubble function and acceleration parameter are presented explicitly; the scale factor is found in quadratures. Asymptotic analysis of the exact solutions has shown that the Big Rip, Little Rip, Pseudo Rip regimes can be realized with the specific choice of guiding parameters of the model. We show that the Coincidence problem can be solved if we consider the memory effect associated with the interactions in the Dark Sector of the universe.

Keywords: Dark Energy; Dark Matter; memory

1. Introduction

Dark Matter (DM) and Dark Energy (DE) play the key roles in all modern cosmological scenaria (see, e.g., [1–17], and references therein for the history of problem, for main ideas and mathematical details). The DM and DE interact by the gravitational field, thus creating the space-time background for various astrophysical and cosmological events. In addition, according to the general view, the direct (non-gravitational) DM/DE coupling exists. One of the motivation of this idea is connected with the so-called Coincidence Problem [18–20]), which is based on the fact that the ratio between DE and DM energy densities is nowadays of the order $\frac{73}{23}$, while at the Planck time this ratio was of the order 10^{-95}, if one uses for calculations the energy density, associated with the cosmological constant (see, e.g., the review [21] for details of estimations). Clearly, the non-gravitational interactions between the DE and DM, or for short, interactions in the Dark Sector of the Universe, could start up the self-regulation procedure thus eliminating the initial disbalance. There are several models, which describe the DE/DM coupling (see, e.g., [21–24]). The most known models are phenomenological; they operate with the so-called kernel of interaction, the function $Q(t)$, which is linear in the energy densities of the DE and DM with coefficients proportional to the Hubble function [21]. In the series of works [25–28] the DE/DM interaction is modeled on the base of relativistic kinetic theory with an assumption that DE acts on the DM particles by the gradient force of the Archimedean type. In [29,30] the DE/DM interactions are considered in terms of extended electrodynamics of continua. In [31] the kernel $Q(t)$ was reconstructed for the case, when the cosmological expansion is described by the hybrid scale factor, composed using both: power-law and exponential functions.

In this work, we present the function $Q(t)$, the kernel of DE/DM interaction, in the integral form, using the analogy with classical theory of fading memory. The appropriate mathematical formalism is based on the theory of linear Volterra operators [32]; the corresponding integrand contains the difference of the DE and DM energy densities. The kernel of interaction vanishes if the DE and DM energy densities coincide; when these quantities do not coincide, the kernel of interaction acts as the source in the balance equations providing the procedure of self-regulation. However, in contrast to the known local phenomenological representations of the interaction kernels, the value of the source-function $Q(t)$ in the model, which includes the Volterra integrals, is predetermined by whole prehistory of the Universe evolution. As the result of modeling, we see that the ratio between the DE and DM energy densities tends asymptotically to some theoretically predicted value, which can be verified using the cosmological observations.

The paper is organized as follows. In Section 2 we recall the main elements of the phenomenological approach to the Universe evolution filled by interacting DE and DM. In Section 3 we formulate the model with kernel of the Volterra type, derive the integro-differential equations describing the Universe evolution, and obtain the so-called key equation, which is the differential equation of the Euler type of the third order in ordinary derivatives for the DE energy density. In Section 4 we classify the exact solutions with respect to the sign of the discriminant of the characteristic polynomial. In Section 5 we consider three examples of explicit analysis of the Universe evolution in the proposed model, and distinguish two exact solutions indicated as bounce and super-inflation, respectively. Section 6 contains discussion and conclusions.

2. Phenomenological Approach to the Problem of Interactions in the Dark Sector of the Universe

First of all, we would like to recall how do the phenomenological elements appear in the theory of interactions in the Dark Sector of the Universe. We consider the well-known two-fluid model, which describes the so-called Dark Fluid joining the DE and DM ; in this model the baryonic matter remains out of consideration.

2.1. *Two-Fluid Model in the Einstein Theory of Gravity*

The master equations for the gravity field

$$R^{ik} - \frac{1}{2}g^{ik}R - \Lambda g^{ik} = \kappa \left[T^{ik}_{(\mathrm{DE})} + T^{ik}_{(\mathrm{DM})}\right] \tag{1}$$

are considered to be derived from the Hilbert-Einstein action functional. Here R^{ik} is the Ricci tensor; R is the Ricci scalar; Λ is the cosmological constant; $T^{ik}_{(\mathrm{DE})}$ and $T^{ik}_{(\mathrm{DM})}$ are the stress-energy tensors of the DE and DM , respectively. These tensors can be algebraically decomposed using the Landau-Lifshitz scheme of definition of the fluid macroscopic velocity:

$$T^{ik}_{(\mathrm{DE})} = WU^iU^k + \mathcal{P}^{ik}\,, \quad T^{ik}_{(\mathrm{DM})} = EV^iV^k + \Pi^{ik}\,. \tag{2}$$

Here U^i and V^i are the timelike velocity four-vectors, the eigen-vectors of the corresponding stress-energy tensors:

$$U_iT^{ik}_{(\mathrm{DE})} = WU^k\,, \quad V_kT^{ik}_{(\mathrm{DM})} = EV^i\,. \tag{3}$$

The corresponding eigen-values, the scalars W and E are the energy density scalars of DE and DM, respectively. The quantities $\mathcal{P}^{ik}$ and Π^{ik} are the pressure tensors of the DE and DM; they are orthogonal to the velocity four-vectors:

$$U_i\mathcal{P}^{ik} = 0\,, \quad V_k\Pi^{ik} = 0\,. \tag{4}$$

The Bianchi identity provides the sum of the DE and DM stress-energy tensors to be divergence free:

$$\nabla_k \left[T^{ik}_{(\mathrm{DE})} + T^{ik}_{(\mathrm{DM})}\right] = 0\,. \tag{5}$$

This means that there exists a vector field Q^i, which possesses the property

$$\nabla_k T^{ik}_{(\mathrm{DE})} = Q^i = -\nabla_k T^{ik}_{(\mathrm{DM})} \,. \tag{6}$$

Until now we did not use the phenomenological assumptions; only the next step, namely the modeling of the vector field Q^i is the essence of the phenomenological approach, which describes the DE/DM interactions.

2.2. Description of the DE/DM Coupling in the Framework of an Isotropic Homogeneous Cosmological Model

When one deals with the spatially isotropic homogeneous cosmological model the key elements of the theory of DE and DM coupling can be simplified essentially. First of all, one uses the metric

$$ds^2 = dt^2 - a^2(t)\left[dx^2 + dy^2 + dz^2\right] , \tag{7}$$

with the scale factor $a(t)$ depending on the cosmological time; one assumes that the energy-density scalars also depend on time only, $W(t)$, $E(t)$. Second, the eigen four-vectors U^i and V^i coincide and are of the form $U^i = V^i = \delta^i_0$. Third, the pressure tensors happen to be reduced to the Pascal-type scalars $P(t)$ and $\Pi(t)$:

$$\mathcal{P}^{ik} = -P\Delta^{ik} , \quad \Pi^{ik} = -\Pi\Delta^{ik} , \quad \Delta^{ik} = g^{ik} - U^i U^k \,. \tag{8}$$

The four-vector Q^i now is presented by one scalar function $Q(t)$, since $Q^i = QU^i$ in the spatially isotropic model. The function $Q(t)$ is called in the review [21] by the term *kernel* of interaction. The master equations of the model can be now reduced to the following three ones:

$$3H^2 - \Lambda = \kappa\left[W(t) + E(t)\right] , \tag{9}$$

$$\dot{W} + 3H(W + P) = Q , \tag{10}$$

$$\dot{E} + 3H(E + \Pi) = -Q , \tag{11}$$

where $H(t) \equiv \frac{\dot{a}}{a}$ is the Hubble function, and the dot denotes the derivative with respect to time. The Equation (9) is taken from the Einstein equations; the sum of (10) and (11) gives the total energy conservation law. Also, we use the standard linear equations of state

$$P = (\Gamma - 1)W , \quad \Pi = (\gamma - 1)E , \tag{12}$$

which allow us to focus on the analysis of the set of three equations for three unknown functions W, E and H. The history of modeling of the function $Q(t)$ is well documented in the review [21]; we focus below on a new (rheological-type) model.

3. Rheological-Type Model of the DE/DM Coupling

3.1. Reconstruction of the Kernel $Q(t)$

To reconstruct phenomenologically the interaction kernel $Q(t)$ we use the ansatz based on the following three assumptions.

(i) The function $Q(t)$ is presented by the integral operator of the Volterra type:

$$Q(t) = \int_{t_0}^{t} d\xi K(t,\xi)\left[E(\xi) - W(\xi)\right] . \tag{13}$$

(ii) The Volterra integral contains the difference of the energy density scalars $E(\xi)$ and $W(\xi)$.

(iii) The kernel of the Volterra integral $K(t,\xi)$ has a specific multiplicative form

$$K(t,\xi) = K_0 H(t) H(\xi) \left[\frac{a(\xi)}{a(t)}\right]^{\nu}. \tag{14}$$

Motivation of our choice is the following.

(1) In the context of rheological approach we assume that the state of a fluid system at the present time moment t is predetermined by whole prehistory of its evolution from the starting moment t_0 till to the moment t. More than century ago it was shown, that the mathematical formalism appropriate for description of this idea can be based on the theory of linear integral Volterra operators, which have found numerous applications to the theory of media with memory. We also use this fruitful idea.

(2) Our ansatz is that the interaction between two constituents of the Dark Fluid vanishes, if the DE energy density coincides identically with the DM energy density, $W \equiv E$. When $W \neq E$ the integral mechanism of self-regulation inside the Dark Fluid switches on. For instance, during the cosmological epochs with DE domination, i.e., when $W > E$, the corresponding contribution into the interaction term Q is negative, the rates $\dot{W}$ and $\dot{E}$ obtain negative and positive contributions, respectively (see (10) and (11)); when $W < E$, the inverse process starts thus regulating the ratio between DE and DM energy densities.

(3) For classical models of fading memory the kernel of the Volterra operator is known to be of exponential form $K(t,\xi) = \mathcal{K} \exp \frac{(\xi - t)}{T_0}$, where the parameter T_0 describes the typical time of memory fading, and the quantity $\mathcal{K}$ has the dimensionality $[\text{time}]^{-2}$. When we work with the de Sitter scale factor $a(t) = a(t_0) \exp H_0 t$, we can rewrite the kernel of the Volterra operator as follows:

$$K(t,\xi) = \mathcal{K} \exp \left[\frac{H_0(\xi - t)}{H_0 T_0}\right] = K_0 H_0^2 \left[\frac{a(\xi)}{a(t)}\right]^{\frac{1}{H_0 T_0}}, \tag{15}$$

where the parameter K_0 is dimensionless. This idea inspired us to formulate the ansatz, that not only for the de Sitter spacetime, but for Friedmann - type spacetimes also, we can use the kernel (14) with two additional model parameters, ν and K_0.

3.2. Key Equation of the Model

To analyze the set of coupled Equations (9)–(12), (13) and (14), let us derive the so-called key equation, which contains only one unknown function, W. In our model with the ansatz (14) the unknown functions W and E depend on time through the scale factor, i.e., $W = W(a(t))$, $E = E(a(t))$. Following the standard approach (see, e.g., the review [21]), we introduce new dimensionless variable x instead of t using the definitions

$$x \equiv \frac{a(t)}{a(t_0)}, \quad \frac{d}{dt} = xH(x)\frac{d}{dx}. \tag{16}$$

When the function $H(x)$ is found, the scale factor as the function of cosmological time can be found from the following quadrature:

$$t - t_0 = \int_1^{\frac{a(t)}{a(t_0)}} \frac{dx}{xH(x)}. \tag{17}$$

In these terms three basic master equations take the form

$$3H^2(x) - \Lambda = \kappa \left[W(x) + E(x)\right], \tag{18}$$

$$x\frac{dW}{dx} + 3\Gamma W = K_0 x^{-\nu} \int_1^x dy y^{\nu-1}[E(y) - W(y)]\,, \tag{19}$$

$$x\frac{dE}{dx} + 3\gamma E = K_0 x^{-\nu} \int_1^x dy y^{\nu-1}[W(y) - E(y)]\,. \tag{20}$$

Also, we have the consequence of two last equations:

$$x\frac{d}{dx}(W+E) + 3\,(\Gamma W + \gamma E) = 0\,. \tag{21}$$

The Equation (18) is decoupled from this set; it can be used to find the Hubble function, when $W(x)$ and $E(x)$ are obtained. Two last integro-differential equations can be reduced to the differential ones:

$$x^2 W'' + xW'(\nu + 1 + 3\Gamma) + 3\nu\Gamma W = K_0(E - W)\,, \tag{22}$$

$$x^2 E'' + xE'(\nu + 1 + 3\gamma) + 3\nu\gamma E = K_0(W - E)\,. \tag{23}$$

Here and below the prime denotes the derivative with respect to dimensionless variable x. The next step is the following: we extract $E(x)$ from (22)

$$E(x) = \frac{1}{K_0}\left[x^2 W'' + xW'(\nu + 1 + 3\Gamma) + W(K_0 + 3\nu\Gamma)\right]\,, \tag{24}$$

and put it into (21), thus obtaining the Euler equation of the third order

$$x^3 W''' + (A+3)x^2 W'' + (B+1)xW' + DW = 0\,, \tag{25}$$

where the auxiliary parameters are the following:

$$A = \nu + 3(\Gamma + \gamma)\,, \quad B = A + 2K_0 + 3\nu(\Gamma + \gamma) + 9\Gamma\gamma\,, \quad D = 3\left[K_0(\Gamma + \gamma) + 3\nu\Gamma\gamma\right]. \tag{26}$$

We indicate Equation (25) as the key equation, since when $W(x)$ is found, we obtain $E(x)$ immediately from (24), and then $H(x)$ from (18).

4. Classification of Solutions

4.1. The Scheme of Classification

The characteristic equation for the Euler Equation (25) is the cubic one:

$$\sigma^3 + \sigma^2 A + \sigma(B - A) + D = 0\,. \tag{27}$$

As usual, we reduce the cubic equation to the canonic form

$$\sigma = z - \frac{A}{3} \;\rightarrow\; z^3 + pz + q = 0\,, \tag{28}$$

using the following definitions of the canonic parameters p and q:

$$p = B - A - \frac{1}{3}A^2\,, \quad q = \frac{2}{27}A^3 + \frac{1}{3}A(A - B) + D\,. \tag{29}$$

The discriminant of the cubic Equation (27) with p and q given by (29) is of the form

$$\Delta = \frac{p^3}{27} + \frac{q^2}{4}\,. \tag{30}$$

When $\Delta < 0$, the roots of Equation (27) are real and do not coincide, $\sigma_1 \neq \sigma_2 \neq \sigma_3$. When $\Delta = 0$, the roots are real, but at least two of them coincide, $\sigma_1 \neq \sigma_2 = \sigma_3$ or $\sigma_1 = \sigma_2 = \sigma_3$. When $\Delta > 0$, there is one real root, and a pair of complex conjugated, $\sigma_1, \sigma_{2,3} = \alpha \pm i\beta$. Let us study all these cases in detail.

4.2. Solutions Corresponding to the Negative Discriminant, $\Delta < 0$

4.2.1. The Structure of the Exact Solution

It is the case, when the parameter p is negative, $p < 0$, and $|\frac{q}{2}| \left(\frac{3}{|p|}\right)^{\frac{3}{2}} < 1$, or in more detail

$$\left|\frac{1}{27}A^3 + \frac{1}{6}A(A-B) + \frac{1}{2}D\right| < \left|\frac{1}{3}(B-A) - \frac{1}{9}A^2\right|^{\frac{3}{2}}. \tag{31}$$

All three roots $\sigma_1, \sigma_2, \sigma_3$ are real and they do not coincide:

$$\sigma_1 = -\frac{A}{3} + 2\sqrt{\frac{|p|}{3}} \cos\frac{\varphi}{3}, \quad \sigma_{2,3} = -\frac{A}{3} + 2\sqrt{\frac{|p|}{3}} \cos\left(\frac{\varphi}{3} \pm \frac{2\pi}{3}\right), \tag{32}$$

where the auxiliary angle $0 \leq \varphi \leq \pi$ is defined as follows:

$$\cos\varphi = -\frac{q}{2}\left(\frac{3}{|p|}\right)^{\frac{3}{2}}. \tag{33}$$

In this case, the key equation for the DE energy density scalar $W(x)$ gives power-law solution:

$$W(x) = C_1 x^{\sigma_1} + C_2 x^{\sigma_2} + C_3 x^{\sigma_3}. \tag{34}$$

Using the relationship (24) we obtain immediately the DM energy density scalar $E(x)$

$$\begin{gathered} E(x) = \frac{1}{K_0}\left\{C_1 x^{\sigma_1}\left[\sigma_1^2 + \sigma_1(\nu + 3\Gamma) + (K_0 + 3\nu\Gamma)\right] + \right. \\ \left. C_2 x^{\sigma_2}\left[\sigma_2^2 + \sigma_2(\nu + 3\Gamma) + (K_0 + 3\nu\Gamma)\right] + C_3 x^{\sigma_3}\left[\sigma_3^2 + \sigma_3(\nu + 3\Gamma) + (K_0 + 3\nu\Gamma)\right]\right\}. \end{gathered} \tag{35}$$

The constants of integration C_1, C_2, C_3 can be expressed in terms of presented functions at $t = t_0$, or equivalently, at $x = 1$; they are the solutions of the system:

$$C_1 + C_2 + C_3 = W(1), \tag{36}$$

$$C_1\sigma_1 + C_2\sigma_2 + C_3\sigma_3 = -3\Gamma W(1), \tag{37}$$

$$\begin{gathered} C_1\left[\sigma_1^2 + \sigma_1(\nu + 3\Gamma) + (K_0 + 3\nu\Gamma)\right] + C_2\left[\sigma_2^2 + \sigma_2(\nu + 3\Gamma) + (K_0 + 3\nu\Gamma)\right] + \\ C_3\left[\sigma_3^2 + \sigma_3(\nu + 3\Gamma) + (K_0 + 3\nu\Gamma)\right] = K_0 E(1). \end{gathered} \tag{38}$$

Clearly, the first and third equations are the direct consequences of (34) and (35), respectively; as for the second relationship, we obtain it from (19), when $x = 1$. The Cramer determinant for this system

$$\mathcal{D} = (\sigma_1 - \sigma_2)(\sigma_2 - \sigma_3)(\sigma_3 - \sigma_1) \neq 0 \tag{39}$$

is not equal to zero, thus the system has the unique solution:

$$C_1 = \frac{1}{(\sigma_1 - \sigma_2)(\sigma_1 - \sigma_3)}\left\{W(1)\left[\sigma_2\sigma_3 + 3\Gamma(\sigma_2 + \sigma_3) + 9\Gamma^2 - K_0\right] + K_0 E(1)\right\}, \tag{40}$$

$$C_2 = \frac{1}{(\sigma_2 - \sigma_1)(\sigma_2 - \sigma_3)}\left\{W(1)\left[\sigma_1\sigma_3 + 3\Gamma(\sigma_1 + \sigma_3) + 9\Gamma^2 - K_0\right] + K_0 E(1)\right\}, \tag{41}$$

$$C_3 = \frac{1}{(\sigma_3-\sigma_1)(\sigma_3-\sigma_2)}\left\{W(1)\left[\sigma_1\sigma_2+3\Gamma(\sigma_1+\sigma_2)+9\Gamma^2-K_0\right]+K_0E(1)\right\}. \quad (42)$$

Then, using the Einstein Equation (18) we find the square of the Hubble function:

$$\begin{aligned} H^2(x) = \tfrac{\Lambda}{3} + \tfrac{\kappa}{3K_0}\left\{C_1x^{\sigma_1}\left[2K_0+\sigma_1^2+\sigma_1(\nu+3\Gamma)+3\nu\Gamma\right]+\right. \\ \left. C_2x^{\sigma_2}\left[2K_0+\sigma_2^2+\sigma_2(\nu+3\Gamma)+3\nu\Gamma\right]+C_3x^{\sigma_3}\left[2K_0+\sigma_3^2+\sigma_3(\nu+3\Gamma)+3\nu\Gamma\right]\right\}. \end{aligned} \quad (43)$$

The scale factor $a(t)$ can be now obtained from the integral

$$\sqrt{\frac{\kappa}{3K_0}}(t-t_0) = \int_1^{\frac{a(t)}{a(t_0)}} \frac{dx}{x\sqrt{\frac{K_0\Lambda}{\kappa}+\sum_{j=1}^3 C_jx^{\sigma_j}\left[2K_0+\sigma_j^2+\sigma_j(\nu+3\Gamma)+3\nu\Gamma\right]}}. \quad (44)$$

Generally, this integral cannot be expressed in elementary functions; results of asymptotic analysis are discussed below.

4.2.2. Two Auxiliary Characteristics of the Model and a Scheme of Estimation of the Kernel Parameters

(1) *The acceleration parameter q*

The formula (43) allows us to calculate immediately the acceleration parameter:

$$-q(x) = 1+\frac{x}{2H^2(x)}\frac{dH^2}{dx} = \frac{\frac{K_0\Lambda}{\kappa}+\sum_{j=1}^3 C_jx^{\sigma_j}\left(1+\frac{\sigma_j}{2}\right)\left[2K_0+\sigma_j^2+\sigma_j(\nu+3\Gamma)+3\nu\Gamma\right]}{\frac{K_0\Lambda}{\kappa}+\sum_{j=1}^3 C_jx^{\sigma_j}\left[2K_0+\sigma_j^2+\sigma_j(\nu+3\Gamma)+3\nu\Gamma\right]}. \quad (45)$$

(2) *The DM/DE energy density ratio ω*

For many purposes it is important to have the ratio $\omega(x) = \frac{E(x)}{W(x)}$. Direct calculation gives

$$\omega(x) = \frac{E(x)}{W(x)} = \frac{\sum_{j=1}^3 C_jx^{\sigma_j}\left[\sigma_j^2+\sigma_j(\nu+3\Gamma)+(K_0+3\nu\Gamma)\right]}{K_0\sum_{j=1}^3 C_jx^{\sigma_j}}. \quad (46)$$

Let us assume that the present moment of the cosmological time is $t = T$, and the corresponding value of the dimensionless scale factor is $X = \frac{a(T)}{a(t_0)}$. Also, we use the following estimations for the present time parameters:

$$\omega(X) = \frac{E(X)}{W(X)} \simeq \frac{23}{73}, \quad q(X) = -0.55. \quad (47)$$

Thus, we have two relationships, which link the kernel parameters K_0 and ν with X and other coupling constants:

$$-0.9 = \frac{\sum_{j=1}^3 C_jX^{\sigma_j}\sigma_j\left[2K_0+\sigma_j^2+\sigma_j(\nu+3\Gamma)+3\nu\Gamma\right]}{\frac{K_0\Lambda}{\kappa}+\sum_{j=1}^3 C_jX^{\sigma_j}\left[2K_0+\sigma_j^2+\sigma_j(\nu+3\Gamma)+3\nu\Gamma\right]}, \quad (48)$$

$$\frac{23}{73} = \frac{\sum_{j=1}^3 C_jX^{\sigma_j}\left[\sigma_j^2+\sigma_j(\nu+3\Gamma)+(K_0+3\nu\Gamma)\right]}{K_0\sum_{j=1}^3 C_jX^{\sigma_j}}. \quad (49)$$

We hope to realize the whole scheme of fitting of the model parameters in a special paper.

4.2.3. Admissible Asymptotic Regimes, and Constraints on the Model Parameters

There are three regimes of asymptotic behavior of the presented solutions.

(i) If the maximal real root, say σ_1, is positive and the set of initial data is general, we see that $W \to \infty$, $E \to \infty$ and $H \to \infty$, when $x \to \infty$. The integral in (17) converges at $a(t) \to \infty$, and the scale factor $a(t)$ follows the law $a(t) = a_*(t_* - t)^{-\frac{2}{\sigma_1}}$, and reaches infinity at $t = t_*$. We deal in this case with the so-called Big Rip asymptotic regime, and the Universe follows the catastrophic scenario [11,19]. In particular, when $\sigma_1 > 0$ and $\sigma_2 < 0, \sigma_3 < 0$, according to the Viète theorem, we can definitely say only that $\sigma_1\sigma_2\sigma_3 = -D > 0$, i.e., $K_0(\Gamma + \gamma) + 3\nu\Gamma\gamma < 0$. The asymptotic value of the acceleration parameter is equal to $-q(\infty) = 1 + \frac{\sigma_1}{2}$. The final ratio between the DM and DE energy densities

$$\omega(\infty) = \frac{\sigma_1^2 + \sigma_1(\nu + 3\Gamma) + K_0 + 3\nu\Gamma}{K_0} \tag{50}$$

does not depend on the initial parameters $W(1)$ and/or $E(1)$.

(ii) If the maximal real root, say σ_1, is equal to zero, we see that $D = 0$, and thus

$$K_0(\Gamma + \gamma) + 3\nu\Gamma\gamma = 0\,. \tag{51}$$

In this case, the Hubble function tends asymptotically to constant H_∞, given by

$$H_\infty = \sqrt{\frac{\Lambda}{3} + \frac{\kappa\,(2K_0 + 3\nu\Gamma)}{3K_0\sigma_2\sigma_3}\left\{W(1)\left[\sigma_2\sigma_3 + 3\Gamma(\sigma_2 + \sigma_3) + 9\Gamma^2 - K_0\right] + K_0E(1)\right\}}\,, \tag{52}$$

thus providing the scale factor to be of the exponential form $a(t) \to a_\infty e^{H_\infty t}$; we deal in this case with the Pseudo Rip, or in other words, the late-time Universe of the quasi-de Sitter type. Clearly, the asymptotic value of the function $-q(x)$, given by (45), is $-q(\infty) = 1$. As for the asymptotic value of the quantity $\omega(x)$ (see (46)), it is now equal to $\omega(\infty) = -\frac{\Gamma}{\gamma}$. Since ω is the non-negatively defined quantity, this situation is possible only if the ratio $\frac{\Gamma}{\gamma}$ is non-positive. Thus, the evolution of the ratio $\frac{E(x)}{W(x)}$ starts from the value $\frac{E(1)}{W(1)}$ and finishes with $\left|\frac{\Gamma}{\gamma}\right|$. One can add that, when $\sigma_1 = 0$ and $\sigma_2 < 0, \sigma_3 < 0$, we obtain two supplementary inequalities:

$$A = -(\sigma_2 + \sigma_3) > 0 \;\to \nu + 3(\Gamma + \gamma) > 0\,, \tag{53}$$

$$B - A = \sigma_2\sigma_3 > 0 \;\to 2K_0 + 3\nu(\Gamma + \gamma) + 9\Gamma\gamma > 0\,. \tag{54}$$

These requirements restrict the choice of model parameters.

(iii) If all the roots are negative, we see that $H \to H_0 \equiv \sqrt{\frac{\Lambda}{3}}$, when $x \to \infty$, thus we obtain the classical de Sitter asymptote with $-q(\infty) = 1$. When $\Lambda = 0$, all the roots are negative, and, say, σ_1 is the maximal among them, we see that $W \to 0$, $E \to 0$ at $x \to \infty$. The scale factor behaves asymptotically as the power-law function $a(t) \propto t^{\frac{2}{|\sigma_1|}}$; the acceleration parameter $-q(\infty) = 1 - \frac{|\sigma_1|}{2}$ is positive, when $|\sigma_1| < 2$. In particular, when $\sigma_1 < 0, \sigma_2 < 0, \sigma_3 < 0$, we see that, first, $\sigma_1\sigma_2\sigma_3 = -D < 0$, i.e., $K_0(\Gamma + \gamma) + 3\nu\Gamma\gamma > 0$; second, $A = -(\sigma_1 + \sigma_2 + \sigma_3) > 0$; third, $B - A = \sigma_1\sigma_2 + \sigma_1\sigma_3 + \sigma_3\sigma_2 > 0$.

There are also cases related to the special choice of initial data $W(1)$, $E(1)$, as well as, of the choice of parameters K_0, ν, Γ, γ. For instance, if we deal with the situation indicated as (i) but now $C_1 = 0$ due to specific choice of $W(1)$, $E(1)$, (see (40)), we obtain the situation (ii) or (iii).

4.3. *Solutions Corresponding to the Positive Discriminant, $\Delta > 0$*

Now one root, say σ_1, is real and $\sigma_{2,3}$ are complex conjugated:

$$\sigma_1 = -\frac{A}{3} + (\mathcal{U} + \mathcal{V})\,, \quad \sigma_{2,3} = \alpha \pm i\beta\,, \quad \alpha \equiv -\frac{A}{3} - \frac{1}{2}(\mathcal{U} + \mathcal{V})\,, \quad \beta \equiv \frac{\sqrt{3}}{2}(\mathcal{U} - \mathcal{V})\,, \tag{55}$$

where the auxiliary real parameters $\mathcal{U}$ and $\mathcal{V}$

$$\mathcal{U} \equiv \left[-\frac{q}{2} + \sqrt{\Delta}\right]^{\frac{1}{3}}, \quad \mathcal{V} \equiv \left[-\frac{q}{2} - \sqrt{\Delta}\right]^{\frac{1}{3}} \tag{56}$$

are chosen so that $\mathcal{U}\mathcal{V} = -\frac{p}{3}$. Similarly to the case with negative discriminant, we obtain the DE energy density scalar

$$W(x) = C_1 x^{\sigma_1} + x^{\alpha}\left[C_2 \cos\beta\log x + C_3 \sin\beta\log x\right], \tag{57}$$

the DM energy density

$$\begin{aligned} K_0 E(x) = C_1 x^{\sigma_1}\left[\sigma_1^2 + \sigma_1(\nu + 3\Gamma) + (K_0 + 3\nu\Gamma)\right] + \\ x^{\alpha}\left\{\left[\alpha^2 - \beta^2 + \alpha(\nu + 3\Gamma) + (K_0 + 3\nu\Gamma)\right]\left[C_2\cos\beta\log x + C_3 \sin\beta\log x\right] + \right. \\ \left. \beta(2\alpha + \nu + 3\Gamma)\left[C_3\cos\beta\log x - C_2 \sin\beta\log x\right]\right\}, \end{aligned} \tag{58}$$

where

$$C_1 = \frac{K_0 E(1) + W(1)\left[(\alpha + 3\Gamma)^2 + \beta^2 - K_0\right]}{\left[(\sigma_1 - \alpha)^2 + \beta^2\right]}, \tag{59}$$

$$C_2 = \frac{-K_0 E(1) + W(1)\left[(\sigma_1 - \alpha)^2 - (\alpha + 3\Gamma)^2 + K_0\right]}{\left[(\sigma_1 - \alpha)^2 + \beta^2\right]}, \tag{60}$$

$$C_3 = \frac{K_0(\alpha - \sigma_1)E(1) + W(1)\left\{(\sigma_1 - \alpha)K_0 + (\sigma_1 + 3\Gamma)\left[(\alpha + 3\Gamma)(\alpha - \sigma_1) - \beta^2\right]\right\}}{\beta\left[(\sigma_1 - \alpha)^2 + \beta^2\right]}. \tag{61}$$

The square of the Hubble function can be extracted from the formula

$$\begin{aligned} \tfrac{3K_0}{\kappa}\left[H^2(x) - \tfrac{\Lambda}{3}\right] = C_1 x^{\sigma_1}\left[2K_0 + \sigma_1^2 + \sigma_1(\nu + 3\Gamma) + 3\nu\Gamma\right] + \\ x^{\alpha}\left\{\left[\alpha^2 - \beta^2 + \alpha(\nu + 3\Gamma) + 2K_0 + 3\nu\Gamma\right]\left[C_2\cos\beta\log x + C_3\sin\beta\log x\right] + \right. \\ \left. \beta(2\alpha + \nu + 3\Gamma)\left[C_3\cos\beta\log x - C_2\sin\beta\log x\right]\right\}. \end{aligned} \tag{62}$$

Admissible Asymptotic Regimes

Clearly, all three asymptotic regimes: the Big Rip, Pseudo-Rip, power-law expansion, mentioned above, also can be realized in this submodel. However, three new elements can be added into the catalog of possible regimes.

(i) The first new regime can be indicated as a quasi-periodic expansion; it can be realized when $\sigma_1 = 0$, α is negative, and $H^2_{\infty} > |h|$. The square of the Hubble function can be now rewritten as follows:

$$H^2 \to H^2_{\infty} + hx^{-|\alpha|}\sin\left[\beta\log x + \psi\right]. \tag{63}$$

Asymptotically, the Universe expansion tends to the Pseudo Rip regime; however, this process has quasi-periodic features.

(ii) The second new regime relates to $\sigma_1 = 0$, $\alpha = 0$ and $H^2_{\infty} > |h|$. The square of the Hubble function, the DE and DM energy densities become now periodic functions (see, e.g., (63) with $\alpha = 0$).

(iii) The third regime is characterized by the following specific feature: H^2 takes zero value at finite $x = x_*$. This regime can be effectively realized in two cases: first, when $\sigma_1 = 0$, $\alpha < 0$ and $H^2_{\infty} < |h|$; second, when $\sigma_1 = 0$, $\alpha > 0$. In both cases the size of the Universe is fixed by the specific value of the scale factor $a^* = a(t_0)x_*$.

4.4. Solutions Corresponding to the Vanishing Discriminant, $\Delta = 0$

4.4.1. Two Roots Coincide, $q \neq 0$

It is the case, when all roots are real, but two of them coincide:

$$\sigma_1 = -\frac{A}{3} + 2\left(-\frac{q}{2}\right)^{\frac{1}{3}}, \quad \sigma \equiv \sigma_2 = \sigma_3 = -\frac{A}{3} - \left(-\frac{q}{2}\right)^{\frac{1}{3}}. \tag{64}$$

The DE and DM energy-density scalars contain logarithmic functions

$$W(x) = C_1 x^{\sigma_1} + x^{\sigma}\left[C_2 + C_3 \log x\right], \tag{65}$$

$$\begin{aligned} K_0 E(x) = C_1 x^{\sigma_1}\left[\sigma_1^2 + \sigma_1(\nu + 3\Gamma) + (K_0 + 3\nu\Gamma)\right] + \\ + x^{\sigma}\left\{(C_2 + C_3 \log x)\left[\sigma^2 + \sigma(\nu + 3\Gamma) + (K_0 + 3\nu\Gamma)\right] + C_3\left(2\sigma + \nu + 3\Gamma\right)\right\}, \end{aligned} \tag{66}$$

where the constants of integration are

$$C_1 = \frac{K_0 E(1) + W(1)\left[(\sigma + 3\Gamma)^2 - K_0\right]}{(\sigma_1 - \sigma)^2}, \tag{67}$$

$$C_2 = -\frac{K_0 E(1) + W(1)\left[(\sigma + 3\Gamma)^2 - K_0 - (\sigma_1 - \sigma)^2\right]}{(\sigma_1 - \sigma)^2}, \tag{68}$$

$$C_3 = \frac{K_0 E(1) + W(1)\left[(\sigma + 3\Gamma)(\sigma_1 + 3\Gamma) - K_0\right]}{(\sigma - \sigma_1)}. \tag{69}$$

The square of the Hubble function is presented as follows:

$$\begin{aligned} H^2(x) = \tfrac{\Lambda}{3} + \tfrac{\kappa}{3K_0}\left\{C_1 x^{\sigma_1}\left[\sigma_1^2 + \sigma_1(\nu + 3\Gamma) + (2K_0 + 3\nu\Gamma)\right] + C_2 x^{\sigma}\left[\sigma^2 + \sigma(\nu + 3\Gamma) + (2K_0 + 3\nu\Gamma)\right] + \right. \\ \left. C_3 x^{\sigma}\left[\log x\left(\sigma^2 + \sigma(\nu + 3\Gamma) + (2K_0 + 3\nu\Gamma)\right) + (2\sigma + \nu + 3\Gamma)\right]\right\}. \end{aligned} \tag{70}$$

4.4.2. Three Roots Coincide, $q = 0$

Now all the roots coincide

$$\sigma_1 = \sigma_2 = \sigma_3 = -\frac{A}{3} = \sigma. \tag{71}$$

The DE and DM energy-density scalars, the square of the Hubble function contain logarithmic function and its square

$$W(x) = x^{\sigma}\left[C_1 + C_2 \log x + C_3 \log^2 x\right], \tag{72}$$

$$\begin{aligned} K_0 E(x) = x^{\sigma}\left\{\left(C_1 + C_2 \log x + C_3 \log^2 x\right)\left[\sigma^2 + \sigma(\nu + 3\Gamma) + K_0 + 3\nu\Gamma\right] + \right. \\ \left. (C_2 + 2C_3 \log x)(2\sigma + \nu + 3\Gamma) + 2C_3\right\}, \end{aligned} \tag{73}$$

$$\begin{aligned} H^2(x) = \tfrac{\Lambda}{3} + \tfrac{\kappa}{3K_0} x^{\sigma}\left\{\left(C_1 + C_2 \log x + C_3 \log^2 x\right)\left[\sigma^2 + \sigma(\nu + 3\Gamma) + (2K_0 + 3\nu\Gamma)\right] + \right. \\ \left. (2\sigma + \nu + 3\Gamma)\left(C_2 + 2C_3 \log x\right) + 2C_3\right\}, \end{aligned} \tag{74}$$

$$C_1 = W(1), \quad C_2 = -W(1)(\sigma + 3\Gamma), \quad C_3 = \frac{1}{2}\left\{K_0 E(1) + W(1)[(\sigma + 3\Gamma)^2 - K_0]\right\}. \tag{75}$$

4.4.3. Admissible Asymptotic Regimes

Since the Hubble function contains now the logarithmic terms $\log x$ and $\log^2 x$, a new asymptotic regime, the so-called Little Rip, is possible. In the case of Little Rip we obtain that asymptotically $H(t) \to \infty$ and $a(t) \to \infty$, the infinite values can be reached during the infinite time interval only.

5. Three Examples of Explicit Model Analysis

As a preamble, we would like to recall that the set of model parameters $(\Gamma, \gamma, K_0, \nu, \Lambda)$ is adequate for the procedure of fitting of the acceleration parameter $-q(T) \simeq 0.55$ and of the factor $\frac{E(T)}{W(T)} \simeq \frac{23}{73}$. Nevertheless, we do not perform this procedure in this paper, and do not accompany this procedure by the detailed plots of $q(t)$, $\omega(t)$, $H(t)$, etc. However, we think that for demonstration of analytical capacities of our new model, it is interesting to consider some exact solutions obtained for the set of parameters specifically chosen. Of course, when we introduce the model parameters "by hands", we restrict the time interval, on which the solution is physically motivated and is mathematically adequate. For instance, the super-inflationary solution discussed below can be applicable for the early Universe, but is not appropriate for the late-time period. Nevertheless, the presented exact solutions seem to be intriguing.

5.1. *First Explicit Submodel, $\Delta < 0$, $q = 0$ and $\Lambda = 0$; How Do the Initial Data Correct the Universe Destiny?*

For illustration, let us consider the case with the following set of parameters:

$$\Lambda = 0\,, \quad \Gamma = 0\,, \quad \nu = \frac{3}{2}\gamma\,, \quad K_0 = -\frac{9}{4}\gamma^2\,. \tag{76}$$

Let us recall that for $\Gamma = 0$ according to (12) we obtain $P = -W$, i.e., the pressure typical for the Dark Energy. One deals with the Cold Dark Matter, when $\gamma = 1$; generally, $\gamma \geq 1$. The (26) and (29) yield

$$q = 0\,, \quad \varphi = \frac{\pi}{2}\,, \quad p = -\frac{27}{4}\gamma^2\,, \quad B = A = \frac{9}{2}\gamma\,, \quad D = -\frac{27}{4}\gamma^3\,, \quad \Delta = -\left(\frac{3\gamma}{2}\right)^6 < 0\,. \tag{77}$$

Thus, for $\gamma > 0$ one root of the characteristic equation is positive, and other two are negative:

$$\sigma_1 = \frac{3}{2}\gamma(\sqrt{3}-1) > 0\,, \quad \sigma_2 = -\frac{3}{2}\gamma(\sqrt{3}+1) < 0\,, \quad \sigma_3 = -\frac{3}{2}\gamma < 0\,. \tag{78}$$

The constants of integration are, respectively,

$$C_1 = \frac{1}{6}\left[(2+\sqrt{3})W(1) - E(1)\right]\,, \quad C_2 = \frac{1}{6}\left[(2-\sqrt{3})W(1) - E(1)\right]\,, \quad C_3 = \frac{1}{3}\left[W(1) + E(1)\right]\,. \tag{79}$$

Clearly, there are three principal situations, which correspond to three ranges of values of the initial parameter $\omega(1) = \frac{E(1)}{W(1)}$.

(i) When $\omega(1) = 2 + \sqrt{3}$, i.e., $C_1 = 0$, and the growing mode is deactivated, the DE energy density, DM energy density take, respectively, the form

$$W(x) = \frac{W(1)}{\sqrt{3}} x^{-\frac{3}{2}\gamma}\left[(\sqrt{3}+1) - x^{-\frac{3\sqrt{3}}{2}\gamma}\right] \geq 0\,, \tag{80}$$

$$E(x) = \frac{E(1)}{\sqrt{3}} x^{-\frac{3}{2}\gamma}\left[(\sqrt{3}-1) + x^{-\frac{3\sqrt{3}}{2}\gamma}\right] \geq 0\,. \tag{81}$$

The function $\omega(x) = \frac{E(x)}{W(x)}$, which is given by

$$\omega(x) = (2+\sqrt{3})\frac{\left[(\sqrt{3}-1) + x^{-\frac{3\sqrt{3}}{2}\gamma}\right]}{\left[(\sqrt{3}+1) - x^{-\frac{3\sqrt{3}}{2}\gamma}\right]}\,, \tag{82}$$

is positive and monotonic; it starts from the value $\omega(1)=2+\sqrt{3}$ and tends asymptotically to $\omega(\infty)=1$. In other words, the energy density of the DM tends to the energy density of the DE due to the interaction of the rheological type. The square of the Hubble function is also non-negative:

$$H^2(x) = \frac{\kappa W(1)(\sqrt{3}+1)}{3\sqrt{3}} x^{-\frac{3}{2}\gamma} \left(2 + x^{-\frac{3\sqrt{3}}{2}\gamma}\right) \geq 0\,. \tag{83}$$

The scale factor $a(t)$ can be found from the quadrature:

$$\sqrt{\frac{2\kappa W(1)(\sqrt{3}+1)}{3\sqrt{3}}}(t-t_0) = \int_1^{\frac{a(t)}{a(t_0)}} \frac{dx x^{\frac{3\gamma}{4}-1}}{\sqrt{1+\frac{1}{2}x^{-\frac{3\sqrt{3}}{2}\gamma}}}\,. \tag{84}$$

In the asymptotic regime the scale factor behaves as $a(t) \propto t^{\frac{4}{3\gamma}}$, and the Hubble function $H(t)$ tends to zero as $H(t) \simeq \frac{4}{3\gamma t}$.

(ii) When $0 < \omega(1) < 2+\sqrt{3}$, i.e., $C_1 > 0$, the integral $\int_1^{\infty} \frac{dx}{xH(x)}$ converges, so the scale factor $a(t)$ reaches infinite value at finite value of the cosmological time. The growing mode, which relates to the positive root σ_1, become the leading mode, and we obtain the model of the Big Rip type.

(iii) When $\omega(1) > 2+\sqrt{3}$, i.e., $C_1 < 0$, we obtain the model in which the square of the Hubble function takes zero value at some finite time moment. In other words, the Universe expansion stops, the Universe volume becomes finite.

5.2. Second and Third Explicit Submodels: The Case $\Delta = 0$ and $q = 0$

For illustration we consider the model, in which all three roots coincide and are equal to zero, $\sigma_1 = \sigma_2 = \sigma_3 = 0$. Equivalently, we assume that the characteristic equation takes the form $\sigma^3 = 0$, and, thus, $A = 0$, $B = 0$, $D = 0$. Only one set of model parameters admits such solution, namely

$$\gamma + \Gamma = 0 \quad \nu = 0\,, \quad K_0 = \frac{9}{2}\Gamma^2\,. \tag{85}$$

In particular, this model covers the case, when $\gamma = 1$ and $\Gamma = -1$, i.e., the DM is pressureless, $\Pi = 0$, and the DE pressure is described by the equation of state $P = -2W$. For this set of guiding parameters we obtain

$$W(x) = W(1)\,(1 - 3\Gamma \log x) + \frac{9}{4}\Gamma^2\,[W(1) + E(1)] \log^2 x\,, \tag{86}$$

$$E(x) = E(1)\,(1 - 3\gamma \log x) + \frac{9}{4}\gamma^2\,[W(1) + E(1)] \log^2 x\,. \tag{87}$$

The square of the Hubble function is presented by the formula

$$H^2(x) = \frac{\Lambda}{3} + \frac{\kappa}{3}\left(1 + \frac{9}{2}\Gamma^2 \log^2 x\right) [W(1) + E(1)] + \kappa\gamma\,[W(1) - E(1)] \log x\,, \tag{88}$$

and the scale factor can now be found in elementary functions from the integral

$$(t - t_0) = \int_0^{\log \frac{a(t)}{a(t_0)}} \frac{dz}{\sqrt{\frac{\Lambda}{3} + \frac{\kappa}{3}\left(1 + \frac{9}{2}\Gamma^2 z^2\right) [W(1) + E(1)] + \kappa\gamma\,[W(1) - E(1)]\, z}}\,. \tag{89}$$

However, the integration procedure is faced with two principally different cases, $W(1) + E(1) = 0$, and $W(1) + E(1) \neq 0$. Let us consider them separately.

5.2.1. The Case $W(1)+E(1)=0$: Solution of the Bounce Type

While this case seems to be exotic (one of the energy densities should be negative), it is interesting to study this case in detail. First, we fix that $\gamma > 0$, $W(1) > 0$. Then, integration gives immediately

$$a(t) = a(t_*) \exp\left[\frac{1}{2}\gamma\kappa W(1)(t-t_*)^2\right], \tag{90}$$

where the following auxiliary parameters are introduced

$$a(t_*) = a(t_0) \exp\left[-\frac{\Lambda}{6\kappa W(1)\gamma}\right], \quad t_* = t_0 - \frac{\sqrt{\frac{\Lambda}{3}}}{\kappa W(1)\gamma}. \tag{91}$$

In terms of cosmological time the Hubble function is the linear one:

$$H(t) = \kappa W(1)\gamma(t-t_*). \tag{92}$$

The corresponding acceleration parameter

$$-q(t) = 1 + \frac{1}{\kappa W(1)\gamma(t-t_*)^2} \tag{93}$$

tends to one asymptotically at $t \to \infty$. In the work [25] the solution of this type was indicated as anti-Gaussian solution; also this solution is known as bounce (see, e.g., [12]).

The DE and DM energy densities behave as quadratic functions of cosmological time:

$$\frac{W(t)}{W(1)} = \frac{3}{2}\gamma^2\kappa W(1)(t-t_*)^2 + 1 - \frac{\Lambda}{2\kappa W(1)}, \tag{94}$$

$$\frac{E(t)}{W(1)} = \frac{3}{2}\gamma^2\kappa W(1)(t-t_*)^2 - 1 - \frac{\Lambda}{2\kappa W(1)}. \tag{95}$$

It is interesting to mention that the rates of evolution of the DE and DM energy density scalars coincide:

$$\dot{E}(t) = \dot{W}(t) = 3\gamma^2\kappa W^2(1)(t-t_*). \tag{96}$$

Clearly, both functions: $\omega(t) = \frac{E(t)}{W(t)}$ and $-q(t)$ tend asymptotically to one, $\omega(\infty)=1$, $-q(\infty)=1$. The acceleration parameter $-q(t)$ described by the simple monotonic function (93).

5.2.2. The Case $W(1)+E(1) \neq 0$: Super-Inflationary Solution

For illustration we consider the simple submodel with $\Lambda = 0$, and assume that at $t = t_0$ the DE and DM energy densities coincide, i.e., $E(1) = W(1)$. The integration in (89) yields now

$$\log\frac{a(t)}{a(t_0)} = \frac{\sqrt{2}}{3\gamma}\sinh\left[\gamma(t-t_0)\sqrt{3\kappa W(1)}\right]. \tag{97}$$

This solution is of the super-inflationary type; at $t \to \infty$ it behaves as

$$\frac{a(t)}{a(t_0)} = e^{\frac{1}{3\sqrt{2}\gamma}e^{\sqrt{3\kappa W(1)}\,\gamma\, t}}. \tag{98}$$

It can be indicated as a Little Rip according to the classification given in [11]. Also this solution appears in the model of Archimedean-type interaction between DE and DM [25].

The DE and DM energy densities behave as follows:

$$\frac{W(t)}{W(1)} = \cosh^2\left[\gamma(t-t_0)\sqrt{3\kappa W(1)}\right] + \sqrt{2}\sinh\left[\gamma(t-t_0)\sqrt{3\kappa W(1)}\right], \tag{99}$$

$$\frac{E(t)}{W(1)} = \cosh^2\left[\gamma(t-t_0)\sqrt{3\kappa W(1)}\right] - \sqrt{2}\sinh\left[\gamma(t-t_0)\sqrt{3\kappa W(1)}\right], \tag{100}$$

so, the function $\omega(t) = \frac{E(t)}{W(t)}$ tends asymptotically to one. The Hubble function and acceleration parameter are, respectively

$$H(t) = \sqrt{\frac{2}{3}\kappa W(1)}\cosh\left[\gamma(t-t_0)\sqrt{3\kappa W(1)}\right], \tag{101}$$

$$-q(t) = 1 + \left(\frac{3\gamma}{\sqrt{2}}\right)\frac{\sinh\left[\gamma(t-t_0)\sqrt{3\kappa W(1)}\right]}{\cosh^2\left[\gamma(t-t_0)\sqrt{3\kappa W(1)}\right]}. \tag{102}$$

When we study the time interval $t \geq t_0$, we see that the function $-q(t)$ starts with $-q(t_0) = 1$, reaches the maximum $-q_{(\max)} = 1 + \frac{3\gamma}{2\sqrt{2}}$ and then tends to one asymptotically, $-q(\infty) = 1$.

6. Discussion

We established the model of DE/DM interaction based on the interaction kernel of the Volterra type, as well as, classified and studied the obtained exact solutions. From our point of view, the results are inspiring. Let us explain our optimism.

1. The model of kernel of the DE/DM interaction, which possesses two extra parameters, K_0 and ν, is able to describe many known interesting cosmic scenaria: Big Rip, Little Rip, Pseudo Rip, de Sitter-type expansion; the late-time accelerated expansion of the Universe is the typical feature of the presented model.
2. When $2K_0+3\nu\Gamma \neq 0$, the solution of a new type appears, which is associated with the so-called Effective Cosmological Constant. Indeed, if the standard cosmological constant vanishes, $\Lambda = 0$, we obtain according to (52) that the parameter $H_\infty \neq 0$ plays the role of an effective Hubble constant. It appears as the result of integration over the whole time interval; it can be associated with the memory effect produced by the DE/DM interaction; we can introduce the effective cosmological constant $\Lambda_* \equiv 3H^2_\infty$, which appears just due to the interaction in the Dark sector of the Universe.
3. The regular bounce-type (see (90)) and super-inflationary (see (97)) solutions appear, when the characteristic polynomial of the key equation admits three coinciding roots $\sigma = 0$. Both exact solutions belong to the class of solutions describing the Little Rip scenaria.
4. The model of the DE/DM coupling based on the Volterra-type interaction kernel can solve the Coincidence problem. Indeed, the asymptotic value $\omega(\infty)$ of the function $\omega(x) = \frac{E(x)}{W(x)}$ is predetermined by the choice of parameters K_0 and ν entering the integral kernel (13), (14). Even if the initial value $E(1)$ of the Dark Matter energy density is vanishing, the final value $E(\infty)$ is of the order of the final value $W(\infty)$ due to the integral procedure of energy redistribution, which is described by the Volterra operator (see, e.g., the example (50)). In other words, the DE component of the Dark Fluid transmits the energy to the DM components during the whole evolution time interval, and this action "is remembering" by the Dark Fluid.
5. Optimization of the model parameters K_0, ν, Γ, γ using the observational data is the goal of our next work. However, some qualitative comments concerning the ways to distinguish the models of DE/DM interactions can be done based on the presented work. For instance, when one deals with the standard ΛCDM model, the profile of the energy density associated with the Dark

Energy is considered to be the horizontal straight line; the DM energy density profile decreases monotonically, thus providing the existence of some cross-point at some finite time moment. For this model the time derivative $\dot{W}(t)$ vanishes, so that $\dot{W}(t) = 0$ and $\dot{E}(t) \neq 0$ never coincide. In the model under discussion, the profiles $E(t)$ and $W(t)$ do not cross; these quantities tend to one another asymptotically. As for the rates of evolution, the quantities $\dot{W}(t)$ and $\dot{E}(t)$ can coincide identically (see, e.g., (96)), or can tend to one another asymptotically. In other words, one can distinguish the models of DE/DM interaction if to analyze and compare the rates of evolution of the DE and DM energy density scalars.

Author Contributions: The authors contributed equally to this work.

Funding: Russian Science Foundation (Project No. 16-12-10401).

Acknowledgments: The work was supported by Russian Science Foundation (Project No. 16-12-10401), and, partially, by the Program of Competitive Growth of Kazan Federal University.

Conflicts of Interest: The authors declare no conflict of interest.

Abbreviations

The following abbreviations are used in this manuscript:

DE	Dark Energy
DM	Dark Matter

References

1. Turner, M.S. The dark side of the universe: From Zwicky to accelerated expansion. *Phys. Rep.* **2000**, *333*, 619–635. [CrossRef]
2. Peebles, P.J.E.; Ratra, B. The Cosmological Constant and Dark Energy. *Rev. Mod. Phys.* **2003**, *75*, 559–606. [CrossRef]
3. Sahni, V. Dark Matter and Dark Energy. *Lect. Notes Phys.* **2004**, *653*, 141–180.
4. Copeland, E.J.; Sami, M.; Tsujikawa, S. Dynamics of dark energy. *Int. J. Mod. Phys. D* **2006**, *15*, 1753–1935. [CrossRef]
5. Sahni, V.; Starobinsky, A. Reconstructing Dark Energy. *Int. J. Mod. Phys. D* **2006**, *15*, 2105–2132. [CrossRef]
6. Capozziello, S.; Nojiri, S.; Odintsov, S.D. Unified phantom cosmology: Inflation, dark energy and dark matter under the same standard. *Phys. Lett. B* **2006**, *632*, 597–604. [CrossRef]
7. Nojiri, S.; Odintsov, S.D. Introduction to Modified Gravity and Gravitational Alternative for Dark Energy. *Int. J. Geom. Meth. Mod. Phys.* **2007**, *4*, 115–146. [CrossRef]
8. Frieman, J.; Turner, M.; Huterer, D. Dark Energy and the Accelerating Universe. *Ann. Rev. Astron. Astrophys.* **2008**, *46*, 385–432. [CrossRef]
9. Padmanabhan, T. Dark Energy and Gravity. *Gen. Relat. Gravit.* **2008**, *40*, 529–564. [CrossRef]
10. Bamba, K.; Odintsov, S.D. Inflation and late-time cosmic acceleration in non-minimal Maxwell-$F(R)$ gravity and the generation of large-scale magnetic fields. *JCAP* **2008**, *0804*. [CrossRef]
11. Nojiri, S.; Odintsov, S.D. Unified cosmic history in modified gravity: From F(R) theory to Lorentz non-invariant models. *Phys. Rept.* **2011**, *505*, 59–144. [CrossRef]
12. Nojiri, S.; Odintsov, S.D.; Oikonomou, V.K. Modified Gravity Theories on a Nutshell: Inflation, Bounce and Late-time Evolution. *Phys. Rept.* **2017**, *692*, 1–104. [CrossRef]
13. Bamba, K.; Capozziello, S.; Odintsov, S.D. Dark energy cosmology: The equivalent description via different theoretical models and cosmography tests. *Astrophys. Space Sci.* **2012**, *342*, 155–228. [CrossRef]
14. Del Popolo, A. Non-baryonic dark matter in cosmology. *Int. J. Mod. Phys. D* **2014**, *23*. [CrossRef]
15. Yepes, G.; Gottlober, S.; Hoffman Y. Dark matter in the Local Universe. *New Astron. Rev.* **2014**, *58*, 1–18. [CrossRef]
16. Zurek, K.M. Asymmetric Dark Matter: Theories, signatures, and constraints. *Phys. Rept.* **2014**, *537*, 91–121. [CrossRef]

17. Gleyzes, J.; Langlois, D.; Vernizzi, F. A unifying description of dark energy. *Int. J. Mod. Phys. D* **2015**, *23*, 1443010. [CrossRef]
18. Chimento, L.P.; Jacubi A.S.; Pavon, D.; Zimdahl, W. Interacting quinessence solution to the coincidence problem. *Phys. Rev. D* **2003**, *67*, 083513. [CrossRef]
19. Scherer, R.J. Phantom Dark Energy, Cosmic Doomsday, and the Coincidence Problem. *Phys. Rev. D* **2005**, *71*, 063519. [CrossRef]
20. Velten, H.E.S.; vom Marttens, R.F.; Zimdahl, W. Aspects of the cosmological "coincidence problem". *Eur. Phys. J. C* **2014**, *74*, 3160. [CrossRef]
21. Wang, B.; Abdalla, E.; Atrio-Barandela, F.; Pavon, D. Dark Matter and Dark Energy Interactions: Theoretical Challenges, Cosmological Implications and Observational Signatures. *arXiv* **2016**, arXiv:1603.08299.
22. Farrar, G.R.; Peebles, P.J.E. Interacting dark matter and dark energy. *Astrophys. J.* **2004**, *604*, 1–11. [CrossRef]
23. Zimdahl, W. Interacting dark energy and cosmological equations of state. *Int. J. Mod. Phys. D*, **2005**, *14*, 2319–2326. [CrossRef]
24. Del Campo, S.; Herrera, R.; Pavon, D. Interaction in the Dark Sector. *Phys. Rev. D* **2015**, *91*, 123539. [CrossRef]
25. Balakin, A.B.; Bochkarev, V.V. Archimedean-type force in a cosmic dark fluid. I. Exact solutions for the late-time accelerated expansion. *Phys. Rev. D* **2011**, *83*. [CrossRef]
26. Balakin, A.B.; Bochkarev, V.V. Archimedean-type force in a cosmic dark fluid. II. Qualitative and numerical study of a multistage universe expansion. *Phys. Rev. D* **2011**, *83*. [CrossRef]
27. Balakin, A.B.; Bochkarev, V.V. Archimedean-type force in a cosmic dark fluid. III. Big Rip, Little Rip and Cyclic solutions. *Phys. Rev. D* **2013**, *87*, doi:10.1103/PhysRevD.87.024006. [CrossRef]
28. Balakin, A.B.; Bochkarev, V.V.; Lemos, J.P.S. Light propagation with non-minimal couplings in a two-component cosmic dark fluid with an Archimedean-type force, and unlighted cosmological epochs. *Phys. Rev. D* **2012**, *85*. [CrossRef]
29. Balakin, A.B.; Dolbilova, N.N. Electrodynamic phenomena induced by a dark fluid: Analogs of pyromagnetic, piezoelectric, and striction effects. *Phys. Rev. D* **2014**, *89*. [CrossRef]
30. Balakin, A.B. Electrodynamics of a CosmicDark Fluid. *Symmetry* **2016**, *8*, 56. [CrossRef]
31. Jiménez, J.B.; Rubiera-Garcia, D.; Sáez-Gómez, D.; Salzana, V. Cosmological future singularities in interacting dark energy models. *Phys. Rev. D* **2016**, *94*, 123520. [CrossRef]
32. Brunner, H. *Volterra Integral Equations*; Cambridge University Press: Cambridge, UK, 2017.

symmetry

MDPI

Article

Tsallis, Rényi and Sharma-Mittal Holographic Dark Energy Models in Loop Quantum Cosmology

Abdul Jawad [1], Kazuharu Bamba [2,*], Muhammad Younas [1], Saba Qummer [1] and Shamaila Rani [1]

1 Department of Mathematics, COMSATS University Islamabad, Lahore Campus, Lahore-54000, Pakistan; jawadab181@yahoo.com or abduljawad@cuilahore.edu.pk (A.J.); muhammadyounas@cuilahore.edu.pk (M.Y.); sabaqummer143@gmail.com (S.Q.); drshamailarani@cuilahore.edu.pk (S.R.)

2 Division of Human Support System, Faculty of Symbiotic Systems Science, Fukushima University, Fukushima 960-1296, Japan

* Correspondence: bamba@sss.fukushima-u.ac.jp

Received: 7 October 2018; Accepted: 31 October 2018; Published: 13 Novemebr 2018

Abstract: The cosmic expansion phenomenon is being studied through the interaction of newly proposed dark energy models (Tsallis, Rényi and Sharma-Mittal holographic dark energy (HDE) models) with cold dark matter in the framework of loop quantum cosmology. We investigate different cosmic implications such as equation of state parameter, squared sound speed and cosmological plane (ω_d-ω'_d, ω_d and ω'_d represent the equation of state (EoS) parameter and its evolution, respectively). It is found that EoS parameter exhibits quintom like behavior of the universe for all three models of HDE. The squared speed of sound represents the stable behavior of Rényi HDE and Sharma-Mittal HDE at the latter epoch while unstable behavior for Tsallis HDE. Moreover, ω_d-ω'_d plane lies in the thawing region for all three HDE models.

Keywords: cosmoligical parameters; dark energy models; loop quantum cosmology

1. Introduction

Observational data from type Ia supernovae (SNIa) [1–4], the large scale structure (LSS) [5–8] and the cosmic microwave background (CMB), anisotropies [9–11], tell us that the universe undergoes an accelerated expansion at the present time. This expanding phase of the universe is supported by an unknown component called dark energy (DE) [12–14]. The simplest candidate for DE is the cosmological constant. This model consists of a fluid with negative pressure and positive energy density. The cosmological constant suffers from some problems such as the fine-tuning problem and the coincidence problem [12]. A feasible way to relieve the cosmic coincidence problem is to suppose an interaction between dark matter and DE. The cosmic coincidence problem can also be reduced by the appropriate choice of interaction between dark matter and DE [15–17]. The nature of DE is mysterious and unknown. Therefore, people have suggested various models for DE such as quintessence, tachyon [18], ghost [19], K-essence [20], phantom [21], Chaplygin gas [22], polytropic gas [23,24] and holographic dark energy (HDE) [25–27].

A second approach for understanding this strange component of the universe is gravitational modification in standard theories of gravity which results in modified theories of gravity that involve some invariants depending upon specific features such as torsion, scalars, curvature etc. The several modified theories are $f(R)$ theory [28–30], where f is a general differentiable function of the curvature scalar R, generalized teleparallel gravity, $f(T)$ [31–33] theory, contributing in the gravitational interaction through the torsion scalar T, Brans-Dicke theory, using a scalar field [34], Gauss-Bonnet theory and its modified version involving the Gauss-Bonnet invariant G [35,36], $f(R,T)$

theory where T is the trace of the energy-momentum tensor [37], etc. For recent reviews on modified gravity theories and dark energy problem, see, for instance [14,30,38–41].

The HDE is a promising candidate of DE, which has been studied extensively in the literature. It is based upon the holographic principle [29,42,43] that states the number of degrees of freedom of a system scales with its area instead of its volume. Cohen et al. studied that the DE should obey the holographic principle and constrained by the infrared (IR) cut-off [44]. Li has examined three choices for the IR cutoff as the Hubble horizon, the future event horizon and the particle horizon and also shown that only the future event horizon is able to provide the sufficient acceleration for the universe [45]. Sheykhi [46], developed the HDE model with Hubble horizon and argued that this model possesses the ability to explain the present state of the universe with the help of interaction of DE and cold dark matter (CDM).

Hu and Ling [47] studied the relationship between interacting, HDE and cosmological parameters through observational constraints. They investigated that HDE model is justified with the present observations in the low redshift region. They also tried to reduce the cosmic coincidence problem by taking different possibilities of time rate of change of the ratio of dark matter to HDE densities for a particular choice of interacting term. Ma et al. [48] explored observational signatures of interacting and non-interacting HDE with dark matter. In these models they also observed the big rip singularity in for different parameters by using a lot of recent observational schemes. They also found that the HDE models are slightly compatible with the observations as compared to the ΛCDM model.

In the context of thermodynamics, horizon entropy and DE can be effected by each other. Recently, due to the long-range nature of gravity, the mysterious nature of spacetime and pushed by the fact that the Bekenstein entropy is a non-extensive entropy measure. The generalized entropies, i.e., Tsallis and Rényi entropies have been assigned to the horizons to study the cosmological and gravitational phenomenon.

To study the cosmological and and gravitational phenomena many generalized entropy formalism has been applied but Tsallis and Rényi entropies generates the suitable model of universe. Sharma-Mittal HDE is compatible with universe expansion and whenever it is dominant in cosmos it is stable. Tsallis and Rényi entropies are attributed to the horizon to study the cosmic implications. Bekenstein entropy is also can be obtained by applying Tsallis statistics to the system. However, Tsallis and Rényi entropies can be recovered from Sharma-Mitall entropy by applying appropriate limits [49–51]. Recently, the HDE models such as Tsallis HDE [52] and Rényi HDE [53] and Sharma-Mittal HDE [54], have been studied extensively.

In classical cosmology, an important role is played by inflationary paradigm in understanding the problems of the big-bang model, by considering that the universe undergoes an expansion. However, classical general relativity (GR) fails when spacetime curvature approaches the Planck scale, due to the singularities where all physical quantities become infinite. So, the quantum gravity is considered to be necessary To interpret the circumstances in which classical (GR) breaks down [55]. In the last few decades, loop quantum gravity (LQG) has been widely applied to understand singularities in different black holes and spacetimes. LQG is not a complete theory, nor has its full stability with GR been established yet.

The loop quantum cosmology (LQC) is the application of LQG to the homogenous systems which removes the singularities. It holds the properties of a non-perturbative and background independent quantization of gravity [56]. The theory has numerous physical applications such as black hole physics and others.Recently many DE models have been studied in the context of LQC.

Here, we discuss the cosmological implications of Tsallis HDE, Rényi HDE and Sharma-Mittal HDE in the frame work of loop quantum cosmology (LQC) in the presence of the non-linear interaction between DE and dark matter [57]. This paper is organized as follows. In Section 2, we provides basics of LQC and DE models. Section 3 is devoted to cosmological parameters such as EoS parameter, cosmological plane and squared sound speed for Tsallis HDE, Rényi HDE and Sharma-Mittal HDEmodels. In the last section, we conclude the results.

2. Basic Equations

In these days, DE phenomenon has been discussed in the framework of LQC to describe the quantum effects on the universe. The LQC is the effective and modern application of quantization techniques from loop quantum gravity. In the context of LQC, many DE models have been studied in last few years. In modern cosmology, the cosmic coincidence problem by taking Chaplygin gas into account with dark matter was studied by Jamil et al. [58]. Chakraborty et al. [59], explored the modified Chaplygin gas in LQC. It is also found that with the help of Loop quantum effects one can avoid the future singularities appearing in the standard cosmology. The Friedmann equation in case of LQC [60–63] is given as

$$H^2 = \frac{\rho_{eff}}{3}\left(1-\frac{\rho_{eff}}{\rho_c}\right), \tag{1}$$

where, $H=\frac{\dot{a}}{a}$ is the Hubble parameter and dot represents the derivative of a, with respect to t and $\rho_{eff}=\rho_m+\rho_d$, ρ_m is matter density and ρ_d is DE density. Also, $\rho_c=\frac{\sqrt{3}}{16\pi^2\beta^3G^2h}$ where, β represents the dimensionless Barbero-Immirzi parameter and ρ_c stands for critical loop quantum density [60]. The different future singularities such as big bang and big rip can be avoided in LQC. It is observed that phantom DE with the negative pressure can push the universe towards the big rip singularity where all the physical objects loose the gravitational bounds and finally get dispersed.

We consider the interacting scenario between DE and cold dark matter (CDM) and thus the energy conservation equation turns to the following equations (we refer to the reader to [64,65])

$$\dot{\rho}_m + 3H\rho_m = -Q, \tag{2}$$

$$\dot{\rho}_d + 3H(\rho_d + p_d) = Q. \tag{3}$$

The cosmological evolution of the universe was analyzed by Arevalo and Acero [66], considering a non-linear interaction term of the general form

$$Q = 3dH\rho_{eff}^{a+b}\rho_m^c\rho_d^{-b-c}. \tag{4}$$

In the above equation the powers a, b and c characterize the interaction and d is a positive coupling constant. For $(a,b,c)=(1,-1,1)$ we can get the interaction, $Q=3dH\rho_m$ and for $(a,b,c)=(1,-1,0)$ one can get, $Q=3dH\rho_d$. In this present work we choose the interaction is given by

$$Q = 3dH\left(\frac{\rho_d^2}{\rho_m+\rho_d}\right), \tag{5}$$

this equation correspond to the choice $(a,b,c)=(1,-2,0)$ where, d is the coupling constant. The coupling of the dark matter and DE is a method to describe the evolution of the universe. The coupling constant sign decides the behavior of transformation between DE and dark matter. The positive sign indicates the decomposition of DE into dark matter while the negative sign shows the decomposition into dark matter to DE. However, the choice of positive sign of coupling constant is most favorable according to observational data. The negative sign of coupling parameter should be avoided due to the violation of laws of thermodynamics.

In this present work, we consider the power-law form of scale factor [67,68] as, $a(t)=a_0t^m$ and $H(t)=\frac{m}{t}$ where, $m>0$. At different values of m, we have different phases of the universe

- $0<m<1$ shows the decelerated phase of the universe.
- $m=\frac{2}{3}$, corresponds to the dust dominated era.
- $m=\frac{1}{2}$, leads to the radiation dominated era.
- $m>1$, shows the accelerated phase of the universe.

Next, we discuss the motivation and derivation of DE models.

2.1. Tsallis Holographic Dark Energy (HDE) Model

Li [45], has suggested the mathematical form of HDE as following constraint on its energy density $L^3\rho_d \leq Lm_p{}^2$. This inequality can be written as

$$\rho_d = \frac{3C^2 m_p{}^2}{L^2}, \tag{6}$$

where, $m_p{}^2 = (8\pi G)^{-1}$ represents the reduced Plank mass, C is a dimensionless quantity and L denotes the IR cutoff. HDE density provides the relation between Ultraviolet and IR cutoff. Many IR cutoffs has been presented for explaining the accelerated expansion of the universe for example Hubble, event, particle, Granda- Oliveros, Ricci scalar etc. Tsallis and Cirto [69], studied that the horizon entropy of the black hole can be modified as $S_\delta = \gamma A_\delta$, where δ the is non-additivity parameter, γ is an unknown constant and $A = 4\pi L^2$, represents the area of the horizon. Cohen et al. [44], proposed the mutual relationship between IR (L) cutoff, system entropy (S) and UV (Λ) cut off as

$$L^3\Lambda^3 \leq (S)^{\frac{3}{4}}, \tag{7}$$

which leads to

$$\Lambda^4 \leq \gamma(4\pi)^\delta L^{2\delta-4}, \tag{8}$$

where, Λ^4 is vacuum energy density and $\rho_d \sim \Lambda^4$. So, the Tsallis HDE density [52], is given as

$$\rho_d = BL^{2\delta-4}. \tag{9}$$

Here, B is an unknown parameter and IR cutoff is Hubble radius which is $L = \frac{1}{H}$. The density of Tsallis HDE model using the scale factor is given as

$$\rho_d = B\frac{t^{2\delta-4}}{m^{2\delta-4}}. \tag{10}$$

Inserting the value of ρ_d along with its derivative in Equation (3) it yields expression for pressure

$$p_d = \frac{1}{3}B\frac{t^{2\delta-8}}{m^{4\delta-3}}\left(\frac{-3t^4}{m^{-2\delta-1}} - \frac{6Bm^5 dt^{2\delta}}{\rho_c + \sqrt{\rho_c\left(\rho_c + \frac{12m^2}{t^2}\right)}} - 2m^{2\delta}t^4(\delta-2)\right). \tag{11}$$

2.2. Rényi HDE Model

We consider a system with n, states with probability distribution P_i and satisfies the condition $\Sigma_{i=1}^n P_i = 1$, Rényi and Tsallis entropies are well known parameters of generalized entropy is defined as

$$S = \frac{1}{\delta}\ln \Sigma_{i=1}^n P_i^{1-\delta}, \quad S_T = \frac{1}{\delta}\Sigma_{i=1}^n (P_i^{1-\delta} - P_i), \tag{12}$$

$\delta \equiv 1 - U$, where, U is a real parameter. Now combining above set of equations we find their mutual relation given as

$$S = \frac{1}{\delta}\ln(1 + \delta S_T). \tag{13}$$

In Equation (13), S belongs to the class of most general entropy functions of homogenous system. Recently, it is observed that Bekenstine entropy $S = \frac{A}{4}$, is in fact Tsallis entropy which gives the expression, $S = \frac{1}{\delta}\ln(1 + \delta\frac{A}{4})$, which is the Rényi entropy of the system.

With the help of following assumption $\rho_d dv \propto Tds$ we can get the Rényi HDE density as

$$\rho_d = \frac{3C^2H^2}{8\pi(1+\frac{\delta\pi}{H^2})}. \tag{14}$$

In our case, we suppose $8\pi = 1$ and consider the power-law scale factor we have the following expression for density

$$\rho_d = \frac{3C^2m^4}{t^2(m^2+\delta\pi t^2)}. \tag{15}$$

The pressure for this case is also obtained from Equation (3) with the help of Equation (15)

$$\begin{aligned} p_d &= \frac{C^2m^3}{t^4(m^2+\pi t^2\delta)^2}(-18C^2m^5d - \left(\rho_c + \sqrt{\rho_c\left(\rho_c + \frac{12m^2}{t^2}\right)}\right) \\ &\times t^2\left(m^2(3m-2)+(3m-4)\pi t^2\delta\right)) \left(\rho_c + \sqrt{\rho_c\left(\rho_c + \frac{12m^2}{t^2}\right)}\right)^{-1}. \end{aligned} \tag{16}$$

2.3. *Sharma-Mittal HDE Model*

From the Rényi entropy, we have the generalized entropy content of the system. Using Equation (12) Sharma-Mittal introduced a two parametric entropy and is defined as

$$S_{SM} = \frac{1}{1-r}\left((\Sigma_{i=1}^{n}P_i^{1-\delta})^{1-r/\delta} - 1\right), \tag{17}$$

where r is a new free parameter. We can observe that Rényi and Tsallis entropies can be recovered at the proper limits. In the limit $r \to 1$, Sharma-Mittal entropy becomes Rényi entropy while for $r \to \delta$, it is Tsallis entropy. Using Equation (12), in Equation (17) we have

$$S_{SM} = \frac{1}{R}((1+\delta S_T)^{R/\delta} - 1), \tag{18}$$

here, $R \equiv 1-r$. It has been recently argued that Bekenstine entropy is the proper candidate for Tsallis entropy. It allow us to replace S_T with S_B in above equation we have

$$S_{SM} = \frac{1}{R}((1+\delta\frac{A}{4})^{R/\delta} - 1). \tag{19}$$

The relation between UV (Λ) cutoff, IR (L) cut off and and system horizon (S) is given as $\Lambda^4 \propto \frac{S}{L^4}$ Now, taking $L \equiv \frac{1}{H} = \sqrt{A/4\pi}$, then the the energy density of DE given by

$$\rho_d = \frac{3C^2H^4}{8\pi R}[(1+\frac{\delta\pi}{H^2})^{R/\delta} - 1], \tag{20}$$

here, C^2 is an unknown free parameter. According to our assumptions we get the following expression for energy density

$$\rho_d = \frac{3C^2m^4}{Rt^4}[(1+\frac{t^2\delta\pi}{m^2})^{R/\delta} - 1]. \tag{21}$$

The expression for pressure is obtained as

$$p_d = \frac{C^2 m}{t^8}\left(-2\pi t^6\left(1+\frac{\pi t^2\delta}{m^2}\right)^{-1+R/\delta}+\left(\frac{\left(-1+\left(1+\frac{\pi t^2\delta}{m^2}\right)^{R/\delta}\right)}{R}\right)\right.$$
$$\left.\times\; m^2 t^4(-3m+4)-\frac{18C^2 m^7 d\left(-1+\left(1+\frac{\pi t^2\delta}{m^2}\right)^{R/\delta}\right)^2}{R^2\left(\rho_c+\sqrt{\rho_c\left(\rho_c+\frac{12m^2}{t^2}\right)}\right)}\right). \quad (22)$$

3. Cosmological Parameters

In this section, we will discuss the physical significance of cosmological parameters such as EoS parameter, squared sound speed ${v_s}^2$ and $\omega_d - \omega_d'$ plane.

3.1. EoS Parameter

To obtain EoS parameter we will use the following equation

$$\omega_d = \frac{p_d}{\rho_d}. \quad (23)$$

Here, ρ_d and p_d represents DE density and pressure of HDE model respectively. EoS parameter is used to categorized decelerated and accelerated phases of the universe. The DE dominated phase has following eras:

- $\omega_d = 0$ corresponds to non-relativistic matter.
- $-1 < \omega_d < -\frac{1}{3} \Rightarrow$ quintessence.
- $\omega_d = -1 \Rightarrow$ cosmological constant.
- $\omega_d < -1 \Rightarrow$ phantom.
- In this case $\omega_d > -1$, evolve across the boundary of cosmological constant shows the quintom behavior.

3.1.1. For Tsallis HDE

The EoS parameter for this model is evaluated by using Equations (10) and (11) in Equation (23)

$$\omega_d = \frac{p_d}{\rho_d} = -\frac{2Bm^{-2\delta+4}dt^{2\delta-4}}{\rho_c+\sqrt{\rho_c\left(\rho_c+\frac{12m^2}{t^2}\right)}} - \frac{2(\delta-2)}{3m} - 1. \quad (24)$$

To check the region of the universe, we plot ω_d versus z in Figure 1. The EoS parameter exhibits the quintom-like behavior of the universe as it crosses the phantom barrier for $\delta = 1.3$. However, for other values of δ, it remains in the quintessence region of the universe.

3.1.2. For Rényi HDE

The EoS parameter for Rényi HDE is evaluated by using Equations (15) and (16) in (23) we get the following expression

$$\omega_d = -\left(18C^2 m^5 d+\left(\rho_c+\sqrt{\rho_c\left(\rho_c+\frac{12m^2}{t^2}\right)}\right)t^2\left(m^2(3m-2)\right.\right.$$
$$\left.\left.+\;(3m-4)\pi t^2\delta\right)\right)\times\left(3m\left(\rho_c+\sqrt{\rho_c\left(\rho_c+\frac{12m^2}{t^2}\right)}\right)t^4\delta\left(m^2+\pi\right)\right)^{-1}. \quad (25)$$

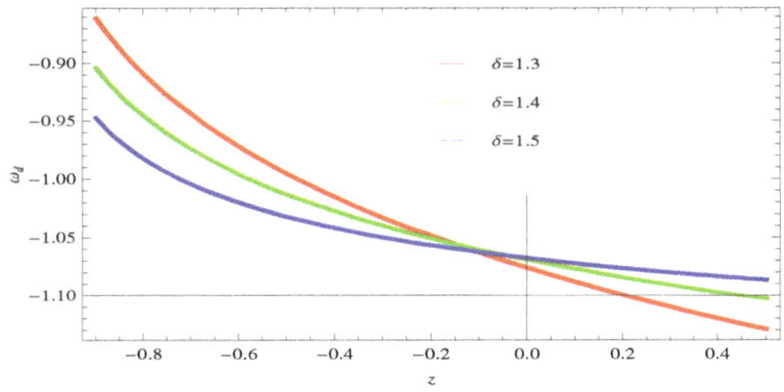

Figure 1. Plot of ω_d versus z for Tsallis HDE at different values of δ. Here $m = 2, \rho_c = 10, B = 2, a_0 = 1, C = 1, d = 1$.

The plot of above parameter versus z is shown in Figure 2. The trajectories of EoS parameter show the transition from phantom region to quintessence region by evolving the vacuum era of the universe. This is called quintom-like nature of the universe.

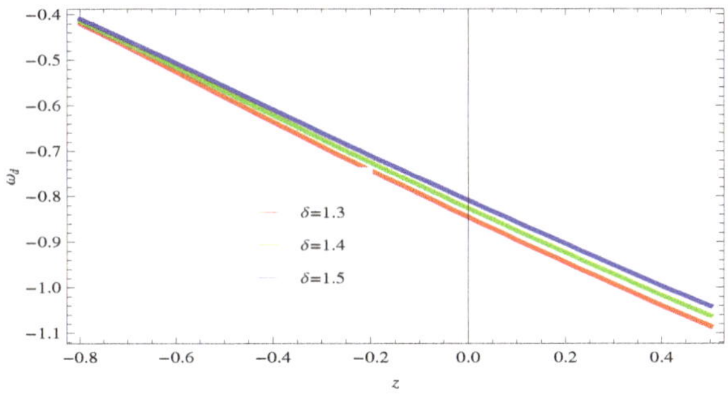

Figure 2. Plot of ω_d versus z for Rényi HDE at different values of δ. Here $m = 2, \rho_c = 10, C = 1, a_0 = 1, d = 1$.

3.1.3. For Sharma-Mittal HDE

The EoS for Sharma-Mittal HDE is obtained by substituting Equations (21) and (22) in Equation (23)

$$\begin{aligned} \omega_d &= R\left(-2\pi t^6 R\left(1+\frac{\pi t^2\delta}{m^2}\right)^{-1+R/\delta} + 4m^2t^4\left(-1+\left(1+\frac{\pi t^2\delta}{m^2}\right)^{R/\delta}\right)\right. \\ &- \left. 3m^3t^4\left(-1+\left(1+\frac{\pi t^2\delta}{m^2}\right)^{R/\delta}\right) - \frac{18C^2m^7d\left(-1+\left(1+\frac{\pi t^2\delta}{m^2}\right)^{R/\delta}\right)^2}{R\left(\rho_c+\sqrt{\rho_c\left(\rho_c+\frac{12m^2}{t^2}\right)}\right)}\right) \\ &\times \left(3m^3t^4\left(-1+\left(1+\frac{\pi t^2\delta}{m^2}\right)^{R/\delta}\right)\right)^{-1}. \end{aligned} \tag{26}$$

In Figure 3, the curves of EoS parameter shows quintom-like behavior of the universe.

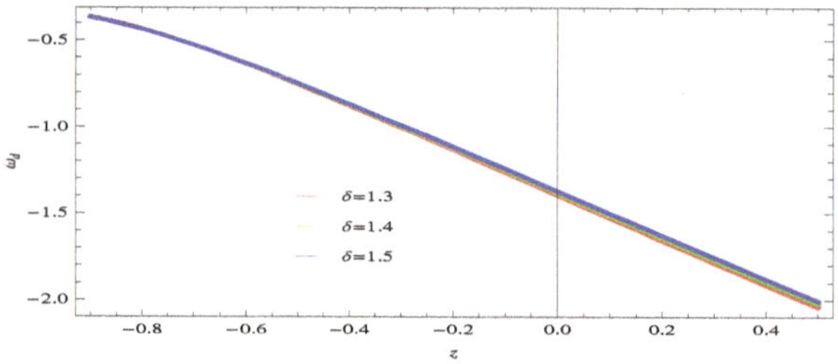

Figure 3. Plot of ω_d versus z for Sharma-Mittal HDE at different values of δ. Here $m = 2, \rho_c = 10$, $C = 1, a_0 = 1, d = 1, R = -2$.

3.2. *Stability Analysis*

To analyze the stability of the HDE models in LQC scenario we evaluate the squared sound speed which is given by

$$v_s^2 = \frac{dp_d}{d\rho_d} = \frac{dp_d/dt}{d\rho_d/dt}. \tag{27}$$

The sign of v_s^2 determines the stability of HDE model. For $v_s^2 > 0$, the model is stable otherwise it is unstable.

3.2.1. For Tsallis HDE

The expression for squared sound speed can be obtained by taking the derivative of Equations (10) and (11) with respect to t, and then substitute in Equation (27) we have

$$\begin{aligned} v_s{}^2 &= \frac{m^{-2\delta-1}}{6t^4(\delta-2)}\left(-\frac{72Bm^7d\sqrt{\rho_c\left(\rho_c+\frac{12m^2}{t^2}\right)}t^{2\delta}}{\left(\rho_c+\sqrt{\rho_c\left(\rho_c+\frac{12m^2}{t^2}\right)}\right)^2(12m^2+\rho_c t^2)}\right. \\ &- \left. 6m^{2\delta+1}t^4(\delta-2) - \frac{24Bm^5dt^{2\delta}(\delta-2)}{\rho_c+\sqrt{\rho_c\left(\rho_c+\frac{12m^2}{t^2}\right)}} - 4m^{2\delta}t^4(\delta-2)^2\right). \end{aligned} \tag{28}$$

Figure 4 shows the graph between $v_s{}^2$ and z. This graph is used to analyze the stability of the Tsallis HDE model under different parametric values. From the figure one can see that $v_s{}^2 < 0$ at the early, present and latter epoch. Hence this model shows unstable behavior at the present, early and latter epoch.

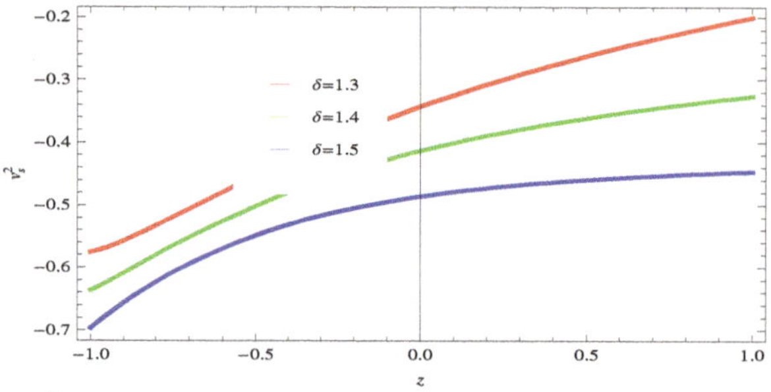

Figure 4. Plot of $v_s{}^2$ versus z for Tsallis HDE at different values of δ. Here $m = 1.1, \rho_c = 10, C = 1$, $B = -1.5, a_0 = 1, d = 1$.

3.2.2. For Rényi HDE

The expression of squared sound speed for Rényi HDE model can be obtained by using Equation (27) is given by

$$
\begin{aligned}
{v_s}^2 &= -\left(2\rho_c\left(t^2\left(12\rho_c m^2+6m^2\sqrt{\rho_c\left(\rho_c+\frac{12m^2}{t^2}\right)}+\rho_c^2t^2+\rho_c t^2\right.\right.\right.\\
&\times \left.\left.\sqrt{\rho_c\left(\rho_c+\frac{12m^2}{t^2}\right)}\right)\left(m^4(-2+3m)+3m^2(-2+3m)\pi t^2\delta\right.\\
&+ \left.2(3m-4)\pi^2t^4\delta^2\right)+18C^2m^5d\left(2\pi t^4\delta\left(\rho_c+\sqrt{\rho_c\left(\rho_c+\frac{12m^2}{t^2}\right)}\right)\right.\\
&+ \left.\left.\left.9m^4+m^2t^2\left(\rho_c+\sqrt{\rho_c\left(\rho_c+\frac{12m^2}{t^2}\right)}+21\pi\delta\right)\right)\right)\right)\left(3mt^4\left(2\pi t^2\delta\right.\right.\\
&+ \left.\left.m^2\right)\left(\rho_c+\sqrt{\rho_c\left(\rho_c+\frac{12m^2}{t^2}\right)}\right)^2\left(m^2+\pi t^2\delta\right)\sqrt{\rho_c\left(\rho_c+\frac{12m^2}{t^2}\right)}\right)^{-1}.
\end{aligned} \tag{29}
$$

In the present model, we significantly investigate the stability analysis of the Rényi HDE model which depends upon the different cosmological parameters. Here we take some specific values $\rho_c = 10, C = 1, d = 1$ for different values of δ. In Figure 5, the curves for ${v_s}^2$ shows the positive behavior for different values of δ at latter epoch which shows the stability the Rényi HDE model at the latter epoch.

3.2.3. For Sharma-Mittal HDE

Using Equation (27) and after some calculations we obtained the expression for squared sound speed which is given by

$$
\begin{aligned}
{v_s}^2 &= \left(2C^2m\left(3\pi t^2(2-m)\left(1+\frac{\pi t^2\delta}{m^2}\right)^{-1+R/\delta}+\frac{2m^2\pi^2t^4(-R+\delta)}{(m^2+\pi t^2\delta)^2}\right.\right.\\
&\times \left.\left(1+\frac{\pi t^2\delta}{m^2}^{R/\delta}\right)+\frac{2m^2}{R}(3m-4)\left(-1+\left(1+\frac{\pi t^2\delta}{m^2}\right)^{R/\delta}\right)\right)\\
&+ \frac{72C^4m^6d}{\left(\rho_cR+R\sqrt{\rho_c\left(\rho_c+\frac{12m^2}{t^2}\right)}\right)^2t^9}\left(\rho_c\left(\pi Rt^2\left(1+\frac{\pi t^2\delta}{m^2}\right)^{-1+R/\delta}\right.\right.\\
&\times \left.\left(1-\left(1+\frac{\pi t^2\delta}{m^2}\right)^{R/\delta}\right)+2m^2\left(-1+\left(1+\frac{\pi t^2\delta}{m^2}\right)^{R/\delta}\right)^2\right)\\
&+ \sqrt{\rho_c\left(\rho_c+\frac{12m^2}{t^2}\right)}\left(\frac{m^2\left(21m^2+2\rho_ct^2\right)\left(-1+\left(1+\frac{\pi t^2\delta}{m^2}\right)^{R/\delta}\right)^2}{12m^2+\rho_ct^2}\right.\\
&+ \left.\left.\left.\pi Rt^2\left(1+\frac{\pi t^2\delta}{m^2}\right)^{-1+R/\delta}\left(1-\left(1+\frac{\pi t^2\delta}{m^2}\right)^{R/\delta}\right)\right)\right)\right)\left(6C^2m^2\pi t^2\right.\\
&\times \left.\left(1+\frac{\pi t^2\delta}{m^2}\right)^{-1+R/\delta}-\frac{12C^2m^4}{R}\left(-1+\left(1+\frac{\pi t^2\delta}{m^2}\right)^{R/\delta}\right)\right)^{-1}.
\end{aligned} \tag{30}
$$

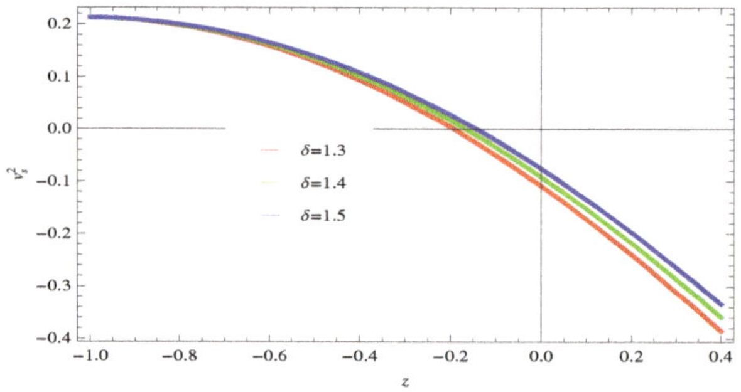

Figure 5. Plot of $v_s{}^2$ versus z for Rényi HDE at different values of δ. Here $m = 1.1, \rho_c = 10, C = 1,$ $a_0 = 1, d = 1$.

To check the stability of the Sharma-Mittal HDE model we plot a graph of $v_s{}^2$ against z. In Figure 6, the curves for $v_s{}^2$ shows the positive behavior for different values of δ at latter epoch which shows the stability the Sharma-Mittal HDE model at the latter epoch.

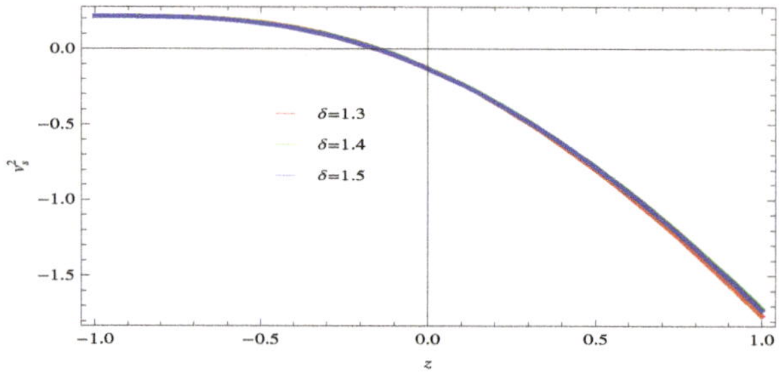

Figure 6. Plot of $v_s{}^2$ versus z for Sharma-Mittal HDE at different values of δ. Here $m = 2.5, \rho_c = 10,$ $C = 1, d = 1, R = -2, 3$ and 4.

3.3. ω_d-ω'_d Plane

Caldwell and Linder [70], proposed the ω_d-ω'_d plane to explain the dynamical property of DE model in quintessence scalar field. Here, ω_d is EoS parameter and ω'_d is its evolutionary form where prime denotes the derivative with respect to $\ln a$. They divided the ω_d-ω'_d plane in two parts, the thawing part ($\omega_d < 0, \omega'_d > 0$) is the region where EoS parameter nearly evolves from $\omega_d < -1$, increases with time while its evolution parameter expresses positive behavior, and the freezing part ($\omega_d < 0, \omega'_d < 0$) is the evolution parameter for EoS parameter remains negative.

3.3.1. For Tsallis HDE

The expression ω'_d for THDE can be obtained by taking the derivative of Equation (24) with respect to $\ln a$ for THDE.

$$\begin{aligned}\omega'_d &= 4\rho_c B m^{-2\delta+3} d t^{2\delta-4}\left(\left(\rho_c+\sqrt{\rho_c\left(\rho_c+\frac{12m^2}{t^2}\right)}\right)(\delta-2)+6m^2(2\delta-3)\right)\\ &\times\left(\left(\rho_c+\sqrt{\rho_c\left(\rho_c+\frac{12m^2}{t^2}\right)}\right)^2\sqrt{\rho_c\left(\rho_c+\frac{12m^2}{t^2}\right)}\right)^{-1}. \qquad (31)\end{aligned}$$

In Figure 7, $\omega_d - \omega'_d$, plane is used to check the region for this Tsallis HDE model. It can be seen that the value of ω'_d decreases as we increase the value of ω_d. We can see that $\omega_d < 0$ and $\omega'_d > 0$ for all values of δ, which corresponds to the thawing region of the universe.

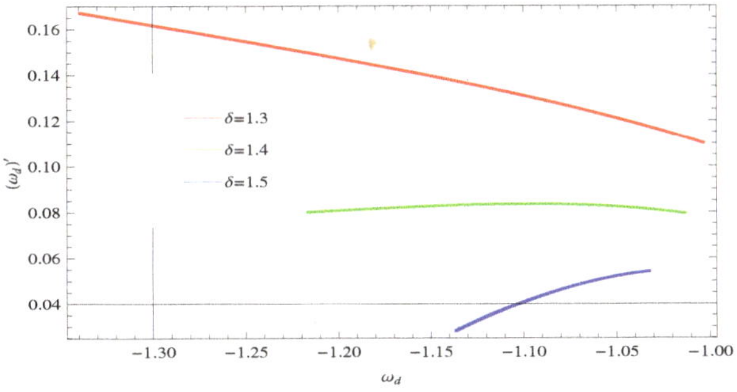

Figure 7. Plot of ω_d versus ω'_d for Tsallis HDE at different values of δ. Here $m = 2, \rho_c = 10, C = 1, B = 2, a_0 = 1, d = 1$.

3.3.2. For Rényi HDE

The expression of ω'_d for Rényi HDE can be obtained by taking the derivative of EoS parameter in Equation (25) with respect to $\ln a$ for Rényi HDE.

$$\begin{aligned} \omega'_d &= 4\rho_c\pi t^2\left(12m^2+\rho_c t^2\right)\delta + 3c^2md\left(6m^4\sqrt{\rho_c\left(\rho_c+\frac{12m^2}{t^2}\right)}+\right. \\ &+ 6m^2\pi(-2\rho_c+3\sqrt{\rho_c\left(\rho_c+\frac{12m^2}{t^2}\right)})\, t^2\delta + \rho_c\pi t^4\delta\,(-\rho_c \\ &+ \left.\sqrt{\rho_c\left(\rho_c+\frac{12m^2}{t^2}\right)}\right)\Bigg) \times \left(3\rho_c\left(12m^2+\rho_c t^2\right)\left(m^2+\pi t^2\delta\right)^2\right)^{-1} \end{aligned} \quad (32)$$

In Figure 8, we find the region on the $\omega_d - \omega'_d$, for the model under consideration. In this plane, the EoS parameter corresponds to the quintessence era, also $\omega_d - \omega'_d$ shows that ($\omega_d < 0, \omega'_d > 0$ which leads to the thawing region.

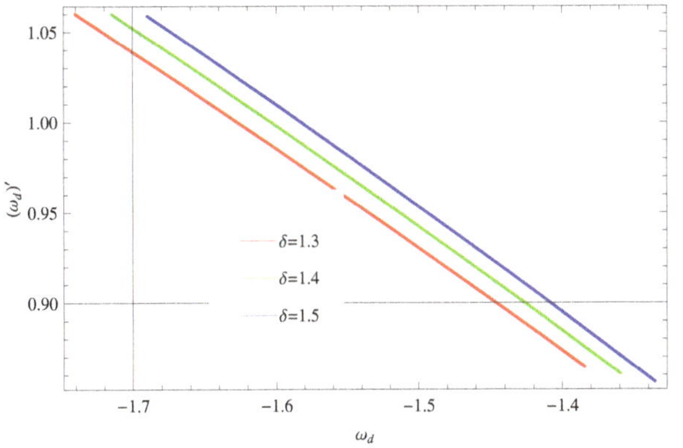

Figure 8. Plot of ω_d versus ω'_d for Rényi HDE at different values of δ. Here $m = 2, \rho_c = 10, a_0 = 1, C = 1, d = 1$.

3.3.3. For Sharma-Mittal HDE

The expression of ω'_d for Sharma-Mittal HDE can be obtained by taking the derivative of Equation (26) with respect to $\ln a$ for Sharma-Mittal HDE.

$$\begin{aligned}
\omega'_d &= -\frac{4\pi R t^2 \left(1+\frac{\pi t^2 \delta}{m^2}\right)^{R/\delta}\left(-\pi R t^2 + m^2\left(-1+\left(1+\frac{\pi t^2\delta}{m^2}\right)^{R/\delta}\right)\right)}{3m^2\left(m^2+\pi t^2\delta\right)^2\left(-1+\left(1+\frac{\pi t^2\delta}{m^2}\right)^{R/\delta}\right)^2} \\
&+ 12C^2 m^3 d\left(\rho_c\left(-2+\frac{\left(2m^2-\pi t^2(R-2\delta)\right)\left(1+\frac{\pi t^2\delta}{m^2}\right)^{R/\delta}}{m^2+\pi t^2\delta}\right)\right. \\
&+ \frac{\sqrt{\rho_c\left(\rho_c+\frac{12m^2}{t^2}\right)}}{\left(12m^2+\rho_c t^2\right)\left(m^2+\pi t^2\delta\right)}\left(-2\left(9m^2+\rho_c t^2\right)\left(m^2+\pi t^2\delta\right)\right. \\
&+ \left.\left.\left(1+\frac{\pi t^2\delta}{m^2}\right)^{R/\delta}\left(18m^4-\rho_c\pi t^4(R-2\delta)+2m^2t^2(\rho_c-6\pi R+9\pi\delta)\right)\right)\right) \\
&\times \left(R\left(\rho_c+\sqrt{\rho_c\left(\rho_c+\frac{12m^2}{t^2}\right)}\right)^2 t^4\right)^{-1}.
\end{aligned} \tag{33}$$

To find out the region of the $\omega_d - \omega'_d$, for the model which is under consideration we construct the $\omega_d - \omega'_d$ plane for different parametric values. In Figure 9 we can see that $\omega_d - \omega'_d$ shows that $(\omega_d < 0, \omega'_d > 0)$ which corresponds to the thawing region.

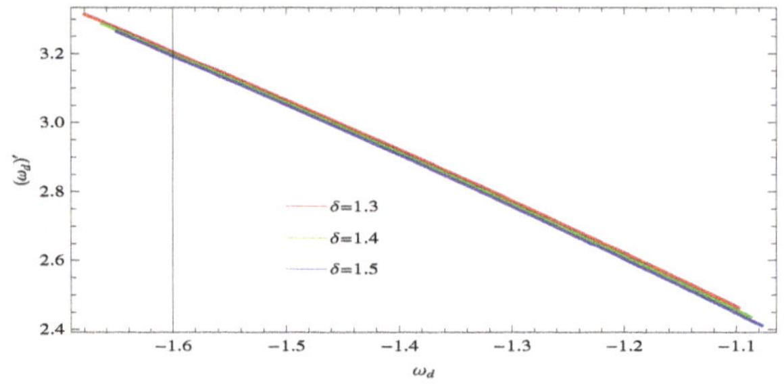

Figure 9. Plot of ω'_d versus ω_d for Sharma-Mittal HDE at different values of δ. Here $m = \frac{2}{3}, \rho_c = 10$, $C = 1, a_0 = 1, d = 1, R = -2$

4. Concluding Remarks

In this paper, cosmological implications with the help of non-linear interaction terms between dark matter and DE models have been discussed in the framework of LQC. For this purpose, we have constructed the EoS parameter, the squared sound speed and $\omega_d - \omega'_d$ plane and discussed their graphical behavior.

- The trajectories of EoS parameter in all three models HDE exhibit the quintom-like nature of the universe as it shows transition of the universe from phantom era (at early and present) towards quintessence era (latter epoch) by evolving phantom barrier.
- To analyze the stability of the Tsallis HDE, Rényi HDE and Sharma-Mittal HDE models we check the graphical behavior of squared sound speed. For Tsallis HDE model, it is observed that $v_s{}^2 < 0$ for all values of z which leads to the instability of this model. On the other hand, for Rényi HDE, the squared speed of sound shows unstable behavior at the early and present epoch while leads

to the stability at the latter epoch. The same behavior of the squared speed of sound has been observed in case of Sharma-Mittal HDE model.

- Also, $\omega_d - \omega'_d$ corresponds to thawing region ($\omega_d < 0$ and $\omega'_d > 0$) for all three models of HDE.

Author Contributions: A.J. has given the idea about this topic and drafted this article, calculations have been done by M.Y. and S.Q. Furthermore, graphs were drawn by S.R. and proofreading has been conducted by K.B.

Funding: This research received no external funding.

Conflicts of Interest: The authors declare no conflict of interest.

References

1. Riess, A.G.; Filippenko, A.V.; Challis, P.; Clocchiattia, A.; Diercks, A.; Garnavich, P.M.; Gilliland, R.L.; Hogan, C.J.; Jha, S.; Kirshner, R.P.; et al. Observational evidence from supernovae for an accelerating Universe and a cosmological constant. *Astron. J.* **1998**, *116*, 1009–1038. [CrossRef]
2. Perlmutter, S.; Aldering, G.; Goldhaber, G.; Knop, R.A.; Nugent, P.; Castro, P.G.; Deustua, S.; Fabbro, S.; Goobar, A.; Groom, D.E.; et al. Measurements of Omega and Lambda from 42 High-Redshift Supernovae. *Astrophys. J.* **1999**, *517*, 565–586. [CrossRef]
3. Bernardis, P.; Ade, P.A.R.; Bock, J.J.; Bond, J.R.; Borrill, J.; Boscaleri, A.; Coble, K.; Crill, B.P.; DeGasperis, G.; Farese, P.C.; et al. A flat Universe from high-resolution maps of the cosmic microwave background radiation. *Nature* **2000**, *404*, 955–959. [CrossRef] [PubMed]
4. Perlmutter, S.; Aldering, G.; Amanullah, R.; Astier, P.; Blanc, G.; Burns, M.S.; Conley, A.; Deustua, S.E.; Doi, M.; Ellis, R.; et al. New Constraints on and w from an Independent Set of 11 High-Redshift Supernovae Observed with the Hubble Space Telescope. *Astrophys. J.* **2003**, *598*, 102–137.
5. Colless, M.; Dalton, G.; Maddox, S.; Sutherland, W.; Norberg, P.; Cole, S.; Hawthrone, J.B.; Bridges, T.; Cannon, R.; Collins, C.; et al. The 2dF Glaxy Redshift Survey: Spectra and redshift. *Mon. Not. R. Astron. Soc.* **2001**, *328*, 1039–1063. [CrossRef]
6. Tegmark, M.; Strauss, M.A.; Blanton, M.R.; Abazajian, K.; Dodelson, S.; Sandvik, H.; Wang, X.; Weinberg, D.H.; Zehavi, I.; Bahcall, N.A.; et al. Cosmological parameters from SDSS and WMAP. *Phys. Rev. D* **2004**, *69*, 103501–103526. [CrossRef]
7. Cole, S.; Percival, W.J.; Peacock, R.A.; Norberg, P.; Baugh, C.M.; Frenk, C.S.; Baldry, I.; Bland-Hawthorn, J.; Bridges, T.; Cannon, R.; et al. The 2dF Galaxy Redshift Survey: Power-spectrum analysis of the final dataset and cosmological implications. *Mon. Not. R. Astron. Soc.* **2005**, *362*, 505–534. [CrossRef]
8. Springel, V.; Frenk, C.S.; White, S.M.D. The large-scale structure of the Universe. *Nature* **2006**, *440*, 1137–1144. [CrossRef] [PubMed]
9. Hanany, S.; Ade, P.; Balbi, A.; Bock, J.; Borrill, J.; Boscaleri, A.; Bernardis, P.; Ferreira, P.G.; Hristov, V.V.; Jaffe, A.H.; et al. MAXIMA-1: A Measurement of the Cosmic Microwave Background Anisotropy on angular scales of 10 arcminutes to 5 degrees. *Astrophys. J. Lett.* **2000**, *545*, L5–L9. [CrossRef]
10. Netterfield, C.B.; Ade, P.A.R.; Bock, J.J.; Bond, J.R.; Borrill, J.; Boscaleri, A.; Coble, K.; Contaldi, C.R.; Crill, B.P.; Bernardis, P.; et al. A measurement by BOOMERANG of multiple peaks in the angular power spectrum of the cosmic microwave background. *Astrophys. J.* **2002**, *571*, 604–614. [CrossRef]
11. Spergel, D.N.; Verde, L.; Peiris, H.V.; Komatsu, E.; Nolta, M.R.; Bennett, C.L.; Halpern, M.; Hinshaw, G.; Jarosik, N.; Kogut, A.; et al. First Year Wilkinson Microwave Anisotropy Probe (WMAP) Observations: Determination of Cosmological Parameters. *Astrophys. J. Suppl.* **2003**, *148*, 175–194. [CrossRef]
12. Roos, M. *Introduction to Cosmology*; John Wiley and Sons: Chichester, UK, 2003.
13. Nojiri, S.; Odintsov, S.D. The new form of the equation of state for dark energy fluid and accelerating universe. *Phys. Lett. B* **2006**, *639*, 144–150. [CrossRef]
14. Bamba, K.; Capozziello, S.; Nojiri, S.; Odintsov, S.D. Dark energy cosmology: The equivalent description via different theoretical models and cosmography tests. *Astrophys. Space Sci.* **2012**, 342, 155–228. [CrossRef]
15. Zimdahl, W.; Pavon, D. Scaling Cosmology. *Gen. Rel. Grav.* **2003**, *35*, 413–422. [CrossRef]
16. Del Campo, S.; Herrera, R.; Olivares, G.; Pavon, D. Interacting models of soft coincidence. *Phys. Rev. D* **2006**, *74*, 023501-9. [CrossRef]
17. Sadjadi, H.M.; Alimohammadi, M. Cosmological coincidence problem in interacting dark energy models. *Phys. Rev. D* **2006**, *74*, 103007-7. [CrossRef]

18. Setare, M.R. The holographic dark energy in non-flat Brans-Dicke cosmology. *Phys. Lett. B* **2007**, *644*, 99–103. [CrossRef]
19. Malekjani, M.; Naderi, T.; Pace, F. Effects of ghost dark energy perturbations on the evolution of spherical overdensities. *MNRAS B* **2015**, *453*, 4148–4158. [CrossRef]
20. Chiba, T.; Okabe, T.; Yamaguchi, M. Kinetically driven quintessence. *Phys. Rev. D* **2000**, *62*, 023511–023518. [CrossRef]
21. Nojiri, S.; Odintsov, S.D. Quantum de Sitter cosmology and phantom matter. *Phys. Lett. B* **2003**, *562*, 147–152. [CrossRef]
22. Pasquier, V.; Moschella, U.; Kamenshchick, A.Y.; Moschella, U.; Pasquier, V. An alternative to quintessence. *Phys. Lett. B* **2001**, *511*, 265–268.
23. Kleidis, K.; Spyrou, N.K. Polytropic dark matter flows illuminate dark energy and accelerated expansion. *Astron. Astrographys.* **2015**, *576*, A23. [CrossRef]
24. Kleidis, K.; Spyrou, N.K. Dark Energy: The Shadowy Reflection of Dark Matter. *Entropy* **2016**, *18*, 94. [CrossRef]
25. Weinberg, S. The cosmological constant problem. *Rev. Mod. Phys.* **1998**, *61*, 1–20. [CrossRef]
26. Copeland, E.J.; Sami, M.; Tsujikawa, S. Dynamics of dark energy. *Int. J. Mod. Phys. D* **2006**, *15*, 1753–1935 . [CrossRef]
27. Padmanabhan, T. Cosmological Constant—The Weight of the Vacuum. *Phys. Rep.* **2003**, *380*, 235–320. [CrossRef]
28. Felice, A.D.; Tsujikawa, S. $f(R)$ Theories. *Living Rev. Relativ.* **2010**, *13*, 3–163. [CrossRef] [PubMed]
29. Sahni, V. The Cosmological Constant Problem and Quintessence. *Class. Quant. Grav.* **2002**, *19*, 3435–3448. [CrossRef]
30. Nojiri, S.; Odintsov, S.D. Unified cosmic history in modified gravity: From $F(R)$ theory to Lorentz non-invariant models. *Phys. Rep.* **2011**, *505*, 59–144. [CrossRef]
31. Linder, E.V. Einstein's other gravity and the acceleration of the Universe. *Phys. Rev. D* **2010**, *81*, 127301–127303. [CrossRef]
32. Ferraro, R.; Fiorini, F. Non-trivial frames for $f(T)$ theories of gravity and beyond. *Phys. Lett. B* **2011**, *702*, 75–80. [CrossRef]
33. Sharif, M.; Rani, S. Wormhole solutions in $f(T)$ gravity with noncommutative geometry. *Phys. Rev. D* **2013**, *88*, 123501–123510. [CrossRef]
34. Brans, C.H.; Dicke, R.H. Mach's Principle and a Relativistic Theory of Gravitation. *Phys. Rev. D* **1961**, *124*, 925–935. [CrossRef]
35. Nojiri, S.; Odintsov, S.D. Modified Gauss-Bonnet theory as gravitational alternative for dark energy. *Phys. Lett. B* **2005**, *631*, 1–6. [CrossRef]
36. Cognola, G.; Elizalde, E.; Nojiri, S.; Odintsov, S.D.; Zerbini, S. Dark energy in modified Gauss-Bonnet gravity: Late-time acceleration and the hierarchy problem. *Phys. Rev. D* **2006**, *73*, 084007–084022. [CrossRef]
37. Harko, T.; Francisco, S.N.; Lobo, F.S.N.; Nojiri, S.; Odintsov, S.D. $f(R,T)$ gravity. *Phys. Rev. D* **2011**, *84*, 024020–024031. [CrossRef]
38. Capozziello, S.; De Laurentis, M. Extended Theories of Gravity. *Phys. Rep.* **2011**, *509*, 167–321. [CrossRef]
39. Capozziello, S.; Faraoni, V. *Beyond Einstein Gravity: A Survey of Gravitational Theories for Cosmology and Astrophysics*, 1st ed.; Springer: Dordrecht, The Netherlands, 2010; Volume 170.
40. Bamba, K.; Odintsov, S. Inflationary cosmology in modified gravity theories. *Symmetry* **2015**, *7*, 220–240. [CrossRef]
41. Nojiri, S.; Odintsov, S.D.; Oikonomou, V.K. Modified Gravity Theories on a Nutshell: Inflation, Bounce and Late-time Evolution. *Phys. Rep.* **2017**, *692*, 1–104. [CrossRef]
42. Susskind, L. The World as a Hologram. *J. Math. Phys.* **1995**, *36*, 6377–6396. [CrossRef]
43. Nojiri, S.; Odintsov, S.D. Unifying phantom inflation with late-time acceleration: Scalar phantom non-phantom transition model and generalized holographic dark energy. *Gen. Rel. Grav.* **2006**, *38*, 1285–1304. [CrossRef]
44. Cohen, A.G.; Kaplan, D.B.; Nelson, A.E. Effective Field Theory, Black Holes, and the Cosmological Constant. *Phys. Rev. Lett.* **1999**, *82*, 4971–4974. [CrossRef]
45. Li, M. A Model of Holographic Dark Energy. *Phys. Lett. B* **2004**, *603*, 1–5. [CrossRef]
46. Sheykhi, A. Holographic scalar field models of dark energy. *Phys. Rev. D* **2011**, *84*, 107302–107306. [CrossRef]

47. Hu, B.; Ling, Y. Interacting dark energy, holographic principle, and coincidence problem. *Phys. Rev. D* **2006**, *73*, 123510–123518. [CrossRef]
48. Ma, Y.Z.; Gong, Y.; Chen, X. Features of holographic dark energy under combined cosmological constraints. *Eur. Phys. J. C* **2009**, *60*, 303–315. [CrossRef]
49. Tsallis, C. Possible generalization of Boltzmann-Gibbs statistics. *J. Stat. Phys.* **1988**, *52*, 479–487. [CrossRef]
50. Rényi, A . On measures of entropy and Information. In *Proceedings of the 4th Berkely Symposium on Mathematics, Statistics and Probability*; University California Press: Berkeley, CA, USA, 1961; Volume 1, pp. 547–561.
51. Sharma, B.D.; Mittal, D.P. New nonadditive measures of entropy for discrete probability distributions. *J. Math. Sci.* **1975**, *10*, 28–40.
52. Tavayef, M.; Sheykhi, A.; Bamba, K.; Moradpour, H. Tsallis holographic dark energy. *Phys. Lett. B* **2018**, *781*, 195–200. [CrossRef]
53. Moradpour, H.; Moosavi, S.A.; Lobo, I.P.; Graca, J.P.M.; Jawad, A.; Salako, I.G. Thermodynamic approach to holographic dark energy and the Renyi entropy. *Eur. Phys. J. C* **2018**, *829*, 78–83. [CrossRef]
54. Jahromi, A.S.; Moosavi, S.A.; Moradpour, H.; Grac, J.P.M.; Lobo, I.P.; Salako, I.G.; Jawad, A. Generalized entropy formalism and a new holographic dark energy model. *Phys. Lett. B* **2018**, *780*, 21–24. [CrossRef]
55. Bojowald, M. Loop Quantum Cosmology. *Living Rev. Relativ.* **2008**, *11*, 4–134. [CrossRef] [PubMed]
56. Rovelli, C. Loop Quantum Gravity. *Living Rev. Relativ.* **2008**, *11*, 5–73. [CrossRef] [PubMed]
57. Oliveros, A. Slow-roll inflation from massive vector fields non-minimally coupled to gravity. *Astrphys. Space Sci.* **2017**, *362*, 1–6. [CrossRef]
58. Jamil, M.; Debnath, U. Interacting modified Chaplygin gas in loop quantum cosmology. *Astrophys. Space Sci.* **2011**, *333*, 3–8. [CrossRef]
59. Chakraborty, S.; Debnath, U.; Ranjit, C. Observational constants of modified Chaplygin gas in loop quantum cosmology. *Eur. Phys. J. C* **2012**, *72*, 2101–2108. [CrossRef]
60. Ashtekar, A. Albert Einstein Century International conference. *AIP Conf. Proc.* **2006**, *861*, 3–14.
61. Ashtekar, A.; Pawlowski, T.; Singh, P. Quantum Nature of the Big Bang: Improved dynamics. *Phys. Rev. D* **2006**, *74*, 084003–084042. [CrossRef]
62. Singh, P. Loop cosmological dynamics and dualities with Randall-Sundrum braneworlds. *Phys. Rev. D* **2006**, *73*, 063508–063516. [CrossRef]
63. Sami, M.; Singh, P.; Tsujikawa, S. Avoidance of future singularities in loop quantum cosmology. *Phys. Rev. D* **2006**, *74*, 043514–043519. [CrossRef]
64. Zimdahl, W.; Pavon, D.; Chimento, L.P. Interacting Quintessence. *Phys. Lett. B* **2001**, *521*, 133–138. [CrossRef]
65. Li, M.; Li, X.; Wang, S.; Wang, Y. Dark Energy. *Commun. Theor. Phys.* **2011**, *56*, 525–604. [CrossRef]
66. Arevalo, F.; Bacalhau, A.P.R.; Zimdahl, W. Cosmological dynamics with non-linear interactions. *Class. Quant. Grav.* **2012**, *29*, 235001–235023 . [CrossRef]
67. Sharif, M.; Zubair, M. Energy Conditions Constraints and Stability of Power Law Solutions in $f(R,T)$ Gravity. *J. Phys. Soc. Jpn.* **2013**, *82*, 014002–014010. [CrossRef]
68. Sharif, M.; Zubair, M. Cosmological reconstruction and stability in $f(R,T)$ gravity. *Gen. Relativ. Gravit.* **2014**, *46*, 1723–1752. [CrossRef]
69. Tsallis, C.; Citro, L.J.L. Black hole thermodynamical entropy. *Eur. Phys. J. C* **2013**, *73*, 2487–2493. [CrossRef]
70. Caldwell, R.R.; Linder, E.V. Limits of Quintessence. *Phys. Rev. Lett.* **2005**, *95*, 141301–141304. [CrossRef] [PubMed]

symmetry

MDPI

Article

Spacetime Symmetry and LemaîTre Class Dark Energy Models

Irina Dymnikova [1,2,*] and Anna Dobosz [2]

1 A.F. Ioffe Physico-Technical Institute, Politekhnicheskaja 26, St. Petersburg 194021, Russia
2 Department of Mathematics and Computer Science, University of Warmia and Mazury, Słoneczna 54, 10-710 Olsztyn, Poland; dobosz@uwm.edu.pl
* Correspondence: irina@uwm.edu.pl; Tel.: +48-601-362-059

Received: 14 December 2018; Accepted: 11 January 2019; Published: 15 January 2019

Abstract: We present the regular cosmological models of the Lemaître class with time-dependent and spatially inhomogeneous vacuum dark energy, which describe relaxation of the cosmological constant from its value powering inflation to the final non-zero value responsible for the present acceleration in the frame of one self-consistent theoretical scheme based on the algebraic classification of stress-energy tensors and spacetime symmetry directly related to their structure. Cosmological evolution starts with the nonsingular non-simultaneous de Sitter bang, followed by the Kasner-type anisotropic expansion, and goes towards the present de Sitter state. Spacetime symmetry provides a mechanism of reducing cosmological constant to a certain non-zero value involving the holographic principle which singles out the special class of the Lemaître dark energy models with the global structure of the de Sitter spacetime. For this class cosmological evolution is guided by quantum evaporation of the cosmological horizon whose dynamics entirely determines the final value of the cosmological constant. For the choice of the density profile modeling vacuum polarization in a spherical gravitational field and the GUT scale for the inflationary value of cosmological constant, its final value agrees with that given by observations. Anisotropy grows quickly at the postinflationary stage, then remains constant and decreases to $A < 10^{-6}$ when the vacuum density starts to dominate.

Keywords: dark energy; spacetime symmetry; de Sitter vacuum

---

1. Introduction

Observational data convincingly testify that our Universe is dominated at above 72% of its density by a dark energy with a negative pressure $p = w\rho$, $w < -1/3$ [1–4] (for a review [5]), with the best fit $w = -1$ corresponding to the cosmological constant λ associated with the vacuum density $\rho_{vac} = (8\pi G)^{-1}\lambda$ [6–11]. However, the Einstein cosmological term $\lambda g_{\mu\nu} = 8\pi G\rho_{vac} g_{\mu\nu}$ cannot be associated with the vacuum dark energy, for two reasons: (i) The quantum field theory estimates ρ_{vac} by $\rho_{Pl} = 5 \times 10^{93}$ g cm^{-3}. Confrontation of this estimate with the observational value $\rho_{obs} \simeq 1.4 \times 10^{-123}\rho_{Pl}$ produces the fine-tuning problem [12]. (ii) A large value $\Lambda = 8\pi G\rho_{vac}$ is needed for powering the inflationary stage of the Universe evolution, the observational data yield its much smaller today value, while the Einstein equations require $\rho_{vac} = const$.

The Planck density ρ_{Pl} provides a natural cutoff on zero-point quantum vacuum fluctuations in QFT which give rise to ρ_{vac} [12], as well as on applicability of classical General Relativity [13]. Proposals to solve the fine-tuning problem typically go beyond the classical General Relativity and involve both its quantization and modifications and extensions on the classical level.

As early as in 1978 Hawking supposed that quantum fluctuations in spacetime topology at small scales may provide a mechanism for vanishing a cosmological constant [14]. The point is that at the very short distances a spacetime itself undergoes uncertainties typical for quantum systems that implies impossibility to determine simultaneously a spacetime geometry and its changes. According to Wheeler, spacetime beyond the Planck scale has foam-like structure and may involve substantial changes in geometry and topology [13] including, in particular, quantum wormholes [15,16].

On the other hand, the effective field theory still valid near the Planck energy scale, and the question why the observed cosmological constant is so small as compared with ρ_{Pl} is equivalent to the question why the universe is accelerating at present [17]. In default of a some fundamental symmetry which would reduce the value of λ to zero, most of proposals were focused on searching for physical mechanisms which could provide its small value today.

The Causal Entropic Principle, which requires disregarding the causally disconnected spacetime regions and maximizing the entropy produced within a causally connected region, has been applied for calculation the expected value for a present cosmological constant in [18] in the reasonable agreement with its observational value. Causality arguments were also used for analysis of relaxation of the cosmological constant to its present value in the context of the eternal inflation in the multiverse and a string landscape [19].

The non-singular model with the curvature square term, a cosmological term and a dilaton field ϕ has been proposed in [20]. In this model the effective potential for a dilaton field demonstrates the proper behavior to provide a successful inflationary stage with a graceful exit in the regime $\phi \to \infty$, and a small value of the vacuum density responsible for the late time acceleration in the regime $\phi \to -\infty$, that appears as the "threshold" for the universe creation [20].

A way to solution of the fine-tuning problem involving the Higgs boson as the most likely candidate for the inflaton field, has been proposed in [17] on the basis of an energy exchange between the inflaton field and a time-dependent Λ. Although the mass of the Higgs boson is much smaller than that needed for an inflaton, the hierarchy problem has been solved by introducing a coupling between $\Lambda(t)$ and the inflaton field, which even in the simplest form, $\mathcal{L}_{int} \sim \phi\Lambda$, can provide a solution of the fine-tuning problem [17].

In the framework of the QFT model with a scalar and a fermion field and a physical cutoff rendering the QFT finite it has been shown that the violation of the Lorentz invariance at the high momentum scale can be made consistent with a suppression of the violation of the Lorentz invariance at the low momenta. The fine tuning required to provide both the suppression and the existence of a light scalar particle in the spectrum has been determined at the one loop level [21].

In the paper [22] it was noted that the extreme smallness of the gravitational fine structure constant $\alpha_g = \sqrt{G\hbar\Lambda/c^3} = 1.91 \times 10^{-61}$ $(\rho_{vac}/\rho_{Pl} = (1/8\pi)\alpha_g)$ may suggest an essentially different structure of a dark energy, which could not be entirely described by a cosmological constant.

The cosmological constant provides the empirically verified explanation for the present accelerated expansion. The ΛCDM model includes a cosmological constant, inflationary initial conditions, cold dark matter, standard radiation and neutrino content and $\Omega_{tot} = 1$, and demonstrates a good agreement with the current cosmological observations [23]. Extensions of the ΛCDM model, abbreviated as $I\Lambda CDM$, involve the vacuum energy interacting with the cold dark matter [24,25] (for a recent review [26]). However, theoretical difficulties including the fine-tuning problem [27,28] enforced looking for alternative models of a dark energy (for a review [29]).

Most of the alternative models introduce a dark energy of a non-vacuum origin which behaves like a cosmological constant when needed. Various models have been developed (for a review [30–33]) and checked out by the cosmography tests [34,35]. Among them phenomenological quintessence models with $-1 < w < -1/3$ [4] and quintom models with two dynamical scalar fields, a canonical field and a phantom field with $w < -1$ [36], supported by symmetry requirements [37], present the dynamically viable dark energy models (for a dynamical analysis of the quintom models [38]).

Modified gravity models (for a review [39]) involve screening mechanisms that are characterized by an effective value of the gravitational coupling G_N for the regions with the different gravitational potentials; the model predictions approach those typical for General Relativity in the regions of a strong gravitational field [29].

A general approach to the viable modified F(R) gravity describing the inflation and the current accelerated expansion has been developed in [40] including investigation of the exponential models of the modified gravity.

The holographic dark energy models [41,42] establish that the QFT ultraviolet cutoff produces the dark energy density $\rho_{DE} = 3C^2 M_{Pl}^2 L^{-2}$ [43], where C is a numerical constant and L indicates the infrared cutoff ([44] and references therein). The scale L^{-1} can be considered as the Hubble scale since the resulting density is comparable with the current dark energy density [45,46]. Another option for L is the particle and future horizons as the generalized holographic dark energy models for a specific $f(R)$ gravity model ([47] and references therein). In [48] a holographic dark energy model has been constructed for the apparent horizon in a curved universe. The holographic models for the particle and future horizons have been analyzed in [49] with the conclusion that the future horizon presents more similarities with the dark energy behavior.

Models with $-1 < w < -1/3$ have been thoroughly tested with using the WMAP (Wilkinson Microwave Anisotropy Probe) - CMB (Cosmic Microwave Background) data, which gave the constraint $w_Q \leq -0.7$ and the best fit $w_Q = -1$ [6–8,11]. CMB measurements [50] together with the BAO (Baryon Acoustic Oscillations) data [51] and SNe (SuperNovae) Ia data [52] confirmed this result giving $w = -1.06 \pm 0.06$ at 68% CL [51].

Model-independent evidence for the character of the dark energy evolution obtained from the BAO data distinguish theories with relaxation of Λ from a large initial value [53].

Relaxation of Λ has been considered in [54] on the basis of the adjustment mechanism proposed in [55] in the model extending General Relativity by adding a class of the invariant terms that reduce an arbitrary initial value of the cosmological constant to a needed value.

A unified description of the dark energy driving both the inflationary stage and the current accelerated expansion is presented in [56] on the basis of a quadratic model of gravity which includes an exponential F(R) gravity contribution with the high-curvature corrections coming from the higher-derivative quantum gravity beyond the one-loop approximation.

Let us note that although the cosmological constant provides the convincing explanation for the observed accelerated expansion, a clear justification of its small value does not exist until now [49].

In this paper we outline our approach to relaxation of a cosmological constant in the frame of the model-independent self-consistent theoretical scheme which makes cosmological constant intrinsically variable. A vacuum dark energy is introduced in general setting suggested by the algebraic classification of stress-energy tensors and directly related to spacetime symmetry that provides a mechanism of reducing a cosmological constant to a certain non-zero value involving the holographic principle which singles out the special class of the Lemaître cosmological models with the global structure of the de Sitter spacetime.

The quantum field theory in a curved spacetime does not contain a unique specification for the vacuum state of a system, and the symmetry of the vacuum expectation value of a stress-energy tensor does not always coincide with the symmetry of the background spacetime ([57] and references therein, for a detailed discussion [58]). What is more important, QFT does not contain an appropriate symmetry to zero out the cosmological vacuum density or to reduce it to a non-zero value.

The key point is that a relevant symmetry does exist in General Relativity as a spacetime symmetry, directly related to the algebraic structure of stress-energy tensors which determine the spacetime geometry as source terms in the Einstein equations. Algebraic classification of stress-energy tensors [59,60] suggests a model-independent definition of a vacuum as a medium by the algebraic structure of its stress-energy tensor and admits the existence of vacua whose symmetry is reduced as compared with the maximally symmetric de Sitter vacuum associated with the Einstein cosmological term $T^i_k = \rho_{vac}\delta^i_k$ ($p = -\rho_{vac}$; $\rho_{vac} =$ const by $T^i_{k;i} = 0$) and responsible for the de Sitter geometry which ensures accelerated expansion, isotropy and homogeneity independently of specific properties of the particular models for ρ_{vac} [61,62]. To make Λ variable it is enough to reduce symmetry of $T^i_k = \rho_{vac}\delta^i_k$ while keeping its vacuum identity ($p_k = -\rho$ for one of two spatial directions) [63,64]. The cosmological constant Λ becomes a time component Λ_t^t of a variable cosmological term $\Lambda_k^i = 8\pi G T_k^i$ [65], which allows a vacuum energy density to be intrinsically dynamical, i.e., time-evolving and spatially inhomogeneous (by virtue of $\Lambda_{k;i}^i = 0$).

A stress-energy tensor T_k^i with a reduced symmetry represents an intrinsically anisotropic vacuum dark fluid which can be evolving and clustering, and provides the unified description of dark energy and dark matter based on the spacetime symmetry [64,66] (for a review [67]). The relevant spherical solutions to the Einstein equations are specified by $T_t^t = T_r^r$ $(p_r = -\rho)$ and belong to the Kerr-Schild class. Regular spherical solutions satisfying the weak energy condition, which implies non-negativity of density as measured by an observer on a time-like curve, have obligatory de Sitter centers, $T_k^i = (8\pi G)^{-1}\Lambda\delta_k^i$. They describe regular cosmological models with time-evolving and spatially inhomogeneous vacuum dark energy, and compact vacuum objects generically related to a dark energy via their de Sitter vacuum interiors: regular black holes, their remnants and self-gravitating vacuum solitons, which can be responsible for observational effects typically related to a dark matter [64,68].

This implies a natural phenomenological inclination with the ΛCDM model: Primordial black hole remnants are considered as (cold) dark matter candidates for more than three decades [69,70]. The problem with singular black holes concerns the existence of the viable products of their evaporation ([71] and references therein). Quantum evaporation of the regular black holes (RBH) involves a 2-nd order phase transition followed by quantum cooling and resulting in thermodynamically stable remnants ([72] and references therein; for a review [73]). Primordial RBH remnants and self-gravitating vacuum solitons appear at the phase transitions in the early universe where they can capture available de Sitter vacuum in their interiors and form graviatoms binding electrically charged particles [74]. Their observational signatures as heavy dark matter candidates generically related to a vacuum dark energy include the electromagnetic radiation whose frequency depends on the scale of the interior de Sitter vacuum, within the range $\sim 10^{11}$ GeV available for observations [68,74]. In graviatoms with the GUT scale interiors, where the baryon and lepton numbers are not conserved, the remnant components of graviatoms can induce the proton decay, which could in principle serve as their additional observational signature in heavy dark matter searches at the IceCUBE experiment [68].

Regular cosmological models with the vacuum dark energy belong to the Lemaître class of cosmological models and are able to describe evolution between different states dominated by the de Sitter vacuum [58]. There exists infinitely many distributions of matter which satisfy $R_{ik} = \Lambda g_{ik}$ and hence model the cosmological constant [75], but in all cases it is ultimately de Sitter vacuum drives the accelerated expansion due to the basic properties of the de Sitter geometry, independently on an underlying particular model for Λ. In a similar way a dynamical vacuum dark energy associated with a variable cosmological term, generates geometries whose basic properties involve in the natural way restoration of the spacetime symmetry asymptotically or/and at certain stages of the universe evolution. Such geometries describe the relaxation of the cosmological constant by the Lemaître class anisotropic cosmological models, which reduce to the isotropic FLRW models at the stages with the spacetime symmetry restored to the de Sitter group. This makes it possible to describe on the common ground the currently observed accelerated expansion and the inflationary stages predicted by the standard model and related to the phase transitions in the universe evolution [76]. Such a model involving the GUT and QCD vacuum scales has been presented in [58] on the basis of the phenomenological density profile with the typical behavior for a cosmological scenario with an inflationary stage followed by decay of the vacuum energy described by the exponential function. The model parameters characterizing the decay rate are uniquely fixed by the requirements of the causality and analyticity. Other model parameters are fixed by the values ρ_{GUT}, ρ_{QCD}, the currently observed density ρ_λ and $\Omega = 1$. The only remaining free parameter, the e-folding number for the first inflation, was estimated by the observational constraints on the CMB anisotropy. It was shown that the Lemaître class cosmological model with the vacuum dark fluid can describe the universe evolution in the frame of one theoretical scheme which fairly well conforms to the basic observational features [58].

The spacetime symmetry provides a mechanism of reducing the cosmological constant to a certain non-zero value. The holographic principle distinguishes the special class of the Lemaître dark energy models, for which the cosmological evolution is guided by the quantum evaporation of the cosmological horizon whose dynamics entirely determines the final value of the cosmological constant.

Here we present the basic features of the Lemaître class dark energy models and outline in some detail the special class of models singled out by the holographic principle. Section 2 presents the spherically symmetric vacuum dark fluid and the generic properties of the spacetime generated by the vacuum dark energy. Section 3 presents the basic equations and the generic properties of the Lemaître class cosmological models with the vacuum dark energy, and the special class of the Lemaître models favored by the holographic principle, including the detailed behavior of the anisotropy parameter in the course of the universe evolution. Section 4 contains summary and discussion.

2. Algebraic Structure of Tress-Energy Tensors for Vacuum Dark Energy and Spacetime Symmetry

Stress-energy tensors of the spherically symmetric vacuum dark fluid have the algebraic structure defined by [63]

$$T_t^t = T_r^r \;\; (p_r = -\rho); \;\; T_\theta^\theta = T_\phi^\phi. \tag{1}$$

The equation of state, following from the conservation equation $T^k_{i;k} = 0$, reads [63,77] $p_r = -\rho$; $p_\perp = -\rho - \frac{r}{2}\rho'$, where $\rho(r) = T_t^t$ is the energy density, $p_r(r) = -T_r^r$ is the radial pressure, and $p_\perp(r) = -T_\theta^\theta = -T_\phi^\phi$ is the transversal pressure for the anisotropic vacuum dark fluid [64]. The stress-energy tensors specified by (1) generate, as the source terms in the Einstein equations, the globally regular spacetimes with the de Sitter centre (replacing the Schwarzschild singularity) provided that the weak energy condition (WEC) is satisfied [78,79]. The spacetime symmetry breaks from the de Sitter group in the origin [78].

The early proposals of replacing a singularity with the de Sitter core were based on the hypotheses of a self-regulation of the geometry by the vacuum polarization effects [77], of the existence of the limiting curvature [80], and of the symmetry restoration at the GUT scale in the course of the gravitational collapse [63,81]. Later a loop quantum gravity and the noncommutative geometry provided arguments in favor of a de Sitter interior in place of a singularity [82–87].

A metric generated by a source term specified by (1) belongs to the Kerr-Schild class [88]

$$ds^2 = g(r)dt^2 - \frac{dr^2}{g(r)} - r^2 d\Omega^2. \tag{2}$$

For the regular spherical solutions with the de Sitter centre WEC leads to monotonical decreasing of the density profile $\rho(r)$ [78]. In the case of two vacuum scales we can thus separate in T_t^t the background vacuum density $\rho_\lambda = (8\pi G)^{-1}\lambda$, introducing $T_t^t(r) = \rho(r) + \rho_\lambda$, where $\rho(r)$ is a dynamical density decreasing from the value at the center $\rho_\Lambda = (8\pi G)^{-1}\Lambda$ to zero at infinity; T_t^t evolves from $(8\pi G)^{-1}(\Lambda + \lambda)$ to $(8\pi G)^{-1}\lambda$ and provides the intrinsic relaxation of a cosmological constant. The metric function [89]

$$g(r) = 1 - \frac{2G\mathcal{M}(r)}{r} - \frac{\lambda r^2}{3}; \quad \mathcal{M}(r) = 4\pi \int_0^r \rho(x) x^2 dx \tag{3}$$

is asymptotically de Sitter with λ as $r \to \infty$ and with $(\Lambda + \lambda)$ as $r \to 0$.

Geometry has three basic length scales: $r_g = 2GM$ $(M = 4\pi \int_0^\infty \rho r^2 dr)$; $r_\Lambda = \sqrt{3/\Lambda}$; $r_\lambda = \sqrt{3/\lambda}$, and an additional length scale $r_* = (r_\Lambda^2 r_g)^{1/3}$ characteristic for the geometry with the de Sitter interior. The relation of r_λ to r_Λ represents the characteristic parameter relating two vacuum scales $q = r_\lambda / r_\Lambda = \sqrt{\Lambda/\lambda} = \sqrt{\rho_\Lambda/\rho_\lambda}$.

Let us note that the characteristic scale $r_* = (r_\Lambda^2 r_g)^{1/3}$, introduced first in [77] as related to a self-regulation of the geometry, appears explicitly in the simple semiclassical model of the vacuum polarization in the spherical gravitational field based on the fact that all fields are involved in a collapse, therefore all of them contribute to the stress-energy tensor and hence to geometry [63,78,90]. The scale r_* defines the zero gravity surface at which the strong energy condition is violated [63,90]. In thermodynamics of regular black holes this scale determines evolution during evaporation [72]. In the Nonlinear Electrodynamics coupled to Gravity the existence of zero gravity surface inside a regular compact object allows to present a certain explanation of the appearance of the minimal length scale in the electromagnetic reaction of the electron-positron annihilation [91].

In the Schwarzschild-de Sitter spacetime the similar scale with the background λ appears in the minimum of the metric function $g(r)$ (with $\mathcal{M}(r) = M = const$) for the Kottler-Trefftz geometry [92], and defines the boundary beyond which there are no bound orbits for test particles [93,94]. This scale plays the fundamental role in the non-linear theories of massive gravity. The original model by Fierz and Pauli described a massive spin-2 particle in the Minkovski space involving five degrees of freedom but was incompatible with the Solar System tests. Nonlinear extensions of this theory, proposed first by Vainshtein [95] include the mechanism of hiding some degrees of freedom and restoring General Relativity below a certain scale r_v called the Vainshtein radius which marks the transition to the regions where the extra degrees of freedom become essential at the large distances ([96,97] and references therein). In the cosmological context the Vainshtein scale r_v involves the background λ and plays the role of the astrophysical scale set by the cosmological constant λ as it was shown in [98] where the general conditions were derived applicable for any theory of massive gravity and responsible for the coincidence of r_v with the relevant scale for the Schwarzshild-de Sitter spacetime obtained in the GR frame with the dynamical metric involving all degrees of freedom. In the context of black hole thermodynamics the basic conditions for an observer in massive gravity [99] provide agreement of the obtained there results with those obtained in GR.

In the considered context of regular spacetimes with the vacuum dark energy the number of the vacuum scales determines the maximal number of the horizons. In accordance with the Einstein equations, the pressure $p_\perp$ is related with the metric function $g(r)$ as

$$8\pi G p_\perp = \frac{g''}{2} + \frac{g'}{r}. \tag{4}$$

It follows that an extremum of $g(r)$ in the region where $p_\perp > 0$, is a minimum. The transversal pressure $p_\perp$ becomes negative in the vicinity of the de Sitter center with the characteristic scale r_Λ and in the asymptotically de Sitter region of large r with the scale r_λ. Then there exists only one region where $p_\perp > 0$. Since $g(r)$ has different signs at $r \to 0$ and $r \to \infty$, the single minimum of $g(r)$ implies the existence of at most 3 zero points of the metric function $g(r)$ and hence 3 horizons of spacetime [100] and five possible configurations shown in Figure 1a [89]. Dependently on the mapping (choice of the observers reference frame) spacetime geometry presents the black (white) hole with three horizons, the internal horizon r_-, the event black (white) hole horizon r_+ and the cosmological horizon r_{++}, two extreme double-horizon states $r_- = r_+$, $r_+ = r_{++}$, and two one-horizon states shown in Figure 1a [100]. Static observers exist in the R-regions $0 \le r < r_-$ and $r_+ < r < r_{++}$.

The T-regions $r > r_{++}$ open to infinity represent in the relevant mapping $r \to T, t \to u$ the regular homogeneous cosmological T-models of the Kantowski-Sachs type with the vacuum dark fluid [101]. Typical features of these models are the existence of a Killing horizon, beginning of the cosmological evolution with a null bang from the Killing horizon, and the existence of a regular static pre-bang region visible to the cosmological observers [101]. The Kantowski-Sachs observers are shown in Figures 1b and 2.

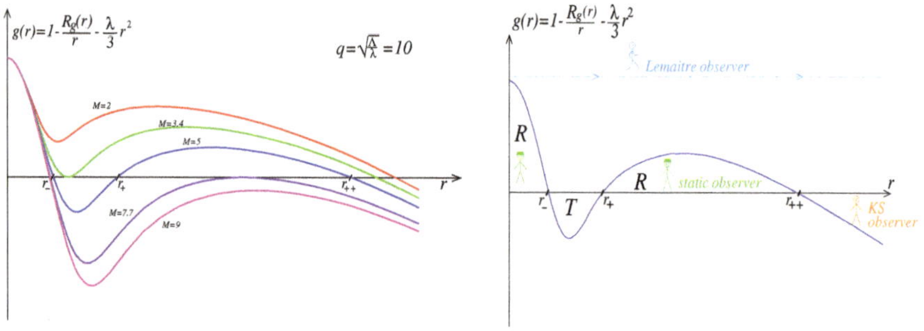

(a) **(b)**

Figure 1. (**a**) Five configurations described by (2). (**b**) Observers in the 3-horizons spacetime.

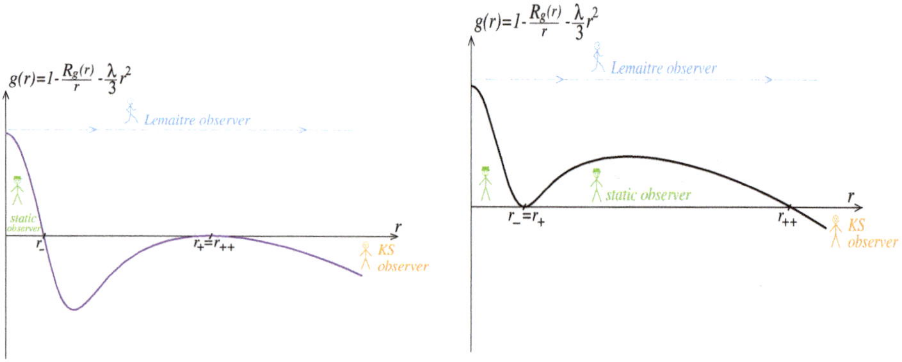

(a) **(b)**

Figure 2. Observers in the double horizon spacetimes: (**a**) $r_+ = r_{++}$, (**b**) $r_- = r_+$.

The Lemaître observers on the time-like radial geodesics, shown in Figures 1b and 2, have to their disposal the whole manifold, $0 \leq r \leq \infty$. Transition to their (geodesic) coordinates (R, τ), where R is the congruence parameter of the geodesic family and τ is the proper time along a geodesic, is given by the matrix relating the mapping $[r, t]$ to the mapping $[R, \tau]$ which reads [100]

$$\frac{\partial t}{\partial \tau} = \frac{E(R)}{g(r)}; \ \frac{\partial r}{\partial R} = \sqrt{E^2(R) - g(r)}; \ \frac{\partial r}{\partial \tau} = \pm \frac{\partial r}{\partial R}; \ \frac{\partial t}{\partial R} = \pm \frac{E^2(R) - g(r)}{E(R) g(r)}. \tag{5}$$

The resulting metric has the form [58,100]

$$ds^2 = d\tau^2 - \frac{[E^2(R) - g(r(R,\tau))]}{E^2(R)} dR^2 - r^2(R,\tau) d\Omega^2. \tag{6}$$

3. The Lemaître Class Models for Relaxing Cosmological Constant

3.1. Basic Equations

The cosmological models dominated by the anisotropic vacuum dark energy belong to the Lemaître class models with the anisotropic fluid and are described by the metric [102]

$$ds^2 = d\tau^2 - e^{2\nu(R,\tau)}dR^2 - r^2(R,\tau)d\Omega^2. \tag{7}$$

The coordinates R, τ are the Lagrange (comoving) coordinates. The function $r(R,\tau)$ is called the luminosity distance. For the metric (7) the Einstein equations read [103]

$$8\pi G p_r = \frac{1}{r^2}\left(e^{-2\nu}r'^2 - 2r\ddot{r} - \dot{r}^2 - 1\right), \tag{8}$$

$$8\pi G p_\perp = \frac{e^{-2\nu}}{r}(r'' - r'\nu') - \frac{\dot{r}\dot{\nu}}{r} - \ddot{\nu} - \dot{\nu}^2 - \frac{\ddot{r}}{r}, \tag{9}$$

$$8\pi G\rho = -\frac{e^{-2\nu}}{r^2}\left(2rr'' + r'^2 - 2rr'\nu'\right) + \frac{1}{r^2}\left(2r\dot{r}\dot{\nu} + \dot{r}^2 + 1\right), \tag{10}$$

$$8\pi G T_t^r = \frac{2e^{-2\nu}}{r}\left(\dot{r}' - r'\dot{\nu}\right) = 0. \tag{11}$$

The component T_t^r zeros out in the comoving reference frame, and the Equation (11) yields [104]

$$e^{2\nu} = \frac{r'^2}{1 + f(R)}, \tag{12}$$

where $f(R)$ is an arbitrary integration function. The dots and primes stand for $\partial/\partial\tau$ and $\partial/\partial R$, respectively. Comparison of the metric (6) with the metric (7) shows that (6) corresponds to (7) with $f(R) = E^2(R) - 1$, and $E^2(R) - g(r) = [r'(R,\tau)]^2 = [\dot{r}(R,\tau)]^2$ [100]. It follows that for the case of the vacuum dark fluid with the anisotropic pressures satisfying (1), generic behavior of the Lemaître class cosmological models is determined by the basic properties of the metric function $g(r)$ in (3).

Putting (12) into (8), we obtain the equation of motion [105]

$$\dot{r}^2 + 2r\ddot{r} + 8\pi G p_r r^2 = f(R). \tag{13}$$

Taking into account that $p_r = -\rho$ for a source term (1), the first integration in (13) gives [100]

$$\dot{r}^2 = \frac{2G\mathcal{M}(r)}{r} + f(R) + \frac{F(R)}{r}. \tag{14}$$

A second arbitrary function $F(R)$ should be chosen equal to zero for the models regular at $r = 0$ since $\mathcal{M}(r) \to 0$ as r^3 for $r \to 0$ where $\rho(r) \to \rho_\Lambda < \infty$. The second integration in (13) gives

$$\tau - \tau_0(R) = \int \frac{dr}{\sqrt{2G\mathcal{M}(r)/r + f(R)}}. \tag{15}$$

The new arbitrary function $\tau_0(R)$ is called the bang-time function [106].

For the expanding models $\dot{r} = \sqrt{E^2(R) - g(r)} = r'$ and hence r is a function of $(R + \tau)$. We can therefore choose $\tau_0(R) = -R$. For the small values of r the Equation (15) reduces to

$$\tau + R = \int \frac{dr}{\sqrt{r^2/r_\Lambda^2 + f(R)}} \tag{16}$$

that corresponds to the de Sitter geometry with $r(R,\tau) = r_\Lambda \cosh\left((\tau + R)/r_\Lambda\right)$ for $f(R) < 0$; $r(R,\tau) = r_\Lambda \exp\left((\tau + R)/r_\Lambda\right)$ for $f(R) = 0$; $r(R,\tau) = r_\Lambda \sinh\left((\tau + R)/r_\Lambda\right)$ for $f(R) > 0$.

3.2. *Basic Features of the Lemaître Cosmological Models With the Vacuum Dark Energy*

For the case $f(R) = 0$ preferred by the observational data ($\Omega = 1$), the cosmological evolution starts from the regular time-like surface $r(R, \tau) = 0$ where the Equation (16) gives [105]

$$r = r_\Lambda e^{(\tau+R)/r_\Lambda}; \;\; e^{2\nu} = r^2/r_\Lambda^2, \tag{17}$$

and the metric (7) takes the FLRW form with the de Sitter scale factor

$$ds^2 = d\tau^2 - r_\Lambda^2 e^{2\tau/r_\Lambda}\left(du^2 + u^2 d\Omega^2\right), \tag{18}$$

where $u = e^{R/r_\Lambda}$. In accordance with (17), it describes a non-singular non-simultaneous de Sitter bang from the surface $r(\tau + R \to -\infty) = 0$ [100,105], as it is shown in Figure 3 which presents the global structure of spacetime for the most general case of 3 horizons. The regions $\mathcal{RC}$ are the regular regions asymptotically de Sitter as $r \to 0$ at the scale of Λ replacing a singularity; the T-regions $\mathcal{WH}$ and $\mathcal{BH}$ represent the white and black holes; the regions $\mathcal{U}$ are the R-regions restricted by the cosmological horizons r_{++}; the regions $\mathcal{CC}$ are the T-regions asymptotically de Sitter with the background λ as $r \to \infty$. The surfaces $\mathcal{J}_-$ and $\mathcal{J}_+$ are the null (photon) boundaries in the past and the future, respectively. The cosmological evolution starts from the regular de Sitter surface $r = 0$ in $\mathcal{RC}_1$ and goes through $\mathcal{WH}$ and $\mathcal{U}_1$ towards $r \to \infty$ in $\mathcal{CC}_1$.

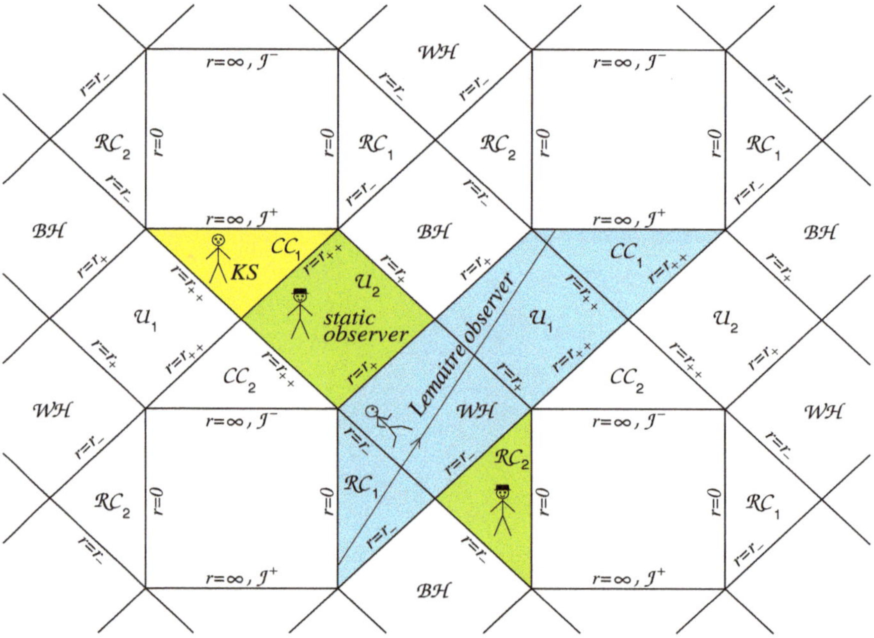

Figure 3. The global structure of the regular spacetime with 3 horizons.

Cosmological evolution is governed by dynamics of pressures. In the spacetime with the de Sitter center the total density monotonically decreases to ρ_λ, hence the total radial pressure $p_r = -\rho$ monotonically increases. Transversal pressure in the case of two vacuum scales evolves from the value $p_\perp = -\rho_\Lambda - \rho_\lambda$ at the inflation to the final value $p_\perp = -\rho_\lambda$, through one maximum in between [100,105]. Typical behavior of pressures (normalized to ρ_Λ) dependently on q is shown in Figure 4; the variable $\tau + R$ is normalized to the GUT time $t_{GUT} = r_\Lambda/c \simeq 0.8 \times 10^{-35}$ s for $M_{GUT} \simeq 10^{15}$ GeV.

The inflationary stage is followed by a strongly anisotropic Kasner-type stage. As follows from the Lemaître metric in the form (6) corresponding to (7) with $f(R) = E^2(R) - 1$, for any function $f(R)$ the expansion in the transversal direction with $\partial_\tau r > 0$ is accompanied by shrinking in the radial direction where $\partial_R |g_{RR}| < 0$ until $dg(r)/dr < 0$ [100]. For $E^2 = 1$ ($f(R) = 0$) the metric at this stage ($r_\Lambda < r \ll r_\lambda$) takes the form [100,105]

$$ds^2 = d\tau^2 - (\tau + R)^{-2/3} K(R) dR^2 - L(\tau + R)^{4/3} d\Omega^2, \tag{19}$$

where $K(R)$ is a smooth regular function and L is a constant, which depend on the specific form of the density profile and hence on the mass function in (3).

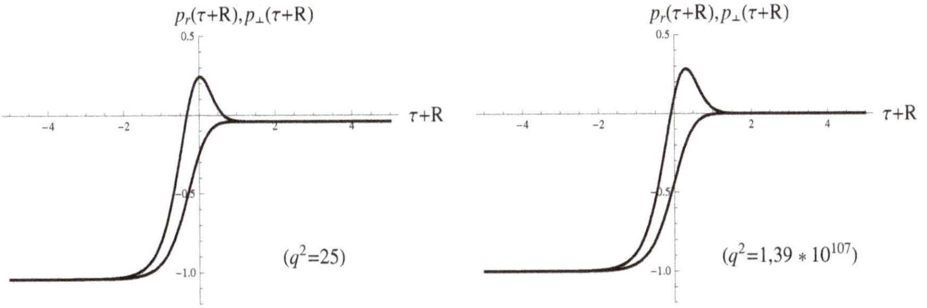

Figure 4. Behavior of p_r (lower curve) and $p_\perp$ (upper curve), at the early and present stages.

The intrinsic anisotropy of the Lemaître class cosmological models is described by the mean anisotropy parameter [107] (for a discussion of the different anisotropy characteristics [108])

$$A = \frac{1}{3H^2} \sum_{i=1}^{3} H_i^2; \quad H_i = \frac{\dot{a}_i(\tau)}{a_i(\tau)}; \quad H = \frac{H_1 + H_2 + H_3}{3}, \tag{20}$$

where $H_i = \dot{a}_i/a_i$ are the directional Hubble parameters corresponding to the scale factors $a_i(\tau)$, and $H = (H_1 + H_2 + H_3)/3$ is the mean Hubble parameter. For the spherically symmetric models with the vacuum dark energy specified by (1), $e^{2\nu} = r'^2$ for $f(R) = 0$, and the scale factors are $a_1 = r'$, $a_2 = a_3 = r$. In terms of the mass function $\mathcal{M}$ the anisotropy parameter takes the form [58,109]

$$A = 2\frac{(\dot{\mathcal{M}}/\mathcal{M}(r) - 3\dot{r}/r)^2}{(\dot{\mathcal{M}}/\mathcal{M} + 3\dot{r}/r)^2}. \tag{21}$$

In the FLRW cosmology the deceleration parameter is introduced on the basis of the Friedmann equations for one scale factor $R(\tau)$. In the case $\Omega = 1$ it reads $q_0 = (\rho + 3p)/2\rho = (1 + 3w)/2$, where w refers to the equation-of-state parameter for an isotropic medium. For $\rho + 3p < 0$ gravitational acceleration becomes repulsive which is responsible for the accelerated expansion. This fact follows directly from the strong energy condition which in general case of an anisotropic medium requires $\rho + \sum p_k \geq 0$. This guarantees, by the Raychaudhuri equation, the attractiveness of gravity [110], and in the cosmological context is responsible for deceleration. Violation of the strong energy condition, $\rho + \sum p_k < 0$, makes gravity repulsive and marks the transition from deceleration to acceleration. In the anisotropic Lemaître cosmology with the vacuum dark energy specified by $T_t^t = T_r^r$ ($p_r = -\rho$), the strong energy condition reads

$$\rho + 2p_\perp \geq 0 \;\rightarrow\; \rho(1 + 2w_\perp) \geq 0. \tag{22}$$

The transition from deceleration to acceleration occurs when $(1+2w_\perp) < 0$. In the Lemaître cosmology describing the unverse evolution from the inflationary beginning to the inflationary end with the possible intermediate inflationary stage(s), the maximal number of acceleration-deceleration-acceleration transitions is determined by the numbers of zeros of the pressure $p_\perp$ which is determined by the number of vacuum (de Sitter) scales. In the case of two vacuum scales $p_\perp$ evolves between two inflationary (negative) pressures and can change its sign twice. It results in not more than two transitions: acceleration-deceleration and deceleration-acceleration. Each additional intermediate inflationary stage can add two zeros of $p_\perp$ [58], and hence not more than two additional transitions, acceleration-deceleration and deceleration-acceleration. In the next subsection we consider acceleration-deceleration-acceleration transitions in the model distinguished by the holographic principle.

3.3. The Lemaître Class Models Singled Out by the Holographic Principle

A family of one-horizon spacetimes with the global structure of the de Sitter spacetime contains the special class distinguished by the holographic principle [111] (which leads to the conjecture that a dynamical system can be entirely determined by the data stored on its boundary [112]) as governed by the quantum evaporation of the cosmological horizon that determines the basic characteristics of the final state in the horizon evaporation for any density profile [113].

Typical behavior of the metric function for this class is shown in Figure 5a. Quantum evaporation of the horizon goes towards decreasing the mass M [72]. In this case it ends up in the triple-horizon state $M = M_{cr}$ which is absolutely thermodynamically stable: Its basic generic features are [113] zero temperature, the infinite positive specific heat capacity, the finite entropy, zero transversal pressure, zero curvature, and the infinite scrambling time (the time needed to thermalize information [114]).

Evolution governed by evaporation goes with increasing entropy from the state $M > M_{cr}$ towards the triple-horizon state $M = M_{cr}$ that satisfies three algebraic equations: $g(r_t) = 0$; $g'(r_t) = 0$; $g''(r_t) = 0$, which determine uniquely the basic parameters: the mass M_t, the triple horizon radius r_t, and $q_t = \sqrt{\rho_\Lambda/\rho_\lambda}$, so that the final non-zero value of the vacuum dark energy density ρ_λ is tightly fixed by the quantum evaporation of the cosmological horizon for a chosen vacuum scale for ρ_Λ [113].

With using the density profile [63]

$$\rho(r) = \rho_\Lambda e^{-r^3/r_\Lambda^2 r_g}, \tag{23}$$

which describes the vacuum polarization in the spherically symmetric gravitational field in a simple semiclassical model [78,90], we obtain [113]

$$M_{cr} = 2.33 \times 10^{56}\,\text{g};\ q_{cr}^2 = 1.37 \times 10^{107};\ r_t = 5.4 \times 10^{28}\,\text{cm}. \tag{24}$$

The cosmological evolution is described by the Lemaître metric (7) which can be written as

$$ds^2 = d\tau^2 - b^2(\tau,R)dR^2 - r^2(\tau,R)d\Omega^2, \tag{25}$$

where the second scale factor $b(\tau,R) \equiv e^{\nu(R,\tau)}$ in accordance with (12). Behavior of two scale factors is shown in Figure 5b for the case $f(R) = 0$ $(\Omega = 1)$ [109]. Distances and times are normalized to $r_* = (r_\Lambda^2 r_g)^{1/3} = 1.26 \times 10^{-7}$ cm, and $t_{GUT} \simeq 0.8 \times 10^{-35}$ s with $M_{GUT} \simeq 10^{15}$ GeV.

Due to the isotropy of pressures (see Figure 4), at the very early and late times the behavior of two scale factors is similar (curves run parallel and differ only by constant), the second stage of the parallel running starts at $r_d \simeq 3 \times 10^{27}$ cm and $(\tau+R)_d \simeq 9{,}5 \times 10^{16}$ s (according to the observational data the vacuum density starts to dominate at the age of about 3×10^9 years). The vacuum density approaches its observed value at the triple horizon r_t at $(\tau+R)_t \simeq 2{,}9 \times 10^{17}$ s. The metric (7) approaches the FLRW form with the de Sitter scale factor $ds^2 = d\tau^2 - r_t^2 e^{2c\tau/r_t}(du^2 + u^2 d\Omega^2)$ where $u = e^{R/r_t}$.

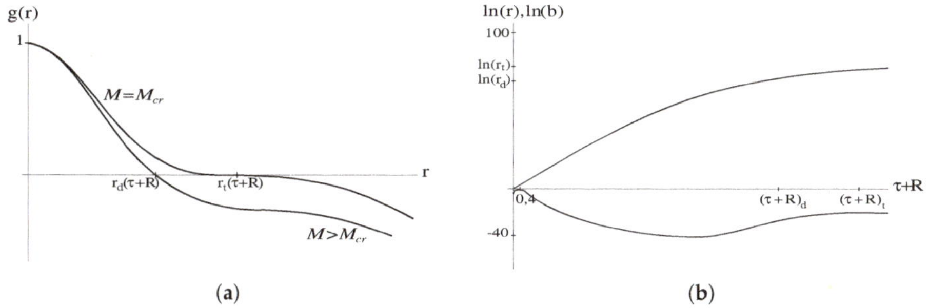

Figure 5. (**a**) Evolution of the metric function during evaporation. (**b**) Behavior of the scale factors: $r(\tau + R)$ (upper curve) and $b(\tau + R)$ (lower curve).

To evaluate the vacuum dark energy density from $q_{cr}^2 = \rho_\Lambda/\rho_\lambda$, we adopt $\rho_\Lambda = \rho_{GUT}$. The Grand Unification scale is estimated as $M_{GUT} \sim 10^{15} - 10^{16}$ GeV, which results in the value for the vacuum density ρ_Λ within the range $1.7 \times 10^{-30} gcm^{-3} - 1.7 \times 10^{-26} gcm^{-3}$, respectively.

The observational value $\rho_{\lambda\ (obs)} \simeq 6.45 \times 10^{-30} g\ cm^{-3}$ [11] corresponds, in the considered context, to $M_{GUT} \simeq 1.4 \times 10^{15}$ GeV which gives $\rho_{GUT} = 8.8 \times 10^{77} gcm^{-3}$, $r_\Lambda = 1.8 \times 10^{-25}$ cm. For this scale q_{cr}^2 gives the value of the present vacuum density ρ_λ in agreement with its observational value [109].

The behavior of two scale factors and of their derivatives at the early stage of evolution is shown in Figures 6 and 7 [115]. The maximum in the scale factor $b(\tau + r)$ at $\tau + R \simeq 0.4\ t_{GUT} = 0.32 \times 10^{-35}$ s corresponds to the maximum of the transversal pressure at $r \simeq 1.5 \times 10^{-7}$ cm in Figure 4.

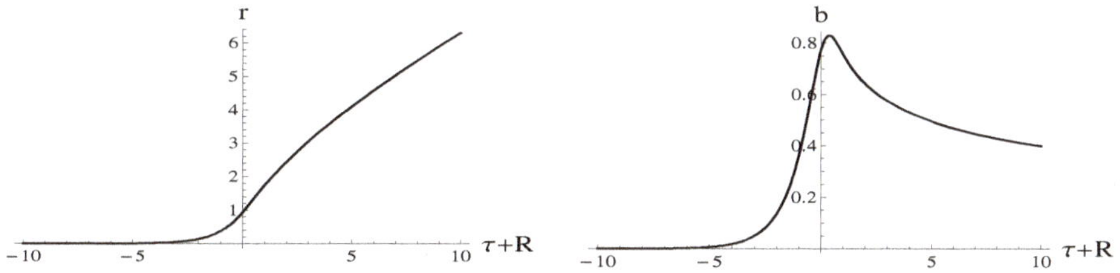

Figure 6. Behavior of the scale factors at the early stage of evolution.

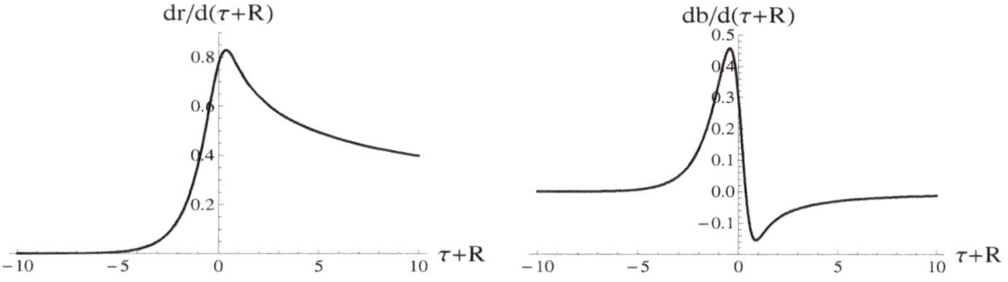

Figure 7. Behavior of the velocities $\dot{r}(\tau + R)$, $\dot{b}(\tau + R)$ at the early stage of evolution.

Behavior of the anisotropy parameter during the whole evolution, shown in Figure 8 [109], was studied numerically with the density profile (23) and the model parameters (24).

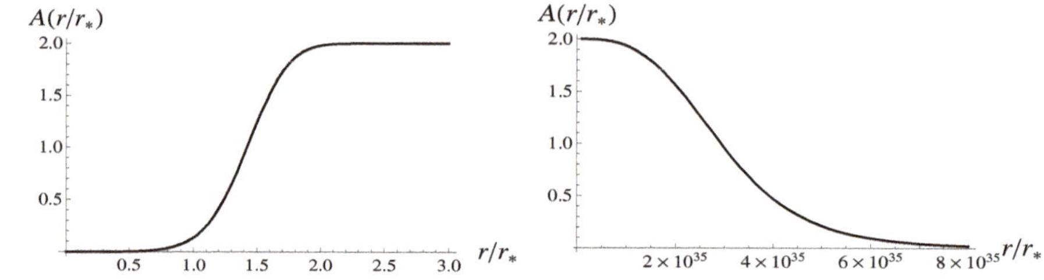

Figure 8. Behavior of the anisotropy parameter at the early and late stages.

The anisotropy develops quickly during the postinflationary stage. At the maximum of $p_\perp$ and $b(\tau + r)$, the anisotropy parameter takes the value $A \simeq 0.4$, grows further achieving $A = 2$ at $r \simeq 2.5 \times 10^{-7}$ cm, and starts to decrease at $r \simeq 6 \times 10^{27}$ cm. The standard estimate $A < 10^{-6}$ is expected to be satisfied already at the recombination epoch, $r \sim 10^{25}$ cm ($z \sim 10^3$). The curve in Figure 7 predicts that the anisotropy starts to satisfy this criterion at approaching $r \sim 10^{28}$ cm when the present vacuum density starts to dominate [109].

In the spacetime with two vacuum scales the pressure $p_\perp$ changes its sign twice [58,100], so that in any triple-horizon model the strong energy condition is violated ($1 + 2w_\perp < 0$ in Equation (22)) in the inflationary stage, and the transition from acceleration to deceleration occurs when satisfaction of the strong energy condition is restored; deceleration changes to acceleration at approaching the current stage of accelerated expansion. The rates of these processes depend essentially on the density profile. For the quickly decreasing density profile (23) the acceleration changes to deceleration after the inflation at $r_1 \simeq 0.4 \times 10^{-7}$ cm, at the essentially anisotropic stage.

4. Summary and Discussion

Responsibility of the spacetime symmetry for reducing the cosmological constant to a certain value in the course of the universe evolution, is suggested on general setting by the algebraic classification for stress-energy tensors which allows for a model-independent definition of a vacuum as a medium and implies the existence of vacua whose symmetry is reduced as compared with the maximally symmetric de Sitter vacuum $p = -\rho$ associated with the Einstein cosmological term. In the spherically symmetric case their stress-energy tensors have the canonical form $T^t_t = T^r_r$ ($p_r = -\rho$) and generate the Lemaître class cosmological models with the anisotropic pressures that allows to describe cosmological evolution by intrinsically dynamical, time-dependent and spatially inhomogeneous vacuum dark energy. In the Lemaître class dark energy models the characteristics of the vacuum dark energy are determined by the algebraic structure of its stress-energy tensor and generically related to the spacetime symmetry. The basic features of the Lemaître class dark energy models are the non-singular non-simultaneous de Sitter bang, followed by the anisotropic Kasner-type stage and directed towards the late-time de Sitter stage, representing the effective relaxation of the cosmological constant from the initial inflationary value Λ to the final late-time inflationary value $\lambda < \Lambda$.

Among these models there is a special class of the one-horizon models distinguished by the holographic principle. The cosmological evolution is governed by the quantum evaporation of the cosmological horizon which determines uniquely the non-zero final value of the cosmological constant in the restoration of the spacetime symmetry to the de Sitter group.

In the case of adopting for the dark energy the density profile representing semi-classically the vacuum polarization in the spherical gravitational field and the GUT scale for the initial vacuum density, its final value appears in agreement with the value given by observations.

Let us note that this special class of models is essentially different from the holographic dark energy models with the isotropic fluid and the postulated density profile. The Lemaître cosmological models are intrinsically anisotropic and describe evolution of the dark energy density represented by the T_0^0 component of the variable cosmological term whose symmetry is reduced as compared with the Einstein cosmological term.

The anisotropy parameter in the special class of the one-horizon Lemaître dark energy models grows quickly during the postinflationary stage, then stays constant and decreases achieving $A < 10^{-6}$ when the present vacuum density starts to dominate. Astronomical observations suggest that our Universe can be deviated from the isotropy. The observed CMB anisotropy, interpreted as a realization of a statistical process originating in the inflationary era [116–118] admit the statistical anisotropy with the confidence level above 99% [116]. The anisotropy has been constrained at the magnitude level of 2–5% by the SNe Ia data [119], and at the level of 4.4% by the Union2 data and the high-redshift gamma-ray bursts [120]. The deviations from the homogeneity also can be confronted with observations. The influence of inhomogeneities on the cosmological distance measurements has been considered in [121].

The Lemaître class cosmological models provide an appropriate tool for the detailed analysis of the anisotropy and inhomogeneity against observations.

Author Contributions: Authors contributed equally to this work.

Funding: This research received no external funding.

Conflicts of Interest: The authors declare no conflict of interest.

References

1. Riess, A.G.; Kirshner, R.P.; Schmidt, B.P.; Jha, S.; Challis, P.; Garnavich, P.M.; Esin, A.A.; Carpenter, C.; Grashius, R. BV RI light curves for 22 type Ia supernovae. *Astron. J.* **1999**, *117*, 707–724. [CrossRef]
2. Perlmutter, S.; Aldering, G.; Goldhaber, G.; Knop, R.A.; Nugent, P.; Castro, P.G.; Deustua, S.; Fabbro, S.; Goobar, A.; Groom, D.E.; et al. Measurements of Ω and Λ from 42 high-redshift supernovae. *Astrophys. J.* **1999**, *517*, 565–586. [CrossRef]
3. Bahcall, N.A.; Ostriker, J.P.; Perlmutter, S.; Steinhardt, P.J. The cosmic triangle: Revealing the state of the universe. *Science* **1999**, *284*, 1481–1488. [CrossRef]
4. Wang, L.; Caldwell, R.R.; Ostriker, J.P.; Steinhardt, P.J. Cosmic Concordance and Quintessence. *Astrophys. J.* **2000**, *530*, 17–35. [CrossRef]
5. Sullivan, M.; Guy, J.; Conley, A.; Regnault, N.; Astier, P.; Balland, C.; Basa, S.; Carlberg, R.G.; Fouchez, D.; Hardin, D.; et al. SNLS3: Constraints on dark energy combining the supernova legacy survey three-year data with other probes. *Astrophys. J.* **2011**, *737*, 102–121. [CrossRef]
6. Corasaniti, P.S.; Copeland, E.J. Constraining the quintessence equation of state with SnIa data and CMB peaks. *Phys. Rev. D* **2002**, *65*, 043004. [CrossRef]
7. Hannestad, S.; Mortsell, E. Probing the dark side: Constraints on the dark energy equation of state from CMB, large scale structure, and type Ia supernovae. *Phys. Rev. D* **2002**, *66*, 063508. [CrossRef]
8. Bassett, B.A.; Kunz, M.; Silk, J.; Ungarelli, C. A late-time transition in the cosmic dark energy? *Mon. Not. R. Astron. Soc.* **2002**, *336*, 1217–1222. [CrossRef]
9. Tonry, J.L.; Schmidt, B.P.; Barris, B.; Candia, P.; Challis, P.; Clocchiatti, A.; Coil, A.L.; Filippenko, A.V.; Garnavich, P.; Hogan, C.; et al. Cosmological results from high-z supernovae. *Astrophys. J.* **2003**, *594*, 1–24. [CrossRef]
10. Ellis, J. Dark matter and dark energy: Summary and future directions. *Philos. Trans. A* **2003**, *361*, 2607–2627. [CrossRef]
11. Corasaniti, P.S.; Kunz, M.; Parkinson, D.; Copeland, E.J.; Bassett, B.A. Foundations of observing dark energy dynamics with the Wilkinson Microwave Anisotropy Probe. *Phys. Rev. D* **2004**, *70*, 083006. [CrossRef]
12. Weinberg, S. The cosmological constant problem. *Rev. Mod. Phys.* **1989**, *61*, 1–23. [CrossRef]
13. Wheeler, J.A.; Ford, K. *Black Holes and Quantum Foam: A Life in Physics*; W.W. Norton and Co.: New York, NY, USA, 1998.
14. Hawking, S. Spacetime foam. *Nucl. Phys. B* **1978**, *144*, 349–362. [CrossRef]
15. Coleman, S. Why There Is Nothing Rather Than Something: A Theory of the Cosmological Constant. *Nucl. Phys. B* **1988**, *310*, 643–668. [CrossRef]

16. Klebanov, I.; Susskind, L.; Banks, T. Wormholes and the cosmological constant. *Nucl. Phys. B* **1989**, *317*, 665–692 . [CrossRef]
17. Feng, C.J.; Li, X.L. Towards a realistic solution of the cosmological constant fine-tuning problem. *Phys. Rev. D* **2014**, *90*, 103009. [CrossRef]
18. Bousso, R.; Harnik, R.; Kribs, G.D.; Perez, G. Predicting the cosmological constant from the causal entropic principle. *Phys. Rev. D* **2007**, *76*, 043513–043530. [CrossRef]
19. Krauss, L.M.; Dent, J.; Starkman, G.D. Late Time Decay of False Vacuum, Measurement, and Quantum Cosmology. *Int. J. Mod. Phys. D* **2009**, *17*, 2501–2505. [CrossRef]
20. Guendelman, E.I. Non singular origin of the Universe and cosmological constant. *Int. J. Mod. Phys. D* **2012**, *20*, 2767–2771. [CrossRef]
21. Cortes, J.L.; Lopez-Sarrion, J. Fine-tuning problem in quantum field theory and Lorentz invariance: A scalar-fermion model with a physical momentum cutoff. *Int. J. Mod. Phys. A* **2017**, *32*, 1750084. [CrossRef]
22. Adler, R.J. Comment on the cosmological constant and a gravitational alpha. *arXiv* **2011**, arXiv:1110.3358.
23. Ade, P.A.R.; Aghanim, N.; Arnaud, M.; Ashdown, M.; Aumont, J.; Baccigalupi, C.; Banday, A.J.; Barreiro, R.B.; Bartlett, J.G.; Bartolo, N.; et al. Planck 2015 results. *Astron. Astrophys.* **2016**, *594*, A13.
24. Li, Y.H.; Zhang, J.F.; Zhang, X. Testing models of vacuum energy interacting with cold dark matter. *Phys. Rev. D* **2016**, *93*, 023002. [CrossRef]
25. Peracanba, J.S.; Perez, J.D.C.; Gomez-Valent, A. Possible signals of vacuum dynamics in the Universe. *Mon. Not. R. Astron. Soc* **2018**, *478*, 4357–4373.
26. Li, H.L.; Feng, L.; Zhang, J.F.; Zhang, X. Models of vacuum energy interacting with cold dark matter: Constraints and comparison. *arXiv* **2018**, arXiv:1812.00319.
27. Sahni, V.; Starobinsky, A.A. The case for a positive cosmological Lambda term. *Int. J. Mod. Phys. D* **2000**, *9*, 373–444. [CrossRef]
28. Bean, R.; Carroll, S.M.; Trodden, M. Insight into dark energy: Interplay between theory and observation. *arXiv* **2005**, arXiv:astro-ph/0510059.
29. Tanabashi, M.; Hagiwara, K.; Hikasa, K.; Nakamura, K.; Sumino, Y.; Takahashi, F.; Tanaka, J.; Agashe, K.; Aielli, G.; Amsler, C.; et al. Dark Energy. *Phys. Rev. D* **2018**, *98*, 030001.
30. Caldwell, R.R.; Kamionkowski, M. The Physics of Cosmic Acceleration. *Ann. Rev. Nucl. Part. Sci.* **2009**, *59*, 397–429. [CrossRef]
31. Copeland E.J. Models of dark energy. In Proceedings of the Invisible Universe International Conference, Paris, France, 29 June–3 July 2009; AIP: New York, NY, USA, 2010; pp. 132–138.
32. Bamba, K.; Capozziello, S.; Nojiri, S.; Odintsov, S.D. Dark energy cosmology: The equivalent description via different theoretical models and cosmography tests. *Astrophys. Space Sci.* **2012**, *342*, 155–228.
33. Rivera, A.B.; Farieta, J.G. Exploring the Dark Universe: Constraint on dynamical dark energy models from CMB, BAO and Growth Rate Measurements. *arXiv* **2016**, arXiv:1605.01984.
34. Aviles, A.; Bravetti, S.; Capozziello, S.; Luongo, O. Precision cosmology with Padé rational approximations: Theoretical predictions versus observational limits. *Phys. Rev. D* **2011**, *90*, 043531. [CrossRef]
35. Capozziello S.; D'Agostino, R.; Luongo, O. Model-independent reconstruction of f(T) teleparallel cosmology. *Gen. Relativ. Gravit.* **2017**, *49*, 141–162. [CrossRef]
36. Cai, Y.F.; Saridakis, E.N.; Setare, M.R.; Xia, J. Quintom Cosmology: Theoretical implications and Observations. *Phys. Rep.* **2010**, *493*, 1–60. [CrossRef]
37. Capozziello, S.; Predipalumbo, E.; Rubano, C.; Scudellaro, P. Noether symmetry approach in phantom quintessence cosmology. *Phys. Rev. D* **2009**, *80*, 104030. [CrossRef]
38. Mishra, S.; Chakraborty, S. Dynamical system analysis of Quintom Dark Energy Model. *Eur. Phys. J. C* **2018**, *78*, 917–923. [CrossRef]
39. Joyce, A.; Lombriser, L.; Schmidt, F. Dark Energy Versus Modified Gravity. *Ann. Rev. Nucl. Part. Sci.* **2016**, *66*, 95–122. [CrossRef]
40. Cognola, G.; Elizalde, E.; Nojiri, S.; Odintsov, S.D.; Sebastiani, L.; Zerbini, S. Class of viable modified f(R) gravities describing inflation and the onset of accelerated expansion. *Phys. Rev. D* **2008**, *77*, 046009. [CrossRef]
41. Horava P.; Minic, D. Probable Values of the Cosmological Constant in a Holographic Theory. *Phys. Rev. Lett.* **2000**, *85*, 1610–1613. [CrossRef] [PubMed]

42. Thomas, S. Holography Stabilizes the Vacuum Energy. *Phys. Rev. Lett.* **2002**, *89*, 81301–81304. [CrossRef] [PubMed]
43. Setare, M.R. Interacting holographic dark energy model in non-flat universe. *Phys. Lett. B* **2006**, *642*, 1–4. [CrossRef]
44. Li, E.K.; Zhang, Y.; Cheng, J.L. Modified holographic Ricci dark energy coupled to interacting relativistic and non-relativistic dark matter in the nonflat universe. *Phys. Rev. D* **2014**, *90*, 083534. [CrossRef]
45. Wang, S.; Wu, J.; Li, M. Holographic Dark Energy. *Phys. Rep.* **2017**, *696*, 1–58. [CrossRef]
46. Cruz, M.; Lepe, S. Holographic approach for dark energy-dark matter interaction in curved FLRW spacetime. *Class. Quantum Gravity* **2018**, *35*, 155013. [CrossRef]
47. Nojiri, S.; Odintsov, S.D. Covariant Generalized Holographic Dark Energy and Accelerating Universe. *Eur. Phys. J. C* **2017**, *77*, 528. [CrossRef]
48. Cruz, M.; Lepe, S. A holographic cut-off inspired in the apparent horizon. *Eur. Phys. J. C* **2018**, *78*, 994. [CrossRef]
49. Cruz M.; Lepe S. Modeling holographic dark energy with particle and future horizons. *arXiv* **2018**, arXiv:1812.06373.
50. Ade, P.A.R.; Aikin, R.W.; Barkats, D.; Benton, S.J.; Bischoff, C.A.; Bock, J.J.; Brevik, J.A.; Buder, I.; Bullock, E.; Dowell, C.D.; et al. Detection of B-Mode Polarization at Degree Angular Scales by BICEP2. *Phys. Rev. Lett.* **2014**, *112*, 241101. [CrossRef] [PubMed]
51. Anderson, L.; Aubourg, E.; Bailey, S.; Beutler, F.; Bhardwaj, V.; Blanton, M.; Bolton, A.S.; Brinkmann, J.; Brownstein, J.R.; Burden, A.; et al. The clustering of galaxies in the SDSS-III Baryon Oscillation Spectroscopic Survey: Baryon acoustic oscillations in the Data Releases 10 and 11 Galaxy samples. *Mon. Not. R. Astron. Soc* **2014**, *441*, 24–62. [CrossRef]
52. Suzuki, N.; Rubin, D.; Lidman, C.; Aldering, G.; Amanullah, R.; Barbary, K.; Barrientos, L.F.; Botyanszki, J.; Brodwin, M.; Connolly, N.; et al. The Hubble Space Telescope Cluster Supernova Survey: Improving the Dark Energy Constraints and Building an Early-Type-Hosted Supernova Sample. *Astrophys. J.* **2012**, *746*, 85–109. [CrossRef]
53. Sahni, V.; Shafiello, A.; Starobinsky, A. A. Model-independent Evidence for Dark Energy Evolution from Baryon Acoustic Oscillations. *Astrophys. J.* **2014**, *793*, L40–L44. [CrossRef]
54. Bauer, F.; Solà, J.; Štefancic, H. Dynamically avoiding fine-tuning the cosmological constant: the "Relaxed Universe". *J. Cosmol. Astropart. Phys.* **2010**, *1012*, 29. [CrossRef]
55. Bauer, F.; Solà, J.; Štefancic, H. Relaxing a large cosmological constant. *Phys. Lett. B* **2009**, *678*, 427–433. [CrossRef]
56. Elizalde E.; Odintsov, S.D.; Sebastiani, L.; Myrzakulov, R. Beyond-one-loop quantum gravity action yielding both inflation and late-time acceleration. *Nucl. Phys. B* **2017**, *921*, 411—435. [CrossRef]
57. Anderson, P.R. Attractor states and infrared scaling in de Sitter space. *Phys. Rev. D* **2000**, *62*, 124019. [CrossRef]
58. Bronnikov, K.A.; Dymnikova, I.; Galaktionov, E. Multihorizon spherically symmetric spacetimes with several scales of vacuum energy. *Class. Quantum Gravity* **2012**, *29*, 095025. [CrossRef]
59. Petrov, A.Z. *Einstein Spaces*; Pergamon Press: Oxford, UK, 1969.
60. Stephani, H.; Kramer, D.; MacCallum, V.; Hoenselaers, C.; Herlt, E. *Exact Solutions of Einstein's Field Equations*; Cambridge University Press: Cambridge, UK, 2003.
61. Gliner, E.B.; Dymnikova, I.G. Nonsingular Friedmann cosmology. *Sov. Astron. Lett.* **1975**, *1*, 93–95.
62. Olive K.A. Inflation. *Phys. Rep.* **1990**, *190*, 307–403. [CrossRef]
63. Dymnikova, I. Vacuum nonsingular black hole. *Gen. Relativ. Gravit.* **1992**, *24*, 235–242. [CrossRef]
64. Dymnikova, I.; Galaktionov, E. Vacuum dark fluid. *Phys. Lett. B* **2007**, *645*, 358–364. [CrossRef]
65. Dymnikova, I. The algebraic structure of a cosmological term in spherically symmetric solutions. *Phys. Lett. B* **2000**, *472*, 33–38. [CrossRef]
66. Dymnikova, I.; Galaktionov, E. Dark ingredients in one drop. *Cent. Eur. J. Phys.* **2011**, *9*, 644–653. [CrossRef]
67. Dymnikova, I. Unification of dark energy and dark matter based on the Petrov classification and space-time symmetry. *Intern. J. Mod. Phys. A* **2016**, *31*, 1641005. [CrossRef]
68. Dymnikova, I.; Khlopov, M. Regular black hole remnants and graviatoms with de Sitter interior as heavy dark matter candidates probing inhomogeneity of early universe. *Int. J. Mod. Phys. D* **2015**, *24*, 1545002. [CrossRef]

69. Polnarev, A.G.; Khlopov, M.Y. Cosmology, primordial black holes, and supermassive particles. *Sov. Phys. Uspekhi* **1985**, *28*, 213–232. [CrossRef]
70. MacGibbon, J.H. Can Planck-mass relics of evaporating black holes close the Universe? *Nature* **1987**, *329*, 308–309. [CrossRef]
71. Dymnikova, I. Regular black hole remnants. In Proceedings of the Invisible Universe International Conference, Paris, France, 29 June–3 July 2009; AIP: New York, NY, USA, 2010; pp. 361–368.
72. Dymnikova, I.; Korpusik, M. Regular black hole remnants in de Sitter space. *Phys. Lett. B* **2010**, *685*, 12–18. [CrossRef]
73. Dymnikova, I. Generic Features of Thermodynamics of Horizons in Regular Spherical Space-Times of the Kerr-Schild Class. *Universe* **2018**, *4*, 63. [CrossRef]
74. Dymnikova, I.; Fil'chenkov, M. Graviatoms with de Sitter Interior. *Adv. High Energy Phys.* **2013**, *2013*, 746894. [CrossRef]
75. Gibbons, G.W. *Phantom Matter and the Cosmological Constant*; DAMTP-2003-19; Cambridge University: Cambridge, UK, 2003; arXiv:hep-th/0302199.
76. Boyanovsky, D.; de Vega, H.J.; Schwarz, D.J. Phase transitions in the early and present universe. *Ann. Rev. Nucl. Part. Sci.* **2006**, *56*, 441–500. [CrossRef]
77. Poisson, E.; Israel, W. Structure of the black hole nucleus. *Class. Quantum Gravity* **1988**, *5*, L201–L205. [CrossRef]
78. Dymnikova, I. The cosmological term as a source of mass. *Class. Quantum Gravity* **2002**, *19*, 725–740. [CrossRef]
79. Dymnikova, I. Spherically symmetric space-time with regular de Sitter center. *Int. J. Mod. Phys. D* **2003**, *12*, 1015–1034. [CrossRef]
80. Frolov, V.P.; Markov, M.A.; Mukhanov, V.F. Black holes as possible sources of closed and semiclosed worlds. *Phys. Rev. D* **1990**, *41*, 383–394. [CrossRef]
81. Dymnikova, I. Internal structure of nonsingular spherical black holes. *Ann. Isr. Phys. Soc.* **1997**, *13*, 422–440.
82. Perez, A. Spin foam models for quantum gravity. *Class. Quantum Gravity* **2003**, *20*, R43–R104. [CrossRef]
83. Rovelli, C. *Quantum Gravity*; Cambridge University Press: Cambridge, UK, 2004.
84. Bonanno, A.; Reuter, M. Spacetime structure of an evaporating black hole in quantum gravity. *Phys. Rev. D* **2006**, *73*, 83005–83017. [CrossRef]
85. Nicolini, P. Noncommutative black holes, the final appeal to quantum gravity: A review. *Int. J. Mod. Phys. A* **2009**, *24*, 1229–1308. [CrossRef]
86. Arraut, I.; Batic, D.; Nowakowski, M. A noncommutative model for a mini black hole. *Class. Quantum Gravity* **2009**, *26*, 245006. [CrossRef]
87. Arraut, I.; Batic, D.; Nowakowski, M. Maximal extension of the Schwarzschild spacetime inspired by noncommutative geometry. *J. Math. Phys.* **2010**, *51*, 022503. [CrossRef]
88. Kerr, R.P.; Schild, A. Some algebraically degenerate solutions of Einstein's gravitational field equations. *Proc. Symp. Appl. Math* **1965**, *17*, 199–207.
89. Dymnikova, I.; Soltysek, B. Spherically symmetric space-time with two cosmological constants. *Gen. Relativ. Gravit.* **1998**, *30*, 1775–1793. [CrossRef]
90. Dymnikova, I. De Sitter-Schwarzschild black hole: Its particlelike core and thermodynamical properties. *Int. J. Mod. Phys. D* **1996**, *5*, 529–540. [CrossRef]
91. Dymnikova, I.; Sakharov, A.; Ulbricht, J. Appearance of a minimal length in e^+e^- annihilation. *Adv. High Energy Phys.* **2014**, *2014*, 707812. [CrossRef]
92. Bażański, S.L.; Ferrari, V. Analytic Extension of the Schwarzschild-de Sitter Metric. *Il Nuovo Cimento B* **1986**, *91*, 126–142. [CrossRef]
93. Balaguera Antolinez, A.; Böhmer, C.G.; Nowakowski, M. Scales set by the Cosmological Constant. *Class. Quantum Gravity* **2006**, *23*, 485–496. [CrossRef]
94. Arraut, I.; Batic, D.; Nowakowski, M. Velocity and velocity bounds in static spherically symmetric metrics, *Cent. Eur. J Phys.* **2011**, *9*, 926–938.
95. Vainshtein, A.I. To the problem of nonvanishing gravitation mass. *Phys. Lett. B* **1972**, *39*, 393–-394. [CrossRef]
96. Chkareuli, G.; Pirtskhalava, D. Vainshtein mechanism in Λ_3-theories. *Phys. Lett. B* **2012**, *713*, 99–103. [CrossRef]

97. Babichev, E.; Deffayet, C. An introduction to the Vainshtein mechanism. *Class. Quantum Gravity* **2013**, *30*, 184001. [CrossRef]
98. Arraut, I. On the Black Holes in alternative theories of gravity: The case of non-linear massive gravity. *Int. J. Mod. Phys. D* **2015**, *24*, 1550022. [CrossRef]
99. Arraut, I. The Astrophysical Scales Set by the Cosmological Constant, Black-Hole Thermodynamics and Non-Linear Massive Gravity. *Universe* **2017**, *3*, 45. [CrossRef]
100. Bronnikov, K.A.; Dobosz, A.; Dymnikova, I. Nonsingular vacuum cosmologies with a variable cosmological term. *Class. Quantum Gravity* **2003**, *20*, 3797–3814. [CrossRef]
101. Bronnikov, K.A.; Dymnikova, I. Regular homogeneous T-models with vacuum dark fluid. *Class. Quantum Gravity* **2007**, *24*, 5803–5816. [CrossRef]
102. Lemaître, G. Evolution of the Expanding Universe. *Proc. Natl. Acad. Sci. USA* **1934**, *20*, 12–17. [CrossRef]
103. Landau, L.D.; Lifshitz, E.M. *Classical Theory of Fields*, 4th ed.; Butterworth-Heinemann: Oxford, UK, 1975.
104. Tolman, R.C. Effect of Inhomogeneity on Cosmological Models *Proc. Natl. Acad. Sc. USA* **1934**, *20*, 169–176. [CrossRef]
105. Dymnikova, I.; Dobosz, A.; Filchenkov, M.; Gromov, A. Universes inside a Λ black hole. *Phys. Lett. B* **2001**, *506*, 351–360. [CrossRef]
106. Olson, D.W.; Silk J. Primordial inhomogeneities in the expanding universe. II—General features of spherical models at late times. *Astrophys. J.* **1979**, *233*, 395–401. [CrossRef]
107. Harko, T.; Mak, M.K. Bianchi Type I universes with dilaton and magnetic fields. *Int. J. Mod. Phys. D* **2002**, *11*, 1171–1189. [CrossRef]
108. Bronnikov, K.A.; Chudayeva, E.N.; Shikin, G.N. Magneto-dilatonic Bianchi-I cosmology: Isotropization and singularity problems. *Class. Quantum Gravity* **2004**, *21*, 3389–3403. [CrossRef]
109. Dymnikova, I.; Dobosz, A.; Sołtysek, B. Lemaître Class Dark Energy Model for Relaxing Cosmological Constant. *Universe* **2017**, *3*, 39. [CrossRef]
110. Wald, R.M. *General Relativity*; University of Chicago Press: Chicago, IL, USA; London, UK, 1984.
111. 't Hooft, G. Dimensional reduction in quantum gravity. *arXiv* **1999**, arXiv:9310026.
112. Susskind, L. The World as a hologram. *J. Math. Phys.* **1995**, *36*, 6377–6396. [CrossRef]
113. Dymnikova, I., Triple-horizon spherically symmetric spacetime and holographic principle. *Int. J. Mod. Phys. D* **2012**, *21*, 1242007–12420016. [CrossRef]
114. Sekino, Y.; Susskind, L. Fast scramblers. *High Energy Phys.* **2008**, *810*, 65–80. [CrossRef]
115. Dymnikova, I.; Dobosz, A.; Soltysek B. Lemaître dark energy model singled out by the holographic principle. *Gravit. Cosmol.* **2017**, *23*, 28–34. [CrossRef]
116. Wiaux, Y.; Vielva, P.; Martinez-Gonzalez, E.; Vandergheynst, P. Global universe anisotropy probed by the alignment of structures in the cosmic microwave background. *Phys. Rev. Lett.* **2006**, *96*, 151303–151306. [CrossRef] [PubMed]
117. Marochnik L.; Usikov, D. Inflation and CMB anisotropy from quantum metric fluctuations. *Gravit. Cosmol.* **2015**, *21*, 118–122. [CrossRef]
118. Sharma, M. Raychaudhuri equation in an anisotropic universe with anisotropic sources. *Gravit. Cosmol.* **2015**, *21*, 252–256. [CrossRef]
119. Chang, Z.; Li, X.; Lin, H.-N.; Wang, S. Constraining anisotropy of the universe from different groups of type-Ia supernovae. *Eur. Phys. J. C* **2014**, *74*, 2821–2829. [CrossRef]
120. Chang, Z.; Li, X.; Lin, H.-N.; Wang, S. Constraining anisotropy of the universe from Supernovae and Gamma-ray Bursts. *Mod. Phys. Lett. A* **2014**, *29*, 1450067. [CrossRef]
121. Nikolaev, A.V.; Chervon, S.V. The effect of universe inhomogeneities on cosmological distance measurements. *Gravit. Cosmol.* **2016**, *22*, 208–216. [CrossRef]

symmetry

MDPI

Article

Cosmological Consequences of New Dark Energy Models in Einstein-Aether Gravity

Shamaila Rani [1], Abdul Jawad [1], Kazuharu Bamba [2,*] and Irfan Ullah Malik [3]

1 Department of Mathematics, COMSATS University Islamabad, Lahore-Campus, Lahore 54000, Pakistan; shamailatoor.math@yahoo.com (S.R.); jawadab181@yahoo.com or abduljawad@cuilahore.edu.pk (A.J.)
2 Division of Human Support System, Faculty of Symbiotic Systems Science, Fukushima University, Fukushima 960-1296, Japan
3 Allied School, Ali Campus, Muhafiz Town Canal Road, Lahore 54000, Pakistan; malik_irfan22@yahoo.com
* Correspondence: bamba@sss.fukushima-u.ac.jp

Received: 16 February 2019; Accepted: 28 March 2019; Published: 8 April 2019

Abstract: In this paper, we reconstruct various solutions for the accelerated universe in the Einstein-Aether theory of gravity. For this purpose, we obtain the effective density and pressure for Einstein-Aether theory. We reconstruct the Einstein-Aether models by comparing its energy density with various newly proposed holographic dark energy models such as Tsallis, Rényi and Sharma-Mittal. For this reconstruction, we use two forms of the scale factor, power-law and exponential forms. The cosmological analysis of the underlying scenario has been done by exploring different cosmological parameters. This includes equation of state parameter, squared speed of sound and evolutionary equation of state parameter via graphical representation. We obtain some favorable results for some values of model parameters

Keywords: Einstein-Aether theory of gravity; dosmological parameters; dark energy models

1. Introduction

Nowadays, it is believed that our universe is undergoing an accelerated expansion with the passage of cosmic time. This cosmic expansion has been confirmed through various observational schemes such as supernova type Ia (SNIa) [1–4] and the cosmic microwave background (CMB) [5–9]. The source behind the expansion of the universe is a mysterious force called dark energy (DE) and its nature is still ambiguous [10–13]. The current Planck data shows that DE accounts for 68.3% of the total energy contents of the universe. The first candidate for describing the DE phenomenon is the cosmological constant but it has fine tuning and cosmic coincidence problems. Due to this reason, different DE models as well as theories of gravity with modifications have been suggested. The dynamical DE models include a family of Chaplygin gas as well as holographic DE models, scalar field models such as K-essence, phantom, quintessence, ghost, etc. [14–26].

One of the DE model is the holographic DE (HDE) model which becomes a favorable technique now-a-days to study the DE mystery. This model is established in the framework of holographic principle which corresponds to the area instead of volume for the scaling of the number of degrees of freedom of a system. This model is an interesting effort in exploring the nature of DE in the framework of quantum gravity. In addition, the HDE model gives the relationship between the energy density of quantum fields in vacuum (as the DE candidate) to the cutoffs (infrared and ultraviolet). Cohen et al. [27] provided a very useful result about the expression of the HDE model density which is based on the vacuum energy of the system. The black hole mass should not be overcome by the maximum amount of the vacuum energy. Taking into account the nature of spacetime along with long term gravity, various entropy formalisms have been used to discuss the gravitational and cosmological setups [28–33]. Recently, some new HDE models are proposed, like Tsallis HDE (THDE) [31], Rényi HDE model (RHDE) [32] and Sharma-Mittal HDE (SMHDE) [33].

The examples of theories with modification setups include $f(R), f(T), f(R,\mathcal{T}), f(G)$ etc., where R shows the Ricci scalar representing the curvature, T means the torsion scalar, $\mathcal{T}$ is the trace of the energy-momentum tensor and G goes as the invariant of Gauss–Bonnet [34–45]. For recent reviews in terms of DE problems including modified gravity theories, see, for instance [46–52]. The Einstein-Aether theory is one of the modified theories of gravity [53,54] and accelerated expansion phenomenon of the universe has also been investigated in this theory [55]. Meng et al. have also discussed the current cosmic acceleration through DE models in this gravity [56,57]. Recently, Pasqua et al. [58] have made versatile studies on cosmic acceleration through various cosmological models in the presence of HDE models.

In the present work, we will develop the Einstein-Aether gravity models in the presence of modified HDE models and well-known scale factors. For these models of modified gravity, we will extract various cosmological parameters. In the next section, we will give a brief review of the Einstein-Aether theory. In Section 3, we present the basic cosmological parameters as well as well-known scale factors. We will discuss the cosmological parameters for modfied HDE models in Sections 4–6. In the last section, we will summarize our results.

2. Einstein-Aether Theory

As our universe is full with many of the natural occurring phenomenons. One of them is transfer of light from one place to another and second is how gravity acts. To explain these kinds of phenomenons, many of the physicists were used the concept of Aether in many of the theories. In modern physics, Aether indicates a physical medium that is spread homogeneously at each point of the universe. Hence, it was considered that it is a medium in space that helps light to travel in a vacuum. According to this concept, a particular static frame reference is provided by Aether and everything has absolute relative velocity in this frame. That is suitable for Newtonian dynamics extremely well. But, when Einstein performed different experiments on optics in his theory of relativity, then Einstein rejected this ambiguity. When CMB was introduced, many of the people took it a modern form of Aether. Gasperini has popularized Einstein-Aether theories [59]. This theory is said to be covariant modification of general relativity in which unit time like vector field(aether) breaks the Lorentz Invariance (LI) to examine the gravitational and cosmological effects of dynamical preferred frame [53]. Following is the action of Einstein-Aether theory [60,61].

$$S = \int d^4x\sqrt{-g}\left(\frac{R}{4\pi G} + L_{EA} + L_m\right), \tag{1}$$

where L_{EA} represents the Lagrangian density for the vector field and L_m indicates Lagrangian density of matter field. Further, g, R and G indicate determinant of the metric tensor $g^{\mu\nu}$, Ricci scalar and gravitational constant respectively. The Lagrangian density for vector field can be written as

$$L_{EA} = \frac{M^2}{16\pi G}F(K) + \frac{1}{16\pi G}\lambda(A^a A_a + 1), \tag{2}$$

$$K = M^{-2}K^{ab}_{cd}\nabla_a A^c \nabla_b A^d c \tag{3}$$

$$K^{ab}_{cd} = c_1 g^{ab} g_{cd} + c_2\delta^a_c\delta^b_d + c_3\delta^a_d\delta^b_c, \quad a,b = 0,1,2,3. \tag{4}$$

where λ represents a Lagrangian multiplier, dimensionless constants are denoted by c_i, M referred as coupling constant parameter and A^a is a tensor of rank one, that is a vector. The function $F(K)$ is any arbitrary function of K. We obtain the Einstein field equations from Equation (1) for the Einstein-Aether theory as follows

$$
\begin{aligned}
G_{ab} &= T_{ab}^{EA} + 8\pi G T_{ab}^{m}, && (5)\\
\nabla_a\left(\frac{dF}{dK}J_b^a\right) &= 2\lambda A_b, && (6)
\end{aligned}
$$

where $J_b^a = -2K_{bc}^{ad}\nabla_d A^c$, T_{ab}^{EA} shows energy momentum-tensor for vector field and T_{ab}^{m} indicates energy-momentum tensor for mater field. These tensors are given as

$$
T_{ab}^{m} = (\rho + p)u_a ub + p g_{ab}, \quad (7)
$$

$$
\begin{aligned}
T_{ab}^{EA} &= \frac{1}{2}\nabla_d\left((J_a^d A_b - J_a^d A_b - J_{(ab)}A^d)\frac{dF}{dK}\right) - Y_{(ab)}\frac{dF}{dK}\\
&+ \frac{1}{2}g_{ab}M^2F + \lambda A_a A_b,
\end{aligned} \quad (8)
$$

where p and ρ represent energy density and pressure of the matter respectively. Furthermore, u_a expresses the four-velocity vector of the fluid and given as $u_a = (1,0,0,0)$ and A_a is time-like unitary vector and is defined as $A_a = (1,0,0,0)$. Moreover Y_{ab} is defined as

$$
Y_{ab} = c_1\Big((\nabla_d A_a)(\nabla^d A_b) - (\nabla_a A_d)(\nabla_a A^d)\Big), \quad (9)
$$

where indices $(a\ b)$ show the symmetry.

The Friedmann equations modified by the Einstein-Aether gravity are given as follows

$$
\begin{aligned}
\epsilon\left(\frac{F}{2K} - \frac{dF}{dK}\right)H^2 + \left(H^2 + \frac{k}{a^2}\right) &= \left(\frac{8\pi G}{3}\right)\rho, && (10)\\
\epsilon\frac{d}{dt}\left(H\frac{dF}{dK}\right) + \left(-2\dot{H} + \frac{2k}{a^2}\right) &= 8\pi G(p+\rho). && (11)
\end{aligned}
$$

Here K becomes $K = \frac{3\epsilon H^2}{M^2}$, where ϵ is a constant parameter. The energy density of Einstein-Aether theory is denoted by ρ_{EA} and called the effective energy density, while the effective pressure in the Einstein-Aether gravity is given by p_{EA}. So, we can rewrite Equations (10) and (11) as

$$
\begin{aligned}
\left(H^2 + \frac{k}{a^2}\right) &= \left(\frac{8\pi G}{3}\right)\rho + \frac{1}{3}\rho_{EA}, && (12)\\
\left(-2\dot{H} + \frac{2k}{a^2}\right) &= 8\pi G(p+\rho) + (\rho_{EA} + p_{EA}), && (13)
\end{aligned}
$$

where

$$
\begin{aligned}
\rho_{EA} &= 3\epsilon H^2\left(\frac{dF}{dK} - \frac{F}{2K}\right), && (14)\\
p_{EA} &= -3\epsilon H^2\left(\frac{dF}{dK} - \frac{F}{2K}\right) - \epsilon\left(\dot{H}\frac{dF}{dK} + H\frac{d\dot{F}}{dK}\right). && (15)\\
&= \rho_{EA} - \frac{\dot{\rho}_{EA}}{3H}. && (16)
\end{aligned}
$$

The equation of state (EoS) parameter for the Einstein-Aether can be obtained by using Equations (14) and (15), and it is given by

$$
\omega_{EA} = \frac{p_{EA}}{\rho_{EA}} = -1 - \frac{\dot{H}\frac{dF}{dK} + H\frac{d\dot{F}}{dK}}{3H^2(\frac{dF}{dK} - \frac{F}{2K})}. \quad (17)
$$

3. Cosmological Parameters

To understand the geometry of the universe, the following are some basic cosmological parameters.

3.1. Equation of State Parameter

In order to categorize the different phases of the evolving universe, the EoS parameter is widely used. In particular, the decelerated and accelerated phases contain DE, DM, radiation dominated eras. This parameter is defined in terms of energy density ρ and pressure p as $\omega = \frac{p}{\rho}$.

- In the decelerated phase, the radiation era $0 < \omega < \frac{1}{3}$ and cold DM era $\omega = 0$ are included.
- The accelerated phase of the universe has following eras: $\omega = -1 \Rightarrow$ cosmological constant, $-1 < \omega < \frac{-1}{3} \Rightarrow$ quintessence and $\omega < -1 \Rightarrow$ phantom era of the universe.

3.2. Squared Speed of Sound

To examine the behavior of DE models, there is another parameter which is known as squared speed of sound. It is denoted by v_s^2 and is calculated by the following formula

$$v_s^2 = \frac{\dot{p}}{\dot{\rho}}. \tag{18}$$

The stability of the model can be checked by this relation. If its graph is showing negative values then we may say that model is unstable and in case of non-negative values of the graph, it represents the stable behavior of the model.

3.3. ω-ω' Plane

There are different DE models which have different properties. To examine their dynamical behavior, we use ω-ω' plane, where prime denotes the derivative with respect to $\ln a$ and subscript Λ indicates DE scenario. This method was developed by Caldwell and Linder [62] and divides ω-ω' plane into two parts. One is the freezing part in which evolutionary parameter gives negative behavior for negative EoS parameter, i.e., $\omega' < 0, \omega < 0$, while for positive behavior of evolutionary parameter corresponding to negative EoS parameter yields the thawing part ($\omega' > 0, \omega < 0$) of the evolving universe.

3.4. Scale Factor

The scale factor is the measure of how much the universe has expanded since a given time. It is represented by $a(t)$. Since the latest cosmic observations have shown that the universe is accelerating so $a(t) > 0$. As Einstein-Aether is one of the modified theory which may produce the accelerated expansion of the universe, by using this theory, we can reconstruct various well-known DE models. In order to do this, we take some modified HDE models such as THDE, RHDE and SMHDE models. Since $F(K)$ is a function that the Einstein-Aether theory contains, which can be determined by comparing the densities with the above DE models. For this purpose, we use some well-known forms of the scale factor, $a(t)$. We consider two forms of scale factors $a(t)$ in terms of power and exponential terms. These are

(i) Power-law form: $a(t) = a_0 t^m$, $m > 0$, where a_0 is a constant which indicates the value of scale factor at present-day [63,64]. From this scale factor, we get $H,\ \dot{H},\ K$ as follows

$$H = \frac{m}{t}, \quad \dot{H} = -\frac{m}{t^2}, \quad K = \frac{3\epsilon m^2}{M^2 t^2}. \tag{19}$$

(ii) Exponential form: $a(t) = e^{\alpha t^{\theta}}$ where α is a positive constant and θ lies between 0 and 1. This scale factor gives

$$H = \alpha\theta t^{\theta-1}, \quad \dot{H} = \alpha\theta(\theta-1)t^{\theta-2}, \quad K = \frac{3\epsilon\alpha^2\theta^2 t^{2(\theta-1)}}{M^2}. \tag{20}$$

4. Reconstruction from the Tsallis Holographic Dark Energy Model

The energy density of THDE model is given by [31]

$$\rho_D = BL^{2\delta-4}, \tag{21}$$

where B is an unknown parameter. Taking into account Hubble radius as IR cutoff L, that is $L = \frac{1}{H}$, we have

$$\rho_D = BH^{-2\delta+4}. \tag{22}$$

In order to construct a DE model in the framework of Einstein-Aether gravity with THDE model, we compare the densities of both models, (i.e., $\rho_{EA} = \rho_D$). This yields

$$\frac{dF}{dK} - \frac{F}{2K} = \frac{B}{3\epsilon}H^{-2\delta+2}, \tag{23}$$

which results the following form

$$F(K) = \frac{2BM^{-2\delta+2}K^{-\delta+2}}{(-2\delta+3)(3\epsilon)^{-\delta+2}} + C_1\sqrt{K}. \tag{24}$$

Power-law form of scale factor:

Using the expression of $F(K)$ along with Equation (19) in (14), we obtain the energy density and pressure as

$$\begin{aligned} \rho_{EA} &= \frac{m^2\left(3^{\delta}2BK^{\frac{5}{2}-\delta}M^{2-2\delta}\epsilon^{\delta} - 9\epsilon^2C_1\right)}{6K^{\frac{3}{2}}t^2\epsilon}, \\ p_{EA} &= m\left(4BM^{-2\delta}\epsilon^{\delta}\left(-3^{\delta}K^{1-\delta}M^2(4-9m-2\delta+6m\delta) - 12M^2(-2\right.\right. \\ &+ \left.\delta)(-1+\delta)\left(\frac{m^2\epsilon}{M^2t^2}\right)^{1-\delta}\right)(-3+2\delta)^{-1} + \left(\frac{9(-1+6m)\epsilon^2}{K^{\frac{3}{2}}} + \frac{3\sqrt{3}}{m^4}\right. \\ &\times \left.\left. M^4t^4\sqrt{\frac{m^2\epsilon}{M^2t^2}}\right)C_1\right)(36t^2\epsilon)^{-1}. \end{aligned} \tag{25}$$

Using these expressions of energy density and pressure, we find the values of some cosmological parameters in the following. The EoS parameter takes the following form

$$\begin{aligned} \omega_{EA} &= K^{\frac{3}{2}}\left(4BM^{-2\delta}\epsilon^{\delta}\left(-3^{\delta}K^{1-\delta}M^2(4-9m-2\delta+6m\delta) - 12M^2(-2\right.\right. \\ &+ \left.\delta)(-1+\delta)\left(\frac{m^2\epsilon}{M^2t^2}\right)^{1-\delta}\right)(-3+2\delta)^{1} + \left(\frac{9(-1+6m)\epsilon^2}{K^{\frac{3}{2}}} + \frac{3\sqrt{3}M^4}{m^4}\right. \\ &\times \left.\left. t^4\sqrt{\frac{m^2\epsilon}{M^2t^2}}\right)C_1\right)\left(6m\left(23^{\delta}BK^{\frac{5}{2}-\delta}M^{2-2\delta}\epsilon^{\delta} - 9\epsilon^2C_1\right)\right)^{-1}. \end{aligned} \tag{26}$$

The derivative of EoS parameter with respect to $\ln a$ is given by

$$
\begin{aligned}
\omega'_{EA} &= K^{\frac{3}{2}} t \Bigg(\frac{96Bm^2M^{-2\delta}(1-\delta)(-2+\delta)(-1+\delta)\epsilon^{1+\delta}\left(\frac{m^2\epsilon}{M^2t^2}\right)^{-\delta}}{t^3(-3+2\delta)} + \Bigg(-\frac{3\sqrt{3}}{m^2} \\
&\times \frac{M^2t\epsilon}{\sqrt{\frac{m^2\epsilon}{M^2t^2}}} + \frac{12\sqrt{3}M^4t^3\sqrt{\frac{m^2\epsilon}{M^2t^2}}}{m^4}\Bigg)C_1\Bigg)\left(6m^2\left(23^{\delta}BK^{\frac{5}{2}-\delta}M^{2-2\delta}\epsilon^{\delta}-9\epsilon^2C_1\right)\right)^{-1}.
\end{aligned}
\tag{27}
$$

We plot the EoS parameter versus z using the relation $t = \frac{1}{(1+z)^{\frac{1}{m}}}$ taking values of constants as $B = 5,\ M = 5,\ \delta = 1.8,\ \epsilon = 1$ and $C_1 = 2$. We plot ω_{EA} for three different values of scale factor parameter m as $m = 2,\ 3, 4$ as shown in Figure 1. All the three trajectories represent the phantom behavior of the universe related to the redshift parameter. In Figure 2, we plot $\omega'_{EA} - \omega EA$ plane taking same values of the parameters for $-1 \leq z \leq 1$. For $m = 2$ and 3, the evolving EoS parameter shows negative behavior with respect to negative EoS parameter which indicates the freezing region of the universe. The trajectory of ω'_{EA} for $m = 4$ represents the positive behavior for negative EoS parameter and expresses the evolving universe in thawing region of the universe.

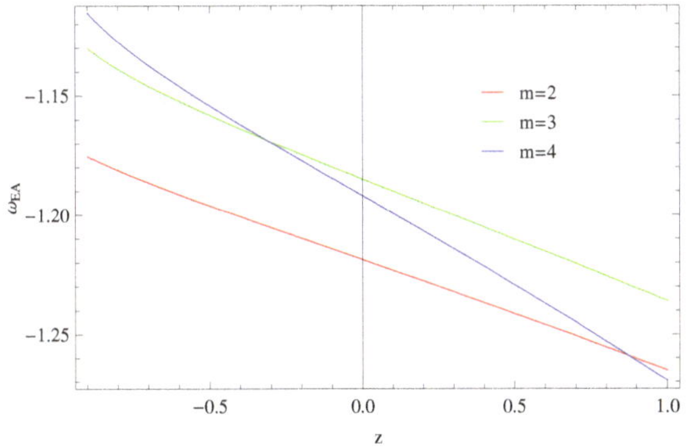

Figure 1. Plot of ω_{EA} versus z taking power-law scale factor for the Tsallis holographic dark energy (THDE) model.

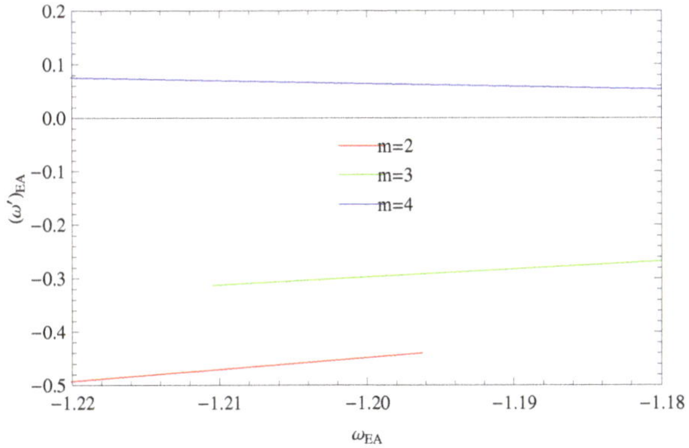

Figure 2. Plot of $\omega'_{EA} - \omega_{EA}$ taking the power-law scale factor for the THDE model.

Also the squared speed of sound in the underlying scenario becomes

$$
\begin{aligned}
v_s^2 &= \Bigg(\Big(\frac{m^2\epsilon}{M^2t^2}\Big)^{-\delta}\Bigg(8Bm^4\epsilon^\delta\Big(-12K^{\frac{3}{2}+\delta}m^2(-2+\delta)^2(-1+\delta)\epsilon+3^\delta K^{\frac{5}{2}}M^2 \\
&\times t^2(4-9m-2\delta+6m\delta)\Big(\frac{m^2\epsilon}{M^2t^2}\Big)^\delta\Big)+3K^\delta M^{2\delta}t^2(-3+2\delta)\Big(\frac{m^2\epsilon}{M^2t^2}\Big)^\delta \\
&\times \Big(6(1-6m)m^4\epsilon^2+\sqrt{3}K^{\frac{3}{2}}M^4t^4\sqrt{\frac{m^2\epsilon}{M^2t^2}}\Big)C_1\Bigg)\Bigg)/\Big(12m^5t^2(-3+2\delta) \\
&\times \Big(-23^\delta BK^{\frac{5}{2}}M^2\epsilon^\delta+9K^\delta M^{2\delta}\epsilon^2C_1\Big)\Big).
\end{aligned}
\tag{28}
$$

Figure 3 shows the plot of v_s^2 versus z to check the behavior of the Einstein-Aether model for the THDE and power-law scale factor for same values of parameters. The trajectories represent the negative behavior of the model which indicated the instability of the model.

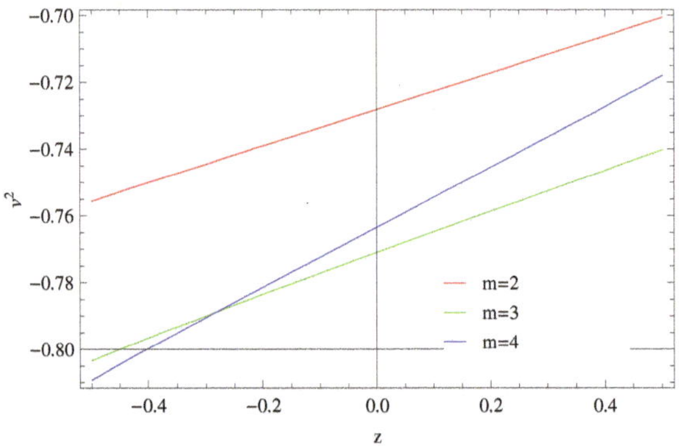

Figure 3. Plot of v_s^2 versus z taking power-law scale factor for the THDE model.

Exponential form of scale factor:

Following the same steps for the exponential form of the scale factor, we get energy density and pressure as

$$
\begin{aligned}
\rho_{EA} &= \frac{t^{-2+2\theta}\alpha^2\theta^2\Big(23^\delta BK^{\frac{5}{2}-\delta}M^{2-2\delta}\epsilon^\delta-9\epsilon^2C_1\Big)}{6K^{\frac{3}{2}}\epsilon}, \\
p_{EA} &= \frac{1}{12\epsilon}t^{-2+\theta}\alpha\theta\Bigg(\frac{2t^\theta\alpha\theta\Big(-23^\delta BK^{\frac{5}{2}-\delta}M^{2-2\delta}\epsilon^\delta+9\epsilon^2C_1\Big)}{K^{\frac{3}{2}}}-\epsilon^2(-1 \\
&+ \theta)\Bigg(8BM^{-2\delta}(-2+\delta)\epsilon^{-2+\delta}\Big(3^\delta K^{1-\delta}M^2-6M^2(-1+\delta)\Big(\frac{t^{-2+2\theta}}{M^2} \\
&\times \alpha^2\epsilon\theta^2\Big)^{1-\delta}\Big)\Big(3(-3+2\delta)\Big)^{-1}+\Bigg(-\frac{3}{K^{\frac{3}{2}}}+\frac{\sqrt{\frac{t^{-2+2\theta}\alpha^2\epsilon\theta^2}{M^2}}}{\alpha^4\epsilon^2\theta^4}\sqrt{3}M^4 \\
&\times t^{4-4\theta}\Bigg)C_1\Bigg)\Bigg).
\end{aligned}
\tag{29}
$$

Now, by using the above density and pressure, we obtain the EoS parameter and its derivative for the Einstein-Aether gravity as follows

$$\omega_{EA} = \Bigg(K^{\frac{3}{2}}t^{-\theta}\Bigg(\frac{2t^{\theta}\alpha\theta\Big(-23^{\delta}BK^{\frac{5}{2}-\delta}M^{2-2\delta}\epsilon^{\delta}+9\epsilon^{2}C_1\Big)}{K^{\frac{3}{2}}}-\epsilon^{2}(-1+\theta)$$
$$\times \Bigg(8BM^{-2\delta}(-2+\delta)\epsilon^{-2+\delta}\Bigg(3^{\delta}K^{1-\delta}M^{2}-6M^{2}(-1+\delta)\Bigg(\frac{t^{-2+2\theta}\alpha^{2}}{M^{2}}$$
$$\times \epsilon\theta^{2}\Bigg)^{1-\delta}\Bigg)\Big(3(-3+2\delta)\Big)^{-1}+\Bigg(-\frac{3}{K^{\frac{3}{2}}}+\frac{\sqrt{3}M^{4}t^{4-4\theta}\sqrt{\frac{t^{-2+2\theta}\alpha^{2}\epsilon\theta^{2}}{M^{2}}}}{\alpha^{4}\epsilon^{2}\theta^{4}}\Bigg)$$
$$\times C_1\Bigg)\Bigg)\Bigg)\Big/\Bigg(2\alpha\theta\Big(23^{\delta}BK^{\frac{5}{2}-\delta}M^{2-2\delta}\epsilon^{\delta}-9\epsilon^{2}C_1\Big)\Bigg),$$

$$\omega'_{EA} = \frac{1}{6\alpha^{2}\theta\Big(23^{\delta}BK^{\frac{5}{2}-\delta}M^{2-2\delta}\epsilon^{\delta}-9\epsilon^{2}C_1\Big)}K^{\frac{3}{2}}t^{-2(1+\theta)}(-1+\theta)\Bigg(8BK^{-\delta}$$
$$\times M^{-2\delta}(-2+\delta)\epsilon^{\delta}\Bigg(\frac{t^{-2+2\theta}\alpha^{2}\epsilon\theta^{2}}{M^{2}}\Bigg)^{-\delta}\Bigg(-6K^{\delta}t^{2\theta}\alpha^{2}(-1+\delta)\epsilon(2+2\delta$$
$$\times (-1+\theta)-\theta)\theta+3^{\delta}KM^{2}t^{2}\Bigg(\frac{t^{-2+2\theta}\alpha^{2}\epsilon\theta^{2}}{M^{2}}\Bigg)^{\delta}\Bigg)\Big(-3+2\delta\Big)^{-1}-3$$
$$\times t^{2-4\theta}\Bigg(3t^{4\theta}\alpha^{4}\epsilon^{2}\theta^{5}+\sqrt{3}K^{\frac{3}{2}}M^{4}t^{4}(3-4\theta)\sqrt{\frac{t^{-2+2\theta}\alpha^{2}\epsilon\theta^{2}}{M^{2}}}\Bigg)C_1\Bigg(K^{\frac{3}{2}}\alpha^{4}$$
$$\times \theta^{5}\Bigg)^{-1}\Bigg).$$

The squared speed of sound for the second form of the scale factor is given by

$$v_s^2 = K^{\frac{3}{2}}t^{-\theta}\Bigg(8BM^{-2\delta}\epsilon^{\delta}\Bigg(-6M^{2}(-2+\delta)(-1+\delta)(4+2\delta(-1+\theta)-3\theta)$$
$$\times \Bigg(\frac{t^{-2+2\theta}\alpha^{2}\epsilon\theta^{2}}{M^{2}}\Bigg)^{1-\delta}-3^{\delta}K^{1-\delta}M^{2}\Bigg((-2+\delta)(-2+\theta)+3t^{\theta}\alpha(-3+2$$
$$\times \delta)\theta\Bigg)\Bigg)(-3+2\delta)^{-1}+3t^{-4\theta}\Bigg(3t^{4\theta}\alpha^{4}\epsilon^{2}(-2+\theta)\theta^{4}+36t^{5\theta}\alpha^{5}\epsilon^{2}\theta^{5}+\sqrt{3}$$
$$\times K^{\frac{3}{2}}M^{4}t^{4}\sqrt{\frac{t^{-2+2\theta}\alpha^{2}\epsilon\theta^{2}}{M^{2}}}(-1+2\theta)\Bigg)C_1(K^{\frac{3}{2}}\alpha^{4}\theta^{4})^{-1}\Bigg)\Bigg(12\alpha\theta\Bigg(23^{\delta}B$$
$$\times K^{\frac{5}{2}-\delta}M^{2-2\delta}\epsilon^{\delta}-9\epsilon^{2}C_1\Bigg)\Bigg).$$

Figure 4 represents the graph of the EoS parameter versus z for the exponential form of the scale factor taking $B = 5 = M$, $\delta = 1.8$, $\epsilon = 1, C_1 = -0.5$, $\theta = 0.5$ and scale factor parameter $\alpha = 2, 3, 4$. This parameter represents the phantom behavior of the universe for $\alpha = 3$ and after a transition from quintessence to phantom era for $\alpha = 2$. For $\alpha = 4$, the trajectory of the EoS parameter corresponds to the Λ-CDM model $\omega_{EA} = -1$. In Figure 5, the graph is plotted between ω'_{EA} and ω_{EA}. The graph represents initially freezing region and then indicates the thawing region of the evolving universe. As we increase the value of α, the trajectories indicate the thawing region only. However, the graph of v_s^2 versus z as shown in Figure 6 shows the unstable behavior.

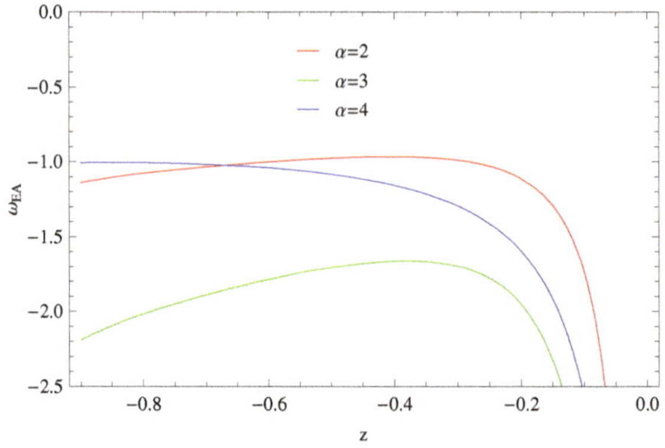

Figure 4. Plot of ω_{EA} versus z, taking an exponential scale factor for the THDE model.

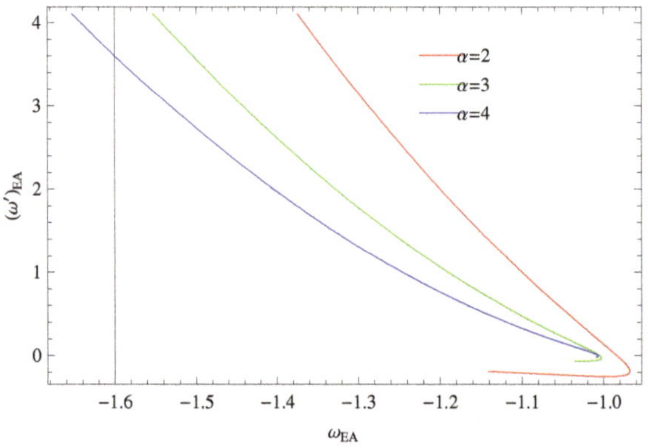

Figure 5. Plot of $\omega'_{EA} - \omega_{EA}$, taking an exponential scale factor for the THDE model.

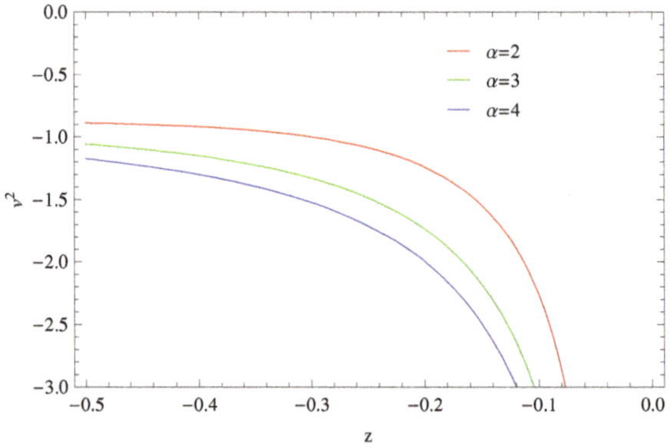

Figure 6. Plot of v_s^2 versus z, taking an exponential scale factor for the THDE model.

5. Reconstruction from Rényi Holographic Dark Energy Model

The energy density of the RHDE model is [33]

$$\rho_D = \frac{3C^2 L^{-2}}{8\pi\left(1 + \frac{\delta\pi}{H^2}\right)}. \tag{30}$$

For the Hubble horizon, it takes the form

$$\rho_D = \frac{3C^2H^2}{8\pi\left(1+\frac{\delta\pi}{H^2}\right)}. \tag{31}$$

Now we compare the Einstein-Aether model energy density with the RHDE model density (i.e., $\rho_{EA} = \rho_D$) in order to get the reconstructed equation,

$$\frac{dF}{dK} - \frac{F}{2K} = \frac{C^2}{8\pi\epsilon(1+\frac{\delta\pi}{H^2})}. \tag{32}$$

The solution of this equation is given by

$$F(K) = \frac{C^2}{4\pi M\epsilon}\left(KM - \sqrt{3K\delta\epsilon\pi}\arctan\left(\frac{M\sqrt{K}}{\sqrt{3\epsilon\pi\delta}}\right)\right) + C_2\sqrt{K}. \tag{33}$$

Power-law form of the scale factor:

Inserting all the corresponding values into Equations (14) and (15), we get density and pressure of the Einstein-Aether gravity model as follows

$$\begin{aligned}
\rho_{EA} &= \frac{1}{8\pi t^2}3C^2m^2\left(1-\frac{3\sqrt{\delta\epsilon\pi}\sqrt{K\delta\epsilon\pi}}{\sqrt{K}\left(KM^2+3\delta\epsilon\pi\right)}\right). \\
p_{EA} &= \frac{1}{24\pi t^5}C^2m\left(9mt^3\left(-1+\frac{3\sqrt{\delta\epsilon\pi}\sqrt{K\delta\epsilon\pi}}{\sqrt{K}\left(KM^2+3\delta\epsilon\pi\right)}\right)-\frac{1}{M^4}\epsilon\left(3M^3\right.\right. \\
&\times\ t^3\left(M\left(-2+\frac{3\sqrt{\delta\epsilon\pi}\sqrt{K\delta\epsilon\pi}}{\sqrt{K}\left(KM^2+3\delta\epsilon\pi\right)}\right)+\frac{\mathrm{ArcTan}\left(\frac{\sqrt{K}M}{\sqrt{3}\sqrt{\delta\epsilon\pi}}\right)}{K}\sqrt{3}\right. \\
&\times\ \left.\sqrt{K\delta\epsilon\pi}-\frac{C_2}{\sqrt{K}}\right)(\epsilon)^{-1}+\left(m^2\left(-3m^2t\delta^2\epsilon^2\pi^2\left(m^2-t^2\delta\pi\right)-3\right.\right. \\
&\times\ m\delta^{3/2}\epsilon^2\pi^{\frac{3}{2}}\left(m^2+t^2\delta\pi\right)^2\mathrm{ArcTan}\left(\frac{m}{t\sqrt{\delta}\sqrt{\pi}}\right)+\sqrt{3}Mt\sqrt{\delta\epsilon\pi}m\epsilon \\
&\times\ \left.\left.\sqrt{\frac{\delta\pi}{M^2t^2}}\left(m^2+t^2\delta\pi\right)^2C_2\right)\right)/\left(\left(\frac{m^2\epsilon}{M^2t^2}\right)^{3/2}\sqrt{\delta\epsilon\pi}\sqrt{\frac{m^2\delta\epsilon^2\pi}{M^2t^2}}\left(m^2\right.\right. \\
&+\ \left.\left.\left.t^2\delta\pi\right)^2\right)\right)\right).
\end{aligned}$$

In this case, the EoS parameter takes the form

$$
\begin{aligned}
\omega_{EA} &= \Bigg(9mt^3\bigg(-1+\frac{3\sqrt{\delta\epsilon\pi}\sqrt{K\delta\epsilon\pi}}{\sqrt{K}\Big(KM^2+3\delta\epsilon\pi\Big)}\bigg)-\frac{1}{M^4}\epsilon\bigg(3M^3t^3\bigg(M\bigg(-2\\
&+ \frac{3\sqrt{\delta\epsilon\pi}\sqrt{K\delta\epsilon\pi}}{\sqrt{K}\Big(KM^2+3\delta\epsilon\pi\Big)}\bigg)+\frac{\sqrt{3}\sqrt{K\delta\epsilon\pi}\mathrm{ArcTan}\Big(\frac{\sqrt{K}M}{\sqrt{3}\sqrt{\delta\epsilon\pi}}\Big)}{K}-\frac{C_2}{\sqrt{K}}\bigg)\\
&\times (\epsilon)^{-1}+\bigg(m^2\bigg(-3m^2t\delta^2\epsilon^2\pi^2\Big(m^2-t^2\delta\pi\Big)-3m\delta^{\frac{3}{2}}\epsilon^2\pi^{\frac{3}{2}}\Big(m^2+t^2\\
&\times \delta\pi\Big)^2\mathrm{ArcTan}\Big(\frac{m}{t\sqrt{\delta}\sqrt{\pi}}\Big)+\sqrt{3}Mt\sqrt{\delta\epsilon\pi}\sqrt{\frac{m^2\delta\epsilon^2\pi}{M^2t^2}}\Big(m^2+t^2\delta\pi\Big)^2\\
&\times C_2\bigg)\bigg)/\bigg(\Big(\frac{m^2\epsilon}{M^2t^2}\Big)^{\frac{3}{2}}\sqrt{\delta\epsilon\pi}\sqrt{\frac{m^2\delta\epsilon^2\pi}{M^2t^2}}\Big(m^2+t^2\delta\pi\Big)^2\bigg)\bigg)\Bigg)/\bigg(9t^3\bigg(m\\
&- \frac{3m\sqrt{\delta\epsilon\pi}\sqrt{K\delta\epsilon\pi}}{\sqrt{K}\Big(KM^2+3\delta\epsilon\pi\Big)}\bigg)\bigg),
\end{aligned}
$$

and ω'_{EA} is given as follows

$$
\begin{aligned}
\omega'_{EA} &= 3\Bigg(6-\frac{4\sqrt{\frac{m^2\delta\epsilon^2\pi}{M^2t^2}}}{\sqrt{\frac{m^2\epsilon}{M^2t^2}}\sqrt{\delta\epsilon\pi}}+m\bigg(-9+\frac{27\sqrt{\delta\epsilon\pi}\sqrt{K\delta\epsilon\pi}}{\sqrt{K}\Big(KM^2+3\delta\epsilon\pi\Big)}\bigg)+\sqrt{\delta\epsilon\pi}\\
&\times \Bigg(-\frac{9\sqrt{K\delta\epsilon\pi}}{\sqrt{K}\Big(KM^2+3\delta\epsilon\pi\Big)}+\frac{4m^2\sqrt{\frac{m^2\epsilon}{M^2t^2}}\Big(m^4+4m^2t^2\delta\pi+t^4\delta^2\pi^2\Big)}{\sqrt{\frac{m^2\delta\epsilon^2\pi}{M^2t^2}}\Big(m^2+t^2\delta\pi\Big)^3}\Bigg)\\
&+ \frac{4t\sqrt{\delta}\sqrt{\pi}\sqrt{\frac{m^2\delta\epsilon^2\pi}{M^2t^2}}\mathrm{ArcTan}\Big(\frac{m}{t\sqrt{\delta}\sqrt{\pi}}\Big)}{m\sqrt{\frac{m^2\epsilon}{M^2t^2}}\sqrt{\delta\epsilon\pi}}-\frac{3\sqrt{3}\sqrt{K\delta\epsilon\pi}\mathrm{ArcTan}\Big(\frac{\sqrt{K}M}{\sqrt{3}\sqrt{\delta\epsilon\pi}}\Big)}{KM}\Bigg)\\
&+ \frac{\Bigg(9-\frac{4\sqrt{3}\sqrt{K}}{\sqrt{\frac{m^2\epsilon}{M^2t^2}}}\Bigg)C_2}{\sqrt{K}M}\Bigg(9m\bigg(m-\frac{3m\sqrt{\delta\epsilon\pi}\sqrt{K\delta\epsilon\pi}}{\sqrt{K}\Big(KM^2+3\delta\epsilon\pi\Big)}\bigg)\Bigg)^{-1}.
\end{aligned}
$$

The expression for squared speed of sound turns out as

$$
\begin{aligned}
v_s^2 &= -3\Bigg(-4+\frac{\sqrt{\frac{m^2\delta\epsilon^2\pi}{M^2t^2}}}{\sqrt{\frac{m^2\epsilon}{M^2t^2}}\sqrt{\delta\epsilon\pi}}+m\left(6-\frac{18\sqrt{\delta\epsilon\pi}\sqrt{K\delta\epsilon\pi}}{\sqrt{K}\left(KM^2+3\delta\epsilon\pi\right)}\right)+\sqrt{\delta\epsilon\pi}\\
&\times\left(\frac{6\sqrt{K\delta\epsilon\pi}}{\sqrt{K}\left(KM^2+3\delta\epsilon\pi\right)}-\frac{m^2\sqrt{\frac{m^2\epsilon}{M^2t^2}}\left(m^4+4m^2t^2\delta\pi+11t^4\delta^2\pi^2\right)}{\sqrt{\frac{m^2\delta\epsilon^2\pi}{M^2t^2}}\left(m^2+t^2\delta\pi\right)^3}\right.\\
&-\left.\frac{t\sqrt{\delta}\sqrt{\pi}\sqrt{\frac{m^2\delta\epsilon^2\pi}{M^2t^2}}\mathrm{ArcTan}\left(\frac{m}{t\sqrt{\delta}\sqrt{\pi}}\right)}{m\sqrt{\frac{m^2\epsilon}{M^2t^2}}\sqrt{\delta\epsilon\pi}}+\frac{2\sqrt{3}\sqrt{K\delta\epsilon\pi}\mathrm{ArcTan}\left(\frac{\sqrt{K}M}{\sqrt{3}\sqrt{\delta\epsilon\pi}}\right)}{KM}\right)\\
&+\frac{\left(-6+\frac{\sqrt{3}\sqrt{K}}{\sqrt{\frac{m^2\epsilon}{M^2t^2}}}\right)C_2}{\sqrt{K}M}\left(18\left(m-\frac{3m\sqrt{\delta\epsilon\pi}\sqrt{K\delta\epsilon\pi}}{\sqrt{K}\left(KM^2+3\delta\epsilon\pi\right)}\right)\right)^{-1}.
\end{aligned}
$$

The plot of the EoS parameter is shown in Figure 7 with respect to z. All the trajectories of the EoS parameter represent the quintessence phase of the universe. Figure 8 shows the graph of $\omega_{EA}-\omega'_{EA}$ plane for same range of z. The trajectories of ω'_{EA} describe the negative behavior for all $\omega_{EA}<0$ give the freezing region of the universe. To check the stability of the underlying model, Figure 9 shows the unstable behavior of the model. However, for $m=2$, we get some stable points for $z<-0.475$.

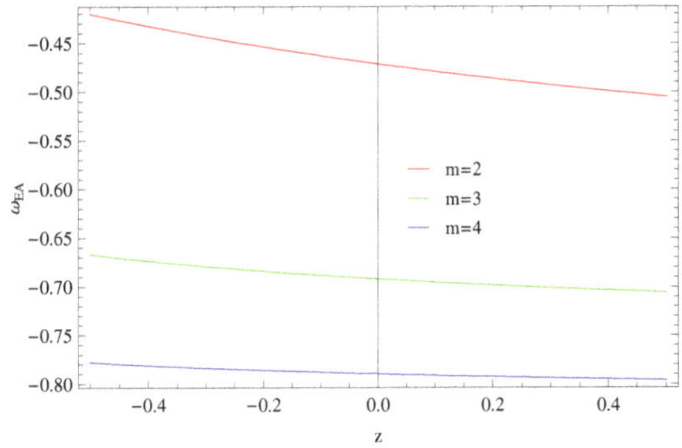

Figure 7. Plot of ω_{EA} versus z, taking a power-law scale factor for the Rényi holographic dark matter (RHDE) model.

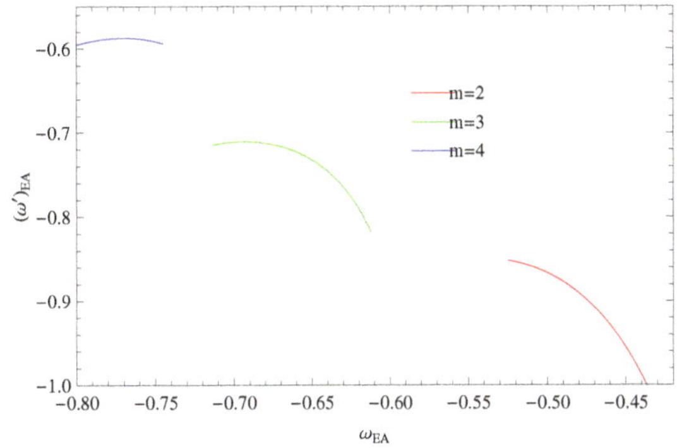

Figure 8. Plot of $\omega'_{EA}-\omega_{EA}$, taking a power-law scale factor for the RHDE model.

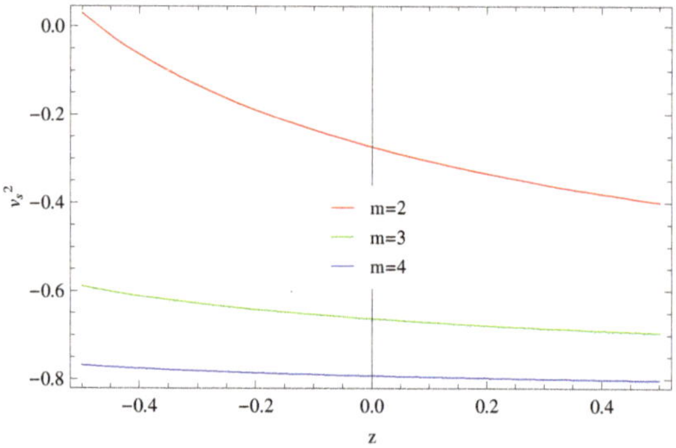

Figure 9. Plot of v_s^2 versus z, taking a power-law scale factor for the RHDE model.

Exponential form of scale factor:

Taking into account second scale factor Equation (15) along with $F(K)$, we get the following energy density and pressure

$$\rho_{EA} = \frac{3C^2 t^{-2+2\theta}\alpha^2\theta^2\Big(-3\sqrt{\delta\epsilon\pi}\sqrt{K\delta\epsilon\pi}+\sqrt{K}\Big(KM^2+3\delta\epsilon\pi\Big)\Big)}{8\sqrt{K}\pi\Big(KM^2+3\delta\epsilon\pi\Big)} \tag{34}$$

$$\begin{aligned} p_{EA} &= \frac{1}{24\pi}C^2 t^{-3+\theta}\alpha\theta\Bigg(9t^{1+\theta}\alpha\theta\Bigg(-1+\frac{3\sqrt{\delta\epsilon\pi}\sqrt{K\delta\epsilon\pi}}{\sqrt{K}\Big(KM^2+3\delta\epsilon\pi\Big)}\Bigg)-\tfrac{1}{M^2}\epsilon(-1 \\ &+ \theta)\Bigg(3Mt\Bigg(M\Bigg(2-\frac{3\sqrt{\delta\epsilon\pi}\sqrt{K\delta\epsilon\pi}}{\sqrt{K}\Big(KM^2+3\delta\epsilon\pi\Big)}\Bigg)-\frac{\mathrm{ArcTan}\Big(\frac{\sqrt{K}M}{\sqrt{3}\sqrt{\delta\epsilon\pi}}\Big)}{K} \\ &\times \sqrt{3}\sqrt{K\delta\epsilon\pi}+\frac{C_2}{\sqrt{K}}\Bigg)(\epsilon)^{-1}+\Bigg(\delta\pi\Bigg(3t^{\theta}\alpha\delta^{\frac{3}{2}}\epsilon^2\theta\pi^{\frac{3}{2}}\Bigg(t^{1+\theta}\alpha\sqrt{\delta}\theta\sqrt{\pi} \\ &\times \Big(t^{2\theta}\alpha^2\theta^2-t^2\delta\pi\Big)+\Big(t^{2\theta}\alpha^2\theta^2+t^2\delta\pi\Big)^2\mathrm{ArcTan}\Bigg(\frac{t^{-1+\theta}\alpha\theta}{\sqrt{\delta}\sqrt{\pi}}\Bigg)\Bigg) \\ &- \sqrt{3}Mt\sqrt{\delta\epsilon\pi}\sqrt{\frac{t^{-2+2\theta}\alpha^2\delta\epsilon^2\theta^2\pi}{M^2}}\Big(t^{2\theta}\alpha^2\theta^2+t^2\delta\pi\Big)^2 C_2\Bigg)\Bigg)/\Bigg(\alpha\theta \\ &\times \sqrt{\frac{t^{-2+2\theta}\epsilon}{M^2}}(\delta\epsilon\pi)^{\frac{3}{2}}\sqrt{\frac{t^{-2+2\theta}\alpha^2\delta\epsilon^2\theta^2\pi}{M^2}}\Big(t^{2\theta}\alpha^2\theta^2+t^2\delta\pi\Big)^2\Bigg)\Bigg)\Bigg). \end{aligned} \tag{35}$$

The EoS parameter is obtained from the above energy density and pressure. This parameter with its derivative are given by

$$\begin{aligned}
\omega_{EA} &= \left(\sqrt{K}t^{-1-\theta}\left(KM^2+3\delta\epsilon\pi\right)\left(9t^{1+\theta}\alpha\theta\left(-1+\frac{3\sqrt{\delta\epsilon\pi}\sqrt{K\delta\epsilon\pi}}{\sqrt{K}\left(KM^2+3\delta\epsilon\pi\right)}\right)\right.\right.\\
&- \tfrac{1}{M^2}\epsilon(-1+\theta)\left(3Mt\left(M\left(2-\frac{3\sqrt{\delta\epsilon\pi}\sqrt{K\delta\epsilon\pi}}{\sqrt{K}\left(KM^2+3\delta\epsilon\pi\right)}\right)-\sqrt{3}\sqrt{K\delta\epsilon\pi}\right.\right.\\
&\times \left.\left.\frac{\mathrm{ArcTan}\left(\frac{\sqrt{K}M}{\sqrt{3}\sqrt{\delta\epsilon\pi}}\right)}{K}+\frac{C_2}{\sqrt{K}}\right)(\epsilon)^{-1}+\left(\delta\pi\left(3t^{\theta}\alpha\delta^{\frac{3}{2}}\epsilon^2\theta\pi^{\frac{3}{2}}\left(t^{1+\theta}\alpha\sqrt{\delta}\right.\right.\right.\\
&\times \left.\left.\theta\sqrt{\pi}\left(t^{2\theta}\alpha^2\theta^2-t^2\delta\pi\right)+\left(t^{2\theta}\alpha^2\theta^2+t^2\delta\pi\right)^2\mathrm{ArcTan}\left(\frac{t^{-1+\theta}\alpha\theta}{\sqrt{\delta}\sqrt{\pi}}\right)\right)\\
&- \left.\left.\sqrt{3}Mt\sqrt{\delta\epsilon\pi}\sqrt{\frac{t^{-2+2\theta}\alpha^2\delta\epsilon^2\theta^2\pi}{M^2}}\left(t^{2\theta}\alpha^2\theta^2+t^2\delta\pi\right)^2C_2\right)\right)\Big/\left(\alpha\theta\sqrt{\epsilon}\right.\\
&\times \left.\left.\left.\left.\sqrt{\frac{t^{-2+2\theta}}{M^2}}(\delta\epsilon\pi)^{\frac{3}{2}}\sqrt{\frac{t^{-2+2\theta}\alpha^2\delta\epsilon^2\theta^2\pi}{M^2}}\left(t^{2\theta}\alpha^2\theta^2+t^2\delta\pi\right)^2\right)\right)\right)\right)\Big/\left(9\right.\\
&\times \left.\alpha\theta\left(-3\sqrt{\delta\epsilon\pi}\sqrt{K\delta\epsilon\pi}+\sqrt{K}\left(KM^2+3\delta\epsilon\pi\right)\right)\right).
\end{aligned} \tag{36}$$

$$\begin{aligned}
\omega'_{EA} &= \left(t^{-5\theta}(-1+\theta)\left(KM^2+3\delta\epsilon\pi\right)\left(3\left(-\frac{3\sqrt{K}Mt^{3\theta}\alpha^3\theta^4\sqrt{\delta\epsilon\pi}\sqrt{K\delta\epsilon\pi}}{KM^2+3\delta\epsilon\pi}\right.\right.\right.\\
&+ \frac{1}{\epsilon^2\left(t^{2\theta}\alpha^2\theta^2+t^2\delta\pi\right)^3}KM^3t^{2+\theta}\alpha\theta\sqrt{\frac{t^{-2+2\theta}\alpha^2\epsilon\theta^2}{M^2}}\left(2t^{6\theta}\alpha^6\epsilon\theta^7\alpha\theta\right.\\
&\times \sqrt{\frac{t^{-2+2\theta}\epsilon}{M^2}}+t^6\delta^2\pi^2\left(2\delta\epsilon\theta\sqrt{\frac{t^{-2+2\theta}\alpha^2\epsilon\theta^2}{M^2}}\pi+(1-2\theta)\sqrt{\delta\epsilon\pi}\alpha\epsilon\theta\right.\\
&\times \left.\sqrt{\frac{t^{-2+2\theta}\delta\pi}{M^2}}\right)+2t^{4+2\theta}\alpha^2\delta\theta^2\pi\left(3\delta\epsilon\theta\sqrt{\frac{t^{-2+2\theta}\alpha^2\epsilon\theta^2}{M^2}}\pi-4(-1\right.\\
&+ \left.\theta)\sqrt{\delta\epsilon\pi}\sqrt{\frac{t^{-2+2\theta}\alpha^2\delta\epsilon^2\theta^2\pi}{M^2}}\right)+t^{2+4\theta}\alpha^4\theta^4\left(6\delta\epsilon\theta\sqrt{\frac{t^{-2+2\theta}\alpha^2\epsilon\theta^2}{M^2}}\pi\right.\\
&+ \left.\left.(-1+2\theta)\sqrt{\delta\epsilon\pi}\sqrt{\frac{t^{-2+2\theta}\alpha^2\delta\epsilon^2\theta^2\pi}{M^2}}\right)\right)-\delta^{3/2}\pi^{3/2}\left(\sqrt{3}t^{3\theta}\alpha^3\epsilon^{3/2}\theta^4\right.\\
&\times \sqrt{K\delta\epsilon\pi}\mathrm{ArcTan}\left(\frac{\sqrt{K}M}{\sqrt{3}\sqrt{\delta}\sqrt{\epsilon}\sqrt{\pi}}\right)+KM^3t^3(1-2\theta)\sqrt{\frac{t^{-2+2\theta}\alpha^2\epsilon\theta^2}{M^2}}\\
&\times \left.\left.\sqrt{\frac{t^{-2+2\theta}\alpha^2\delta\epsilon^2\theta^2\pi}{M^2}}\mathrm{ArcTan}\left(\frac{t^{-1+\theta}\alpha\theta}{\sqrt{\delta}\sqrt{\pi}}\right)\right)((\delta\epsilon\pi)^{3/2})^{-1}\right)+\alpha\theta\left(3\right.\\
&\times \left.\left.\sqrt{K}t^{3\theta}\alpha^2\epsilon\theta^3-\sqrt{3}KM^2t^{2+\theta}\sqrt{\frac{t^{-2+2\theta}\alpha^2\epsilon\theta^2}{M^2}}(-1+2\theta)\right)C_2(\epsilon)^{-1}\right)\\
&\times \left.\right)\Big/\left(9\sqrt{K}M\alpha^5\theta^5\left(-3\sqrt{\delta\epsilon\pi}\sqrt{K\delta\epsilon\pi}+\sqrt{K}\left(KM^2+3\delta\epsilon\pi\right)\right)\right).
\end{aligned} \tag{37}$$

The correspond expression for v_s^2 is given by

$$
\begin{aligned}
v_s^2 \;=\;& \Bigg(t^{-\theta}\Big(KM^2+3\delta\epsilon\pi\Big)\Bigg(3\Bigg(\frac{3\sqrt{K}M\Big(-2+\theta+6t^{\theta}\alpha\theta\Big)\sqrt{\delta\epsilon\pi}\sqrt{K\delta\epsilon\pi}}{KM^2+3\delta\epsilon\pi} \\
&+ KM\Bigg(4-\frac{t^{6}\delta^{3}\pi^{3}\sqrt{\frac{t^{-2+2\theta}\alpha^2\delta\epsilon^2\theta^2\pi}{M^2}}}{\sqrt{\frac{t^{-2+2\theta}\alpha^2\epsilon\theta^2}{M^2}}\sqrt{\delta\epsilon\pi}\Big(t^{2\theta}\alpha^2\theta^2+t^2\delta\pi\Big)^3}+t^{6\theta}\alpha^6\theta^6\sqrt{\delta\epsilon\pi} \\
&\times \frac{\sqrt{\frac{t^{-2+2\theta}\alpha^2\delta\epsilon^2\theta^2\pi}{M^2}}}{M^2\Big(\frac{t^{-2+2\theta}\alpha^2\epsilon\theta^2}{M^2}\Big)^{3/2}\Big(t^{2\theta}\alpha^2\theta^2+t^2\delta\pi\Big)^3}+8M^2t^6\sqrt{\frac{t^{-2+2\theta}\alpha^2\epsilon\theta^2}{M^2}} \\
&\times (\delta\epsilon\pi)^{\frac{3}{2}}\frac{\sqrt{\frac{t^{-2+2\theta}\alpha^2\delta\epsilon^2\theta^2\pi}{M^2}}}{\epsilon^3\Big(t^{2\theta}\alpha^2\theta^2+t^2\delta\pi\Big)^3}+2\theta\Bigg(-1-3t^{\theta}\alpha-4M^2t^6\alpha\theta(\delta\epsilon\pi)^{\frac{3}{2}} \\
&\times \sqrt{\frac{t^{-2+2\theta}\epsilon}{M^2}}\frac{\sqrt{\frac{t^{-2+2\theta}\alpha^2\delta\epsilon^2\theta^2\pi}{M^2}}}{\epsilon^3\Big(t^{2\theta}\alpha^2\theta^2+t^2\delta\pi\Big)^3}\Bigg)\Bigg)+Kt^{-1+\theta}\alpha\theta\sqrt{\delta\epsilon\pi}\alpha\epsilon\theta\sqrt{\frac{t^{-2+2\theta}\delta\pi}{M^2}} \\
&\times \frac{\mathrm{ArcTan}\Big(\frac{t^{-1+\theta}\alpha\theta}{\sqrt{\delta}\sqrt{\pi}}\Big)}{M\sqrt{\delta}\Big(\frac{t^{-2+2\theta}\alpha^2\epsilon\theta^2}{M^2}\Big)^{3/2}\sqrt{\pi}}+\sqrt{3}(-2+\theta)\sqrt{K\delta\epsilon\pi}\,\mathrm{ArcTan}\Big(\frac{\sqrt{K}M}{\sqrt{3}\sqrt{\delta\epsilon\pi}}\Big)\Bigg) \\
&+ \sqrt{K}\Bigg(6-3\theta-\frac{\sqrt{3}\sqrt{K}}{\sqrt{\frac{t^{-2+2\theta}\alpha^2\epsilon\theta^2}{M^2}}}\Bigg)C_2\Bigg)\Bigg)\Bigg/\Bigg(18\sqrt{K}M\alpha\theta\Bigg(-3\sqrt{\delta\epsilon\pi}\sqrt{K\delta\epsilon\pi} \\
&+ \sqrt{K}\Big(KM^2+3\delta\epsilon\pi\Big)\Bigg)\Bigg).
\end{aligned}
\tag{38}
$$

Figure 10 represents the graph of the EoS parameter versus z for the RHDE model taking an exponential form of the scale factor. For $\alpha = 2$, initially the trajectory expresses the transition from decelerated phase to accelerated phase and then crosses the phantom divide line and gives the phantom phase of the universe. For higher values of the α, that is for $\alpha = 3, 4$, the trajectories of the EoS parameter represents the quintessence phase. In Figure 11, we plot the graph of evolution parameter of EoS versus EoS parameter which gives the freezing region of the universe. Figure 12 shows the graph of v_s^2 for stability analysis of the model. Initially the graph gives the stability and then for decreasing z, the model becomes unstable. As we increase the value of α, the trajectories give more stable points.

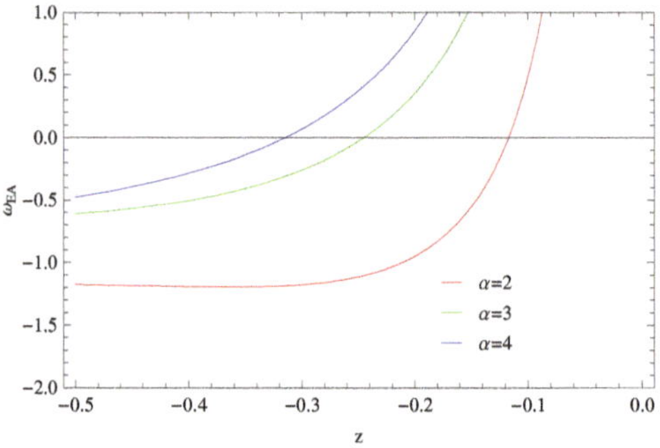

Figure 10. Plot of ω_{EA} versus z, taking an exponential scale factor for the RHDE model.

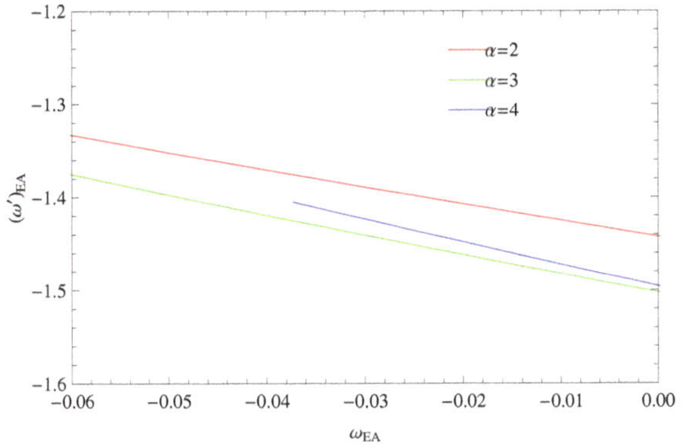

Figure 11. Plot of $\omega'_{EA} - \omega_{EA}$ taking an exponential scale factor for the RHDE model.

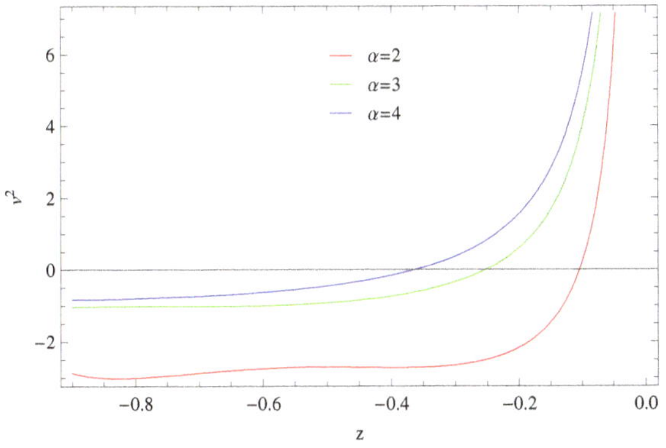

Figure 12. Plot of v_s^2 versus z, taking an exponential scale factor for the RHDE model.

6. Reconstruction from the Sharma-Mittal Holographic Dark Energy Model

Sharma-Mittal introduced two parametric entropy and defined it as [32]

$$S_{SM} = \frac{1}{1-r}\left((\Sigma_{i=1}^{n} P_i^{1-\delta})^{1-r/\delta} - 1\right), \tag{39}$$

where r is a new free parameter. The expression of the SMHDE model for the Hubble horizon is given by

$$\rho_D = \frac{3\epsilon H^4}{8\pi R}\left((1+\frac{\delta\pi}{H^2})^{\frac{R}{\delta}} - 1\right). \tag{40}$$

By comparing the energy densities of the SMHDE model and Einstein-Aether gravity model, we find

$$\frac{dF}{dK} - \frac{F}{2K} = \frac{KM^2}{24\epsilon\pi R}\left(\left(1+\frac{3\epsilon\pi\delta}{KM^2}\right)^{\frac{R}{\delta}} - 1\right), \tag{41}$$

which leads us to the following solution

$$F(K) = \frac{K^2M^2\left(-1 + {}_2F_1(-\frac{3}{2}, -\frac{R}{\delta}, \frac{-1}{2}, \frac{-3\pi\delta\epsilon}{KM^2})\right)}{36\pi R\epsilon} + C_3\sqrt{K}. \tag{42}$$

Power-law form of scale factor:

For this scale factor, we obtain

$$\rho_{EA} = \frac{Km^2M^2\Big(-1+\Big(1+\frac{3\pi\delta\epsilon}{KM^2}\Big)^{\frac{R}{\delta}}\Big)}{8\pi Rt^2}. \tag{43}$$

$$\begin{aligned}
p_{EA} &= \frac{1}{72t^4}m\Big(\frac{1}{\pi R\Big(m^2+\pi t^2\delta\Big)}\Big(3m^2\Big(-8\Big(m^2+\pi t^2\delta\Big) \\
&+ 3\Big(1+\frac{\pi t^2\delta}{m^2}\Big)^{\frac{R}{\delta}}\Big(3m^2+\pi t^2(-2R+3\delta)\Big)\Big)\epsilon+KM^2 \\
&\times t^2\Big(m^2+\pi t^2\delta\Big)\Big(-4+9m-3(-1+3m)\Big(1+\frac{3\pi\delta\epsilon}{KM^2}\Big)^{\frac{R}{\delta}}\Big) \\
&- \Big(m^2+\pi t^2\delta\Big)\Big(3m^2\epsilon{}_2F_1\Big(-\frac{3}{2},-\frac{R}{\delta},-\frac{1}{2},-\frac{\pi t^2\delta}{m^2}\Big) \\
&- KM^2t^2{}_2F_1\Big(-\frac{3}{2},-\frac{R}{\delta},-\frac{1}{2},-\frac{3\pi\delta\epsilon}{KM^2}\Big)\Big)\Big) \\
&- \frac{12t^2\epsilon\Big(-3+\frac{\sqrt{3}\sqrt{K}}{\sqrt{\frac{m^2\epsilon}{M^2t^2}}}\Big)C_3}{\sqrt{K}}\Big).
\end{aligned} \tag{44}$$

The cosmological parameters are given by

$$\begin{aligned}
\omega_{EA} &= \frac{1}{9KmM^2t^2\Big(-1+\Big(1+\frac{3\pi\delta\epsilon}{KM^2}\Big)^{\frac{R}{\delta}}\Big)}\pi R\Big(\frac{1}{\pi R\Big(m^2+\pi t^2\delta\Big)}\Big(3m^2 \\
&\times \Big(-8\Big(m^2+\pi t^2\delta\Big)+3\Big(1+\frac{\pi t^2\delta}{m^2}\Big)^{\frac{R}{\delta}}\Big(3m^2+\pi t^2(-2R+3\delta)\Big)\Big) \\
&\times \epsilon+KM^2t^2\Big(m^2+\pi t^2\delta\Big)\Big(-4+9m-3(-1+3m)\Big(1+\frac{3\pi\delta\epsilon}{KM^2}\Big)^{\frac{R}{\delta}} \\
&\times \Big)-\Big(m^2+\pi t^2\delta\Big)\Big(3m^2\epsilon{}_2F_1\Big(-\frac{3}{2},-\frac{R}{\delta},-\frac{1}{2},-\frac{\pi t^2\delta}{m^2}\Big) \\
&- KM^2t^2{}_2F_1\Big(-\frac{3}{2},-\frac{R}{\delta},-\frac{1}{2},-\frac{3\pi\delta\epsilon}{KM^2}\Big)\Big)\Big)-12t^2\epsilon \\
&\times \frac{\Big(-3+\frac{\sqrt{3}\sqrt{K}}{\sqrt{\frac{m^2\epsilon}{M^2t^2}}}\Big)C_3}{\sqrt{K}}\Big),
\end{aligned} \tag{45}$$

$$\begin{aligned}
\omega'_{EA} &= m^4\epsilon\Big(-3\Big(1+\frac{\pi t^2\delta}{m^2}\Big)^{\frac{R}{\delta}}\Big(5m^4+2m^2\pi t^2(-3R+5\delta)+\pi^2t^4\Big(4R^2 \\
&- 10R\delta+5\delta^2\Big)\Big)-\Big(m^2+\pi t^2\delta\Big)^2\Big(-16 \\
&+ {}_2F_1\Big(-\frac{3}{2},-\frac{R}{\delta},-\frac{1}{2},-\frac{\pi t^2\delta}{m^2}\Big)\Big)\Big)-4\sqrt{3}M^2\pi Rt^4 \\
&\times \Big(m^2+\pi t^2\delta\Big)^2\sqrt{\frac{m^2\epsilon}{M^2t^2}}C_3\Big(3Km^4M^2t^2\Big(m^2+\pi t^2\delta\Big)^2\Big(-1+\Big(1 \\
&+ \frac{3\pi\delta\epsilon}{KM^2}\Big)^{\frac{R}{\delta}}\Big)\Big),
\end{aligned} \tag{46}$$

$$
\begin{aligned}
v_s^2 &= \frac{1}{18K^{\frac{3}{2}}m^3M^2t^2\left(m^2+\pi t^2\delta\right)^2\left(-1+\left(1+\frac{3\pi\delta\epsilon}{KM^2}\right)^{R/\delta}\right)}\Bigg(\sqrt{K}m^2 \\
&\times \left(3m^2\left(-32\left(m^2+\pi t^2\delta\right)^2+3\left(1+\frac{\pi t^2\delta}{m^2}\right)^{R/\delta}\left(11m^4+2m^2\pi t^2\right.\right.\right. \\
&\times \left.\left.\left.(-5R+11\delta)+\pi^2t^4\left(4R^2-14R\delta+11\delta^2\right)\right)\right)\epsilon-2KM^2t^2\left(m^2\right. \\
&+ \left.\pi t^2\delta\right)^2\left(4-9m+3(-1+3m)\left(1+\frac{3\pi\delta\epsilon}{KM^2}\right)^{R/\delta}\right)-\left(m^2+\pi t^2\delta\right)^2 \\
&\times \left(3m^2\epsilon_2F_1\left(-\frac{3}{2},-\frac{R}{\delta},-\frac{1}{2},-\frac{\pi t^2\delta}{m^2}\right)-2KM^2t^2\right. \\
&\times \left.\left.{}_2F_1\left(-\frac{3}{2},-\frac{R}{\delta},-\frac{1}{2},-\frac{3\pi\delta\epsilon}{KM^2}\right)\right)\right)+12\pi Rt^2\left(m^3\right. \\
&+ \left.m\pi t^2\delta\right)^2\epsilon\left(6-\frac{\sqrt{3}\sqrt{K}}{\sqrt{\frac{m^2\epsilon}{M^2t^2}}}\right)C_3\Bigg),
\end{aligned} \tag{47}
$$

We plot the EoS parameter for the SMHDE model with respect to the redshift parameter as shown in Figure 13 for the power-law scale factor. For $m = 3$ and 4, the trajectories represent the transition from quintessence to phantom phase while $m = 2$ indicates the phantom era throughout for z. The plot of this parameter with its evolution parameter is given in Figure 14, which shows the freezing region of the evolving universe. However, for higher values of m, we may get thawing region ($\omega'_{EA} > 0$). Figure 15 gives the graph of squared speed of sound versus redshift. The trajectory for $m = 2$ shows the stability of the model as redshift parameter decreases while other trajectories describe the unstable behavior of the model.

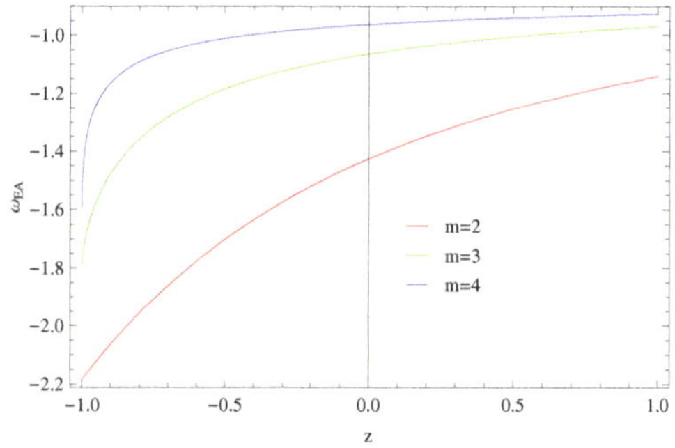

Figure 13. Plot of ω_{EA} versus z, taking a power-law scale factor for the Sharma-Mittal holographic dark matter (SMHDE) model.

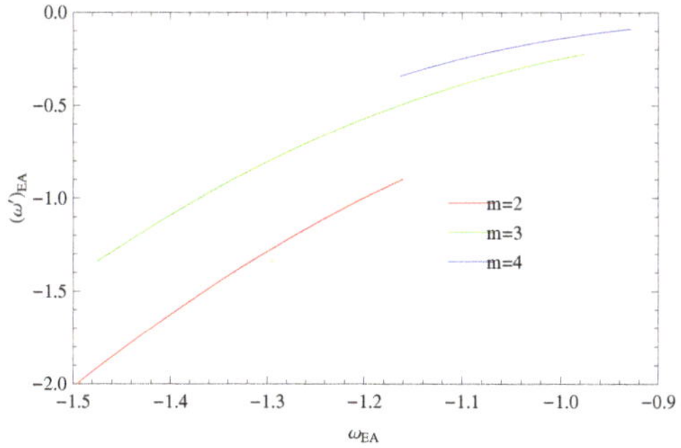

Figure 14. Plot of $\omega'_{EA} - \omega_{EA}$, taking a power-law scale factor for the SMHDE model.

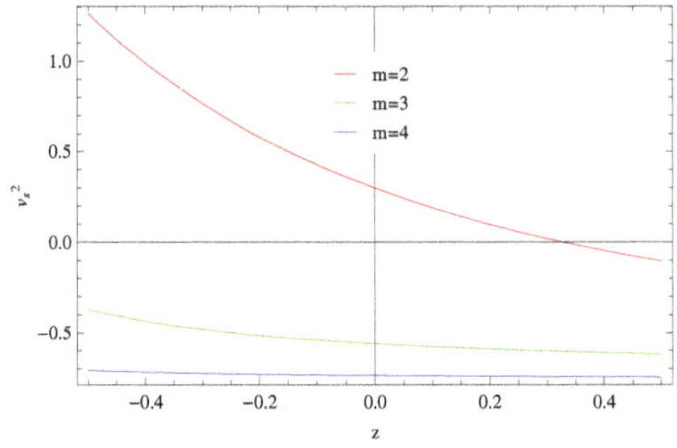

Figure 15. Plot of v_s^2 versus z, taking a power-law scale factor for the SMHDE model.

Exponential form of scale factor:

Following the same steps, we obtain the following expressions for energy density, pressure and parameters for exponential scale factor. These are:

$$\rho_{EA} = \frac{KM^2 t^{-2+2\theta}\alpha^2\Big(-1+\Big(1+\frac{3\pi\delta\epsilon}{KM^2}\Big)^{R/\delta}\Big)\theta^2}{8\pi R}, \tag{48}$$

$$\begin{aligned} p_{EA} &= \frac{1}{24}t^{-4+\theta}\alpha\theta\bigg(-\frac{3KM^2t^{2+\theta}\alpha\Big(-1+\Big(1+\frac{3\pi\delta\epsilon}{KM^2}\Big)^{R/\delta}\Big)\theta}{\pi R}-\epsilon(-1+\theta)\\ &\times \bigg(t^2\bigg(K^{3/2}M^2\bigg(-4+3\Big(1+\frac{3\pi\delta\epsilon}{KM^2}\Big)^{R/\delta}\\ &+ {}_2F_1\Big(-\frac{3}{2},-\frac{R}{\delta},-\frac{1}{2},-\frac{3\pi\delta\epsilon}{KM^2}\Big)\bigg)\bigg)+36\pi R\epsilon C_3\bigg)\\ &\times (3\sqrt{K}\pi R\epsilon)^{-1}+t^{2\theta}\alpha^2\theta^2\bigg(-8-3\Big(1+\frac{\pi t^{2-2\theta}\delta}{\alpha^2\theta^2}\Big)^{R/\delta}\Big(\pi t^2(-2R\\ &+ 3\delta)+3t^{2\theta}\alpha^2\theta^2\Big)(\pi t^2\delta+t^{2\theta}\alpha^2\theta^2)^{-1}\\ &+ {}_2F_1\Big(-\frac{3}{2},-\frac{R}{\delta},-\frac{1}{2},-\frac{\pi t^{2-2\theta}\delta}{\alpha^2\theta^2}\Big)(\pi R)^{-1}-4\sqrt{3}\epsilon\\ &\times \frac{C_3}{M^2\Big(\frac{t^{-2+2\theta}\alpha^2\epsilon\theta^2}{M^2}\Big)^{3/2}}\bigg)\bigg)\bigg), \end{aligned} \tag{49}$$

$$\begin{aligned} \omega_{EA} &= \frac{1}{3KM^2\alpha\Big(-1+\Big(1+\frac{3\pi\delta\epsilon}{KM^2}\Big)^{R/\delta}\Big)\theta}\pi R t^{-2-\theta}\bigg(-3KM^2t^{2+\theta}\alpha\bigg(-1\\ &+ \Big(1+\frac{3\pi\delta\epsilon}{KM^2}\Big)^{R/\delta}\bigg)\theta(\pi R)^{-1}-\epsilon(-1+\theta)\bigg(t^2\bigg(K^{3/2}M^2\bigg(-4\\ &+ 3\Big(1+\frac{3\pi\delta\epsilon}{KM^2}\Big)^{R/\delta}+{}_2F_1\Big(-\frac{3}{2},-\frac{R}{\delta},-\frac{1}{2},-\frac{3\pi\delta\epsilon}{KM^2}\Big)\bigg)\bigg)\\ &+ 36\pi R\epsilon C_3\bigg)(3\sqrt{K}\pi R\epsilon)^{-1}+t^{2\theta}\alpha^2\theta^2\bigg(-8-3\Big(1+\frac{\pi t^{2-2\theta}\delta}{\alpha^2\theta^2}\Big)^{R/\delta}\Big(\pi t^2\\ &\times (-2R+3\delta)+3t^{2\theta}\alpha^2\theta^2\Big)(\pi t^2\delta+t^{2\theta}\alpha^2\theta^2)^{-1}\\ &+ {}_2F_1\Big(-\frac{3}{2},-\frac{R}{\delta},-\frac{1}{2},-\frac{\pi t^{2-2\theta}\delta}{\alpha^2\theta^2}\Big)(\pi R)^{-1}-4\sqrt{3}\epsilon\\ &\times \frac{C_3}{M^2\Big(\frac{t^{-2+2\theta}\alpha^2\epsilon\theta^2}{M^2}\Big)^{3/2}}\bigg)\bigg)\bigg), \end{aligned} \tag{50}$$

$$
\begin{aligned}
\omega'_{EA} &= \frac{1}{9KM^2\alpha^2\Big(-1+\Big(1+\frac{3\pi\delta\epsilon}{KM^2}\Big)^{R/\delta}\Big)\theta^2}\pi R t^{-2(1+\theta)}(-1+\theta)\\
&\times \Bigg(\frac{1}{\pi R\Big(\pi t^2\delta+t^{2\theta}\alpha^2\theta^2\Big)^2}\theta\Bigg(KM^2t^2\Bigg(-4+3\Big(1+\frac{3\pi\delta\epsilon}{KM^2}\Big)^{R/\delta}\Bigg)\\
&\times \Big(\pi t^2\delta+t^{2\theta}\alpha^2\theta^2\Big)^2+3t^{2\theta}\epsilon\theta\Big(8(-2+\theta)\Big(\pi t^2\alpha\delta+t^{2\theta}\alpha^3\theta^2\Big)^2-3\alpha^2\\
&\times \Big(1+\frac{\pi t^{2-2\theta}\delta}{\alpha^2\theta^2}\Big)^{R/\delta}\Big(t^{4\theta}\alpha^4\theta^4(-5+2\theta)+2\pi t^{2+2\theta}\alpha^2\theta^2(R(3-2\theta)\\
&+ \delta(-5+2\theta))+\pi^2t^4\Big(2R\delta(5-4\theta)+4R^2(-1+\theta)+\delta^2(-5+2\theta)\Big)\Big)\Big)\\
&+ \Big(\pi t^2\delta+t^{2\theta}\alpha^2\theta^2\Big)^2\Big(KM^2t^2{}_2F_1\Big(-\frac{3}{2},-\frac{R}{\delta},-\frac{1}{2},-\frac{3\pi\delta\epsilon}{KM^2}\Big)\\
&- 3t^{2\theta}\alpha^2\epsilon\theta(-1+2\theta){}_2F_1\Big(-\frac{3}{2},-\frac{R}{\delta},-\frac{1}{2},-\frac{\pi t^{2-2\theta}\delta}{\alpha^2\theta^2}\Big)\Big)\Big)\\
&+ 12t^2\epsilon\Bigg(\frac{3\theta}{\sqrt{K}}+\frac{\sqrt{3}(1-2\theta)}{\sqrt{\frac{t^{-2+2\theta}\alpha^2\epsilon\theta^2}{M^2}}}\Bigg)C_3\Bigg),
\end{aligned}
\tag{51}
$$

$$
\begin{aligned}
v_s^2 &= \frac{1}{18KM^2\alpha\Big(-1+\Big(1+\frac{3\pi\delta\epsilon}{KM^2}\Big)^{R/\delta}\Big)\theta}\pi R t^{-2-\theta}\Bigg(\frac{1}{\pi R\Big(\pi t^2\delta+t^{2\theta}\alpha^2\theta^2\Big)^2}\\
&\times \Bigg(KM^2t^2\Big(\pi t^2\delta+t^{2\theta}\alpha^2\theta^2\Big)^2\Bigg(-8+4\theta+18t^\theta\alpha\theta-3\Big(1+\frac{3\pi\delta\epsilon}{KM^2}\Big)^{R/\delta}\\
&\times \Big(-2+\theta+6t^\theta\alpha\theta\Big)\Bigg)+3t^{2\theta}\epsilon\Big(8(-4+3\theta)\Big(\pi t^2\alpha\delta\theta+t^{2\theta}\alpha^3\theta^3\Big)^2-3\alpha^2\\
&\times \Big(1+\frac{\pi t^{2-2\theta}\delta}{\alpha^2\theta^2}\Big)^{R/\delta}\theta^2\Big(t^{4\theta}\alpha^4\theta^4(-11+8\theta)+2\pi t^{2+2\theta}\alpha^2\theta^2(R(5-4\theta)\\
&+ \delta(-11+8\theta))+\pi^2t^4\Big(2R\delta(7-6\theta)+4R^2(-1+\theta)+\delta^2(-11+8\theta)\Big)\Big)\Big)\\
&- \Big(\pi t^2\delta+t^{2\theta}\alpha^2\theta^2\Big)^2\Big(KM^2t^2(-2+\theta){}_2F_1\Big(-\frac{3}{2},-\frac{R}{\delta},-\frac{1}{2},-\frac{3\pi\delta\epsilon}{KM^2}\Big)\\
&+ 3t^{2\theta}\alpha^2\epsilon\theta^2{}_2F_1\Big(-\frac{3}{2},-\frac{R}{\delta},-\frac{1}{2},-\frac{\pi t^{2-2\theta}\delta}{\alpha^2\theta^2}\Big)\Big)\Bigg)+12t^2\epsilon C_3\\
&\times \frac{\Bigg(6-3\theta-\frac{\sqrt{3}\sqrt{K}}{\sqrt{\frac{t^{-2+2\theta}\alpha^2\epsilon\theta^2}{M^2}}}\Bigg)}{\sqrt{K}}\Bigg).
\end{aligned}
\tag{52}
$$

For the exponential scale factor for the SMHDE model, the plot of the EoS parameter in Figure 16 represents the phantom behavior initially, but converges to the cosmological constant behavior for $\alpha = 2, 3$, as z decreases. For $\alpha = 4$, the EoS parameter gives the phantom behavior. Figure 17 represents the graph of the $\omega'_{EA} - \omega_{EA}$ plane, which shows the positive behavior of ω'_{EA} versus negative ω_{EA} expressing thawing region of the universe. The squared speed of sound graph gives unstable behavior of the SMHDE model in the framework of the Einstein-Aether theory of gravity as shown in Figure 18.

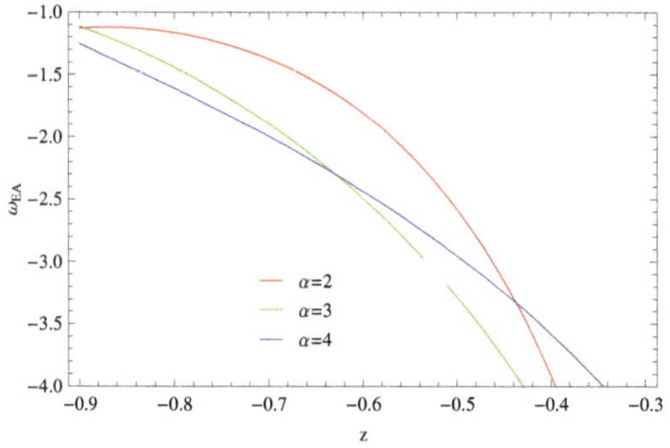

Figure 16. Plot of ω_{EA} versus z taking exponential scale factor for SMHDE model.

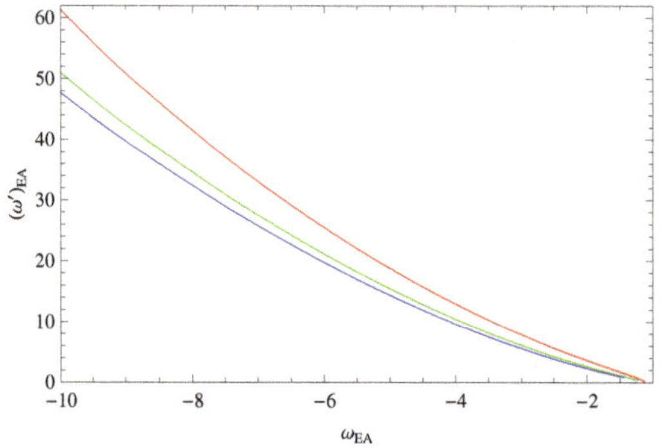

Figure 17. Plot of $\omega'_{EA} - \omega_{EA}$ taking exponential scale factor for SMHDE model.

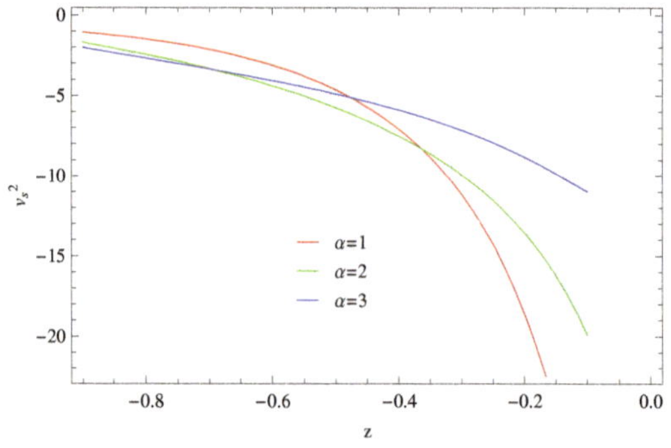

Figure 18. Plot of v_s^2 versus z taking exponential scale factor for SMHDE model.

7. Summary

In this work, we have discussed Einstein-Aether gravity and utilized its effective density and pressure. We have developed the Einstein-Aether models by using some holographic dark energy models. In the presence of a free function $F(K)$, we have treated the affective density and pressure as DE. From the modified HDE models such as the THDE, RHDE and SMHDE models, we have formed the unknown function $F(K)$ for the Einstein-Aether theory by considering the power-law form and exponential forms of scale factor. We have discussed some cosmological parameters, like the EoS parameter with its evolutionary parameter and squared speed of sound to check the stability of the reconstructed models for this theory.

The remaining results have been summarized as follows:

EoS parameter for power-law scale factor:

- THDE $\Rightarrow$ phantom behavior,
- RHDE $\Rightarrow$ quintessence phase,
- SMHDE $\Rightarrow$ transition from quintessence to phantom phase for $m = 3, 4$, phantom era for $m = 2$.

EoS parameter for exponential scale factor:

- THDE $\Rightarrow$ transition from quintessence to phantom era for $\alpha = 2$, phantom behavior for $\alpha = 3$, ΛCDM model for $\alpha = 4$,
- RHDE $\Rightarrow$ phantom phase for $\alpha = 2$, quintessence phase for $\alpha = 3, 4$
- SMHDE $\Rightarrow$ cosmological constant behavior for $\alpha = 2, 3$, phantom behavior for $m = 4$.

ω'-ω plane for power-law scale factor:

- THDE $\Rightarrow$ freezing region for $m = 2, 3$, thawing region for $m = 4$,
- RHDE $\Rightarrow$ freezing region,
- SMHDE $\Rightarrow$ freezing region.

ω'-ω plane for exponential scale factor:

- THDE $\Rightarrow$ freezing region to thawing region,
- RHDE $\Rightarrow$ freezing region,
- SMHDE $\Rightarrow$ thawing region.

Squared speed of sound for power-law scale factor:

- THDE $\Rightarrow$ unstable,
- RHDE $\Rightarrow$ unstable,
- SMHDE $\Rightarrow$ stable for $m = 2$, unstable for $m = 3, 4$.

Squared speed of sound for exponential scale factor:

- THDE $\Rightarrow$ unstable,
- RHDE $\Rightarrow$ stability for higher values and instability for lower values,
- SMHDE $\Rightarrow$ unstable.

It is mentioned here that for $m = 2$ for the power-law form of the scale factor in the case of the SMHDE model, we obtain a phantom region with stable behavior in the freezing region which leads to the most favorable result within the current cosmic expansion scenario.

Author Contributions: A.J. and S.R. proposed and completed the draft of the paper, K.B. done the proof reading and I.U.M. contributed in the mathematical work of the manuscript.

Funding: This research received no external funding.

Acknowledgments: S.R. and A.J. are thankful to the Higher Education Commission, Islamabad, Pakistan for its financial support under the grant No: 5412/Federal/NRPU/R&D/HEC/2016 of NATIONAL RESEARCH PROGRAMME FOR UNIVERSITIES (NRPU). The work of KB was partially supported by the JSPS KAKENHI Grant Number JP 25800136 and Competitive Research Funds for Fukushima University Faculty (18RI009).

Conflicts of Interest: The authors declare no conflict of interest.

References

1. Komatsu, E.; Dunkley, J.; Nolta, M.R.; Bennett, C.L.; Gold, B.; Hinshaw, G.; Jarosik, N.; Larson, D.; Limon, M.; Page, L.; et al. Five-year Wilkinson Microwave Anisotropy probe Observations: Cosmological Interpretation. *Astrophys. J. Suppl.* **2009**, *180*, 330–376. [CrossRef]

2. Nolta, M.R.; Dunkley, J.; Hill, R.S.; Hinshaw, G.; Komatsu, E.; Larson, D.; Page, L.; Spergel, D.N.; Bennett, C.L.; Gold, B.; et al. Five-year Wilkinson Microwave Anisotropy probe Observations: Angular Power Spectra. *Astrophys. J. Suppl.* **2009**, *180*, 296–305. [CrossRef]
3. Bahcall, N.; Ostriker, J.P.; Perlmutter, S.; Steinhardt, P.J. The Cosmic Triangle: Revealing the State of the Universe. *Science* **1999**, *284*, 1481–1488. [CrossRef]
4. Perlmutter, S.; Aldering, G.; Goldhaber, G.; Knop, R.A.; Nugent, P.; Castro, P.G.; Deustua, S.; Fabbro, S.; Goobar, A.; Groom, D.E.; et al. Measurements of Omega and Lambda from 42 High-Redshift Supernovae. *APJ* **1999**, *517*, 565–586. [CrossRef]
5. Perlmutter, S.; Aldering, G.; Valle, M.D.; Deustua, S.; Ellis, R.S.; Fabbro, S.; Fruchter, A.; Goldhaber, G.; Goobar, A.; Groom, D.E.; et al. Discovery of a Supernova Explosion at Half the Age of the Universe and its Cosmological Implications. *Nature* **1998**, *391*, 51–54. [CrossRef]
6. Riess, A.G.; Filippenko, A.V.; Challis, P.; Clocchiattia, A.; Diercks, A.; Garnavich, P.M.; Gilliland, R.L.; Hogan, C.J.; Jha, S.; Kirshner, R.P.; et al. Observational Evidence from Supernovae for an Accelerating Universe and a Cosmological Constant. *Astron. J.* **1998**, *116*, 1009–1038. [CrossRef]
7. Riess, A.G.; Strolger, L.G.; Tonry, J.; Casertano, S.; Ferguson, H.C.; Mobasher, B.; Challis, P.; Filippenko, A.V.; Jha, S.; Li, W.; et al. Type Ia Supernova Discoveries at z > 1 From the Hubble Space Telescope: Evidence for Past Deceleration and Constraints on Dark Energy Evolution. *Astrophys. J.* **2004**, *607*, 665–687. [CrossRef]
8. Boisseau, B.; Esposito-Farese, G.; Polarski, D.; Starobinsky, A.A. Reconstruction of a Scalar-Tensor Theory of Gravity in an Accelerating Universe. *Phys. Rev. Lett.* **2000**, *85*, 2236. [CrossRef] [PubMed]
9. Spergel, D.N.; Bean, R.; Dore, O.; Nolta, M.R.; Bennett, C.L.; Dunkley, J.; Hinshaw, G.; Jarosik, N.; Komatsu, E.; Page, L.; et al. Wilkinson Microwave Anisotropy Probe (WMAP) Three Year Results: Implications for Cosmology. *Astrophys. J. Suppl. Ser.* **2007**, *170*, 377. [CrossRef]
10. Gardner, C.L. Quintessence and the Transition to an Accelerating Universe. *Nucl. Phys. B* **2005**, *707*, 278. [CrossRef]
11. De Leon, J.P. Transition from Decelerated to Accelerated Cosmic Expansion in Braneworld Universes. *Gen. Relativ. Gravit.* **2006**, *38*, 61–81. [CrossRef]
12. Cunha, J.V. Kinematic Constraints to the Transition Redshift from Supernovae Type Ia Union Data. *Phys. Rev. D* **2009**, *79*, 047301. [CrossRef]
13. Roos, M. *Introduction to Cosmology*; John Wiley and Sons: Chichester, UK, 2003.
14. Jawad, A.; Majeed, A. Ghost Dark Energy Models in Specific Modified Gravity. *Astrophy. Space Sci.* **2015**, *356*, 375. [CrossRef]
15. Jawad, A. Cosmological Analysis of Pilgrim Dark Energy in Loop Quantum Cosmology. *Eur. Phys. J. C* **2015**, *75*, 206. [CrossRef]
16. Jawad, A.; Chattopadhyay, S.; Pasqua, A. A Holographic Reconstruction of the Modified $f(R)$ Horava-Lifshitz Gravity with Scale Factor in Power-law form. *Astrophy. Space Sci.* **2013**, *346*, 273. [CrossRef]
17. Jawad, A.; Chattopadhyay, S.; Pasqua, A. Reconstruction of $f(G)$ Gravity with the New Agegraphic Dark-energy Model. *Eur. Phys. J. Plus* **2013**, *128*, 88. [CrossRef]
18. Jawad, A.; Chattopadhyay, S.; Pasqua, A. Reconstruction of $f(\tilde{R})$ models via well-known scale factors. *Eur. Phys. J. Plus* **2014**, *129*, 54. [CrossRef]
19. Jawad, A.; Pasqua, A.; Chattopadhyay, S. Correspondence between $f(G)$ gravity and holographic dark energy via power-law solution. *Astrophy. Space Sci.* **2013**, *344*, 489. [CrossRef]
20. Jawad, A.; Pasqua, A.; Chattopadhyay, S. Holographic reconstruction of $f(G)$ gravity for scale factors pertaining to emergent, logamediate and intermediate scenarios. *Eur. Phys. J. Plus* **2013**, *128*, 156. [CrossRef]
21. Jawad, A. New Agegraphic Pilgrim Dark Energy in $f(T, T_G)$ Gravity. *Astrophy. Space Sci.* **2014**, *353*, 691. [CrossRef]
22. Jawad, A.; Iqbal, A. Modified Cosmology through Renyi and logarithmic Entropies. *Int. J. Geom. Meth. Mod. Phys.* **2018**, *15*, 1850130. [CrossRef]
23. Jawad, A.; Iqbal, A. Cosmological Implications of Non-canonical Scalar Field Model in Fractal Universe. *Phys. Dark Univ.* **2018**, *22*, 16–26. [CrossRef]
24. Iqbal, A.; Jawad, A. Thermodynamics of Ricci-Gauss-Bonnet Dark Energy. *Adv. High Energy Phys.* **2018**, *2018*, 6139430. [CrossRef]
25. Jawad, A.; Bamba, k.; Younas, M.; Qummer, S.; Rani, S. Tsallis, Rényi and Sharma-Mittal Holographic Dark Energy Models in Loop Quantum Cosmology. *Symmetry* **2018**, *10*, 635. [CrossRef]

26. Younas, M.; Jawad, A.; Qummer, S.; Moradpour, H.; Rani, S. Cosmological Implications of the Generalized Entropy Based Holographic Dark Energy Models in Dynamical Chern-Simons Modified Gravity. *Adv. High Energy Phys.* **2019**, *2019*, 1287932. [CrossRef]
27. Kaplan, D.B.; Nelson, A.E. Effective Field Theory, Black Holes, and the Cosmological Constant. *Phys. Rev. Lett.* **1999**, *73*, 4971–4974.
28. Moradpour, H.; Sheykhi, A.; Corda, C.; Salako, I.G. Energy Definition and Dark Energy: A Thermodynamic Analysis. *Phys. Lett. B* **2018**, *783*, 82. [CrossRef]
29. Moradpour, H.; Bonilla, A.; Abreu, E.M.C.; Neto, J.A. Einstein and Rastall Theories of Gravitation in Comparison. *Phys. Rev. D* **2017**, *96*, 123504. [CrossRef]
30. Moradpour, H. Necessity of Dark Energy from Thermodynamic Arguments. *Int. J. Theor. Phys.* **2016**, *55*, 4176. [CrossRef]
31. Tavayef, M.; Sheykhi, A.; Bamba, K.; Moradpour, H. Tsallis holographic dark energy. *Phys. Lett. B* **2018**, *781*, 195. [CrossRef]
32. Moosavi, S.A.; Lobo, I.P.; Morais Graca, J.P.; Jawad, A.; Salako, I.G. Thermodynamic approach to holographic dark energy and the Rényi entropy. *arXiv* **2018**, arXiv:1803.02195.
33. Moosavi, S.A.; Moradpour, H.; Morais Graça, J.P.; Lobo, I.P.; Salako, I.G.; Jawad, A. Generalized entropy formalism and a new holographic dark energy model. *Phys. Lett. B* **2018**, *780*, 21–24.
34. Nojiri, S.; Odintsov, S.D. Modified non-local-$F(R)$ gravity as the key for the inflation and dark energy. *Phys. Lett. B* **2008**, *659*, 821–826. [CrossRef]
35. Li, B.; Barrow, J.D. Cosmology of $f(R)$ gravity in the metric variational approach. *Phys. Rev. D* **2007**, *75*, 084010. [CrossRef]
36. Nojiri, S.; Odintsov, S.D. Modified $f(R)$ gravity consistent with realistic cosmology: From matter dominated epoch to dark energy universe. *Phys. Rev. D* **2006**, *74*, 086005. [CrossRef]
37. Dunsby, P.K.S.; Elizalde, E.; Goswami, R.; Odintsov, S.; Gomez, D.S. On the LCDM universe in $f(R)$ gravity. *Phys. Rev. D* **2010**, *82*, 023519. [CrossRef]
38. Elizalde, E.; Odintsov, S.D.; Pozdeeva, E.O.; Yu, S.; Vernov, J. Cosmological attractor inflation from the RG-improved Higgs sector of finite gauge theory. *J. Cosmo. Astropart. Phys.* **2016**, *1602*, 25. [CrossRef]
39. Abdalla, M.C.B.; Nojiri, S.; Odintsov, S.D. Consistent modified gravity: Dark energy, acceleration and the absence of cosmic doomsday. *Class. Quantum Gravity* **2005**, *22*, L35. [CrossRef]
40. Linder, E.V. Einstein's other gravity and the acceleration of the Universe. *Phys. Rev. D* **2010**, *81*, 127301. [CrossRef]
41. Yerzhanov, K.K. Emergent Universe in Chameleon, $f(R)$ and $f(T)$ Gravity Theories. *arXiv* **2010**, arXiv:1006.3879v1.
42. Nojiri, S.; Odintsov, S.D. Modified Gauss-Bonnet theory as gravitational alternative for dark energy. *Phys. Lett. B* **2005**, *631*, 1–6. [CrossRef]
43. Antoniadis, I.; Rizos, J.; Tamvakis, K. Singularity-free cosmological solutions of the superstring effective action. *Nucl. Phys. B* **1994**, *415*, 497–514. [CrossRef]
44. Horava, P. Membranes at Quantum Criticality. *JHEP* **2009**, *903*, 20. [CrossRef]
45. Brans, C.; Dicke, H. Mach's Principle and a Relativistic Theory of Gravitation. *Phys. Rev.* **1961**, *124*, 925. [CrossRef]
46. Nojiri, S.; Odintsov, S.D. Unified cosmic history in modified gravity: From $F(R)$ theory to Lorentz non-invariant models. *Phys. Rep.* **2011**, *505*, 59–144. [CrossRef]
47. Capozziello, S.; de Laurentis, M. Extended Theories of Gravity. *Phys. Rep.* **2011**, *509*, 167–321. [CrossRef]
48. Faraoni, V.; Capozziello, S. *Beyond Einstein Gravity: A Survey of Gravitational Theories for Cosmology and Astrophysics*; Springer: Dordrecht, The Netherlands, 2011.
49. Bamba, K.; Odintsov, S.D. Inflationary cosmology in modified gravity theories. *Symmetry* **2015**, *7*, 220–240. [CrossRef]
50. Cai, Y.F.; Capozziello, S.; de Laurentis, M.; Saridakis, E.N. $f(T)$ teleparallel gravity and cosmology. *Rep. Prog. Phys.* **2016**, *79*, 106901. [CrossRef]
51. Nojiri, S.; Odintsov, S.D.; Oikonomou, V.K. Modified Gravity Theories on a Nutshell: Inflation, Bounce and Late-time Evolution. *Phys. Rep.* **2017**, *692*, 1–104. [CrossRef]
52. Bamba, K.; Capozziello, S.; Nojiri, S.; Odintsov, S.D. Dark energy cosmology: The equivalent description via different theoretical models and cosmography tests. *Astrophys. Space Sci.* **2012**, *342*, 155–228. [CrossRef]

53. Jacobson, T.; Mattingly, D. Gravity with a dynamical preferred frame. *Phys. Rev. D* **2001**, *64*, 024028. [CrossRef]
54. Jacobson, T.; Mattingly, D. Einstein-aether waves. *Phys. Rev. D* **2004**, *70*, 024003. [CrossRef]
55. Barrow, J.D. Errata for cosmological magnetic fields and string dynamo in axion torsioned spacetime. *Phys. Rev. D* **2012**, *85*, 047503. [CrossRef]
56. Meng, X.; Du, X. Einstein-aether theory as an alternative to dark energy model. *Phys. Lett. B* **2012**, *710*, 493–499. [CrossRef]
57. Meng, X.; Du, X. A Specific Case of Generalized Einstein-aether Theories. *Commun. Theor. Phys.* **2012**, *57*, 227. [CrossRef]
58. Achucarro, A.; Gong, J.O.; Hardeman, S.; Palma, G.A.; Patil, S.P. Features of heavy physics in the CMB power spectrum. *JCAP* **2011**, *1*, 30. [CrossRef]
59. Gasperini, M. Repulsive gravity in the very early Universe. *Gen. Relativ. Gravit.* **1998**, *30*, 1703. [CrossRef]
60. Zlosnik, T.G.; Ferreira, P.G.; Starkman, G.D. Modifying gravity with the Aether: An alternative to Dark Matter. *Phys. Rev. D* **2007**, *75*, 044017. [CrossRef]
61. Zlosnik, T.G.; Ferreira, P.G.; Starkman, G.D. On the growth of structure in theories with a dynamical preferred frame. *Phys. Rev. D* **2008**, *77*, 084010. [CrossRef]
62. Caldwell, R.R.; Linder, E.V. The Limits of Quintessence. *Phys. Rev. Lett.* **2005**, *95*, 141301. [CrossRef]
63. Nojiri, S.; Odintsov, S.D. Introduction to Modified Gravity and Gravitational Alternative for Dark Energy. *Int. J. Geom. Meth. Mod. Phys.* **2007**, *4*, 115–146. [CrossRef]
64. Moradpour, H.; Moosavi, S.A.; Lobo, I.P.; Graca, J.P.M.; Jawad, A.; Salako, I.G. Tsallis, Rényi and Sharma-Mittal Holographic Dark Energy Models in Loop Quantum Cosmology. *Eur. Phys. J. C* **2018**, *829*, 78.

symmetry

MDPI

Article

Cosmological Consequences of a Parametrized Equation of State

Abdul Jawad [1], Shamaila Rani [1], Sidra Saleem [1], Kazuharu Bamba [2,*] and Riffat Jabeen [3]

1 Department of Mathematics, COMSATS University Islamabad Lahore-Campus, Lahore-54000, Pakistan
2 Division of Human Support System, Faculty of Symbiotic Systems Science, Fukushima University, Fukushima 960-1296, Japan
3 Department of Statistics, COMSATS University Islamabad, Lahore-Campus, Lahore-54000, Pakistan
* Correspondence: bamba@sss.fukushima-u.ac.jp

Received: 24 May 2019; Accepted: 16 July 2019; Published: 5 August 2019

Abstract: We explore the cosmic evolution of the accelerating universe in the framework of dynamical Chern–Simons modified gravity in an interacting scenario by taking the flat homogeneous and isotropic model. For this purpose, we take some parametrizations of the equation of state parameter. This parametrization may be a Taylor series extension in the redshift, a Taylor series extension in the scale factor or any other general parametrization of ω. We analyze the interaction term which calculates the action of interaction between dark matter and dark energy. We explore various cosmological parameters such as deceleration parameter, squared speed of sound, Om-diagnostic and statefinder via graphical behavior.

Keywords: dynamical Chern–Simons modified gravity; parametrizations; cosmological parameters

PACS: 95.36.+d; 98.80.-k

1. Introduction

It is believed that in present day cosmology, one of the most important discoveries is the acceleration of the cosmic expansion [1–10]. It is observed that the universe expands with repulsive force and is not slowing down under normal gravity. This unknown force, called dark energy (DE), and is responsible for current cosmic acceleration. In physical cosmology and astronomy, DE is a mysterious procedure of energy which is assumed to pervade all of space which tends to blast the extension of the universe. However, the nature of DE is still unknown which requires further attention [11] (for recent reviews on the so-called geometric DE, i.e., modified gravity theories to explain the late-time cosmic acceleration, see, for example [12–17]). In the standard Λ-cold dark matter (CDM) model of cosmology, the whole mass energy of the cosmos includes 4.9% of usual matter, 26.8% of DM and 68.3% of a mysterious form of energy recognized as dark energy. In astrophysics, DM is an unknown form of matter which appears only participating in gravitational interaction, but does not emit nor absorb light [18]. The nature of DM is still unknown, but its existence is proved by astrophysical observations [19]. The majority of DM is thought to be non-baryonic in nature [20].

In order to explain DE, a large number of models have been suggested such as quintessence [21], quintom [22–24], Chaplygin gas with its modified model [25–27], K-essence [28–30], new agegraphic DE [31,32], holographic DE model [33–35], pilgrim DE model [36–38], Tsallis holographic DE (THDE) [39]. Among all of these, the simplest is the cosmological constant model and this model is compatible with observations [1]. In the cosmological framework, the equation of state (EoS) parameter, ω, gives the relation among energy density and pressure [40]. This is a dimensionless parameter and descrbes the phases of the cosmos [41]. The EoS parameter might be used in Friedmann–Robertson–Walker (FRW)' equations to define the evolution of an isotropic universe filled with a perfect fluid. The EoS parameter governs not only the gravitational properties of DE but also its evolution. The EoS parameter may be a constant or a time dependent function [1]. It is observed that this parameter gives constant ranges by using various observational schemes. For deviating DE, a parametrized formation of ω is supposed. Parametrization may be a Taylor series extension in the redshift, a Taylor series extension in the scale factor or any other parametrization of ω [42–56]. Using different considerations of parametrization, the cosmological parameters can be constrained [57–59].

On the other hand, different modified theories of gravity have been proposed in order to explain cosmic acceleration. The dynamical Chern–Simons modified gravity has been recently proposed [60] which is motivated from string theory and loop quantum gravity [61,62]. In this gravity, Jawad and Rani [63] investigated various cosmological parameters and planes for pilgrim DE models in that FRW universe. Jawad and Sohail [62] explored different cosmological planes as well as parameters for modified DE. Till now, various works have been done on the investigation of cosmic expansion scenario with different cosmological parameters [64–71]. In the present work, we use the constructed models in the frame work of dynamical Chern–Simons modified gravity and investigate the different cosmological parameters such as the deceleration parameter, squared speed of sound, state finder parameters and Om-diagnostic.

This paper is organized as follows: in the next section, we provide the basic cosmological scenario of dynamical Chern–Simons modified gravity and construct the field equations in for flat FRW spacetime. We take interaction scenario for constitutes DE and DM with the help of conservation equations. The holographic DE (HDE) density is used as DE model with Hubble horizon as IR cut-off. In Section 3, we provide the parametrization model of EoS parameter and construct the setup to discuss the cosmic evolution of the universe. Also, we analyze the interaction term for the corresponding parametrizations. In Section 4, we discuss the cosmological parameters such as deceleration, squared speed of sound, Om-diagnostic and statefinder. Last section comprises the results.

2. Dynamical Chern–Simons Modified Gravity

In this section, we describe the dynamical Chern–Simons modified gravity by the following action

$$S = \frac{1}{16\pi G}\int_V d^4x[\sqrt{-g}R + \frac{l}{4}{}^*R^{\rho\sigma\mu\nu}R_{\rho\sigma\mu\nu}\theta - \frac{1}{2}g^{\mu\nu}\nabla_\mu\theta\nabla_\nu\theta + V(\theta)] + S_{mat}, \tag{1}$$

here R is the Ricci scalar, ${}^*R^{\rho\sigma\mu\nu}R_{\rho\sigma\mu\nu}$ is the topological invariant called the pontryagin term, l is the coupling constant, θ is the dynamical variable, S_{mat} is the action of matter and $V(\theta)$ is the potential. Now in case of string theory, we take $V(\theta) = 0$. The variation of Equation (1) corresponding to metric $g_{\mu\nu}$ and scalar field θ, respectively, give the following field equations

$$G_{\mu\nu} + lC_{\mu\nu} = 8\pi G T_{\mu\nu}, \tag{2}$$

$$g^{\mu\nu}\nabla_\mu\nabla_\nu\theta = -\frac{l}{64\pi}{}^*R^{\rho\sigma\mu\nu}R_{\rho\sigma\mu\nu}, \tag{3}$$

where $G_{\mu\nu}$ is known as Einstein tensor and $C_{\mu\nu}$ appears as Cotton tensor which is defined as

$$C_{\mu\nu} = -\frac{1}{2\sqrt{-g}}((\nabla_\rho\theta)\varepsilon^{(\rho\beta\tau^{\mu}_{\nabla_\tau}R^{\nu}_{\beta})} + (\nabla_\sigma\nabla\rho\theta){}^*R^{\rho(\mu\nu)\sigma}. \tag{4}$$

The energy–momentum tensor related to scalar field and matter are given by

$$\hat{T}^{\theta}_{\mu\nu} = \nabla_\mu\theta\nabla_\nu\theta - \frac{1}{2}g_{\mu\nu}\nabla^\rho\theta\nabla_\rho\theta, \tag{5}$$

$$T_{\mu\nu} = (\rho + p)u_\mu u_\nu + p g_{\mu\nu}, \tag{6}$$

here $\hat{T}^{\theta}_{\mu\nu}$ shows the scalar field contribution and $T_{\mu\nu}$ represents the matter contribution while p and ρ indicate the pressure and energy density respectively. Also, $u_{\mu} = (1,0,0,0)$ is the four velocity. In case of flat FRW universe, first Friedmann equation for dynamical Chern–Simons modified gravity becomes

$$H^2 = \frac{1}{3}(\rho_m + \rho_d) + \frac{1}{6}\dot{\theta}^2, \tag{7}$$

where $H = \frac{\dot{a}}{a}$ represents the Hubble parameter, a is the scale factor and dot indicates the derivative with respect to cosmic time, $\rho = \rho_m + \rho_d$ is the effective density and $8\pi G = 1$. We assume $p_m = 0$ then for ordinary matter, the conservation equations are given as

$$\dot{\rho}_m + 3H\rho_m = 0, \tag{8}$$

$$\dot{\rho}_d + 3H(\rho_d + p_d) = 0. \tag{9}$$

For FRW universe the pontryagin term vanishes, so the scalar field in Equation (3) reduces to the following form

$$g^{\mu\nu}\nabla_{\mu}\nabla_{\nu}\theta = g^{\mu\nu}(\partial_{\nu}\partial_{\mu}\theta) = 0. \tag{10}$$

By taking $\theta = \theta(t)$, we get the following equation

$$\ddot{\theta} + 3H\dot{\theta} = 0. \tag{11}$$

The solution of this equation for $\dot{\theta}$ is $\dot{\theta} = ba^{-3}$ where b is an integration constant. Using this solution in Equation (6), we have

$$H^2 = \frac{1}{3}(\rho_m + \rho_d) + \frac{1}{6}b^2a^{-6}. \tag{12}$$

Taking into account the equation of continuity Equation (8), Equation (12) takes the form

$$-2\dot{H} - 3H^2 - \frac{1}{6}b^2a^{-6} = p_d. \tag{13}$$

Equation (12) can be re-written as

$$E^2(z) = \frac{1}{3H_0^2}(\rho_m + \rho_d) + \frac{1}{6H_0^2}b^2a^{-6}, \tag{14}$$

where $E(z) = \frac{H}{H_0}$ is a normalized Hubble parameter, z is the redshift function which is defined as $1 + z = \frac{a_0}{a}$. Interaction is an idea of two way action that occur when two or more objects have effect on each other. The continuity equations for energy densities are defined as

$$\rho_m{}' - \left(\frac{3}{1+z}\right)\rho_m = -\frac{Q}{H_0E(z)(1+z)}, \tag{15}$$

$$\rho_d{}' - 3\left(\frac{1+\omega_d}{1+z}\right)\rho_d = \frac{Q}{H_oE(z)(1+z)}. \tag{16}$$

Here, prime denotes the derivative with respect to the redshift and Q is the interaction term which calculates the action of interaction between the DM and DE. Basically, Q tells about the rate of energy exchange between DM and DE. When $Q > 0$, it means that energy is being converted from DE to DM. For $Q < 0$, the energy is being converted from DM to DE [72]. In the preceding prospectus of DE, HDE is one of the sensational attempts to analyze the nature of DE in the frame of quantum gravity. The HDE is based on holographic principle which states that all information relevant to a physical system inside a spatial region can be observed on its boundary instead of its volume. The relation of ultra-violet UV (Λ) and infra-red IR (L) is introduced by Cohen et al. [73] which plays a key role in the construction of HDE model [74]. The relation is about the energy of vacuum of a system with particular size whose maximal quantity should not be greater than the black hole mass of the similar size. This can be indicated as $L^3\rho_d \leq LM_p^2$, here $M_p^2 = (8\pi G)^{-1}$ and L represents the reduced Planck mass and IR cutoff respectively [75]. From the above inequalities, the HDE density takes the following form

$$\rho_d = 3c^2H_0^2E^2(z), \tag{17}$$

where c is the constant parameter of the dimensionless HDE and describes the expansion of universe and it lies in the interval $0 < c^2 < 1$ and L is taken as Hubble horizon. By inserting Equations (15)–(17) in (14) we get the following expression

$$\omega_d = (1+z)\frac{dE^2(z)}{3c^2E^2(z)dz} - \frac{b^2(1+z)^6}{6c^2H_0^2E^2(z)} - \frac{1}{c^2}, \tag{18}$$

after some calculation we get the following result

$$\frac{dE^2(z)}{dz} - (3+3c^2\omega_d)\left(\frac{E^2(z)}{1+z}\right) = \frac{b^2(1+z)^5}{2H_0^2}. \tag{19}$$

3. Parametrizations of Equation of State Parameter

Parametrization is a process of choosing different parameters and is used for the comparison of two datasets. In cosmological context, the EoS parameter is the relation between energy density and pressure and it helps to classify the accelerated and decelerated phases of the universe. At $\omega = 0$, this parameter corresponds to non-relativistic matter and involves the radiation era $0 < \omega < \frac{1}{3}$ for the accelerated phase of the universe. At $\omega < -1$, $\omega = -1$ and $-1 < \omega < -\frac{1}{3}$ it represents the phantom, cosmological constant and quintessence eras respectively. A parametrized formation of ω is assumed for deviating DE. We construct two different models; one with a constant EoS parameter and other with a dark fluid in the existence of DM [1,76]. By using a function of redshift the variation of EoS parameter can be estimated and many parametrizations have been suggested so far. We use the following parametrizations

$$\omega_{1d} = \omega_0, \tag{20}$$

$$\omega_{2d} = \omega_0 + \omega_1 q. \tag{21}$$

At present, ω_0 is the value of EoS parameter, ω_1 is the parameter of the model that is determined by using the observational data [76] and q is the deceleration parameter. By inserting Equation (20) in (19), we have

$$E^2(z) = \frac{b^2(1+z)^6}{2H_0^2(3-3c^2\omega_0)} + A(1+z)^{3+3c^2\omega_0}, \tag{22}$$

where A is a constant of integration. Similarly by inserting Equation (21) in (19), we get the following result

$$E^2(z) = \frac{b^2}{2H_0^2\left(\frac{3-3c^2\omega_0-8c^2\omega_1}{1-\frac{3}{2}(c^2\omega_1)}\right)}(1+z)^{\frac{6-11c^2\omega_1}{1-\frac{3}{2}(c^2\omega_1)}} + B(1+z)^{\frac{3+3c^2\omega_0-3c^2\omega_1}{1-\frac{3}{2}(c^2\omega_1)}}. \tag{23}$$

Now, we analyze the interaction term Q for the chosen parametrization of EoS parameter. Using Equations (14) and (17), we obtain the energy density of DM in the following form

$$\rho_m = 3H_0^2E^2(z)(1-c^2) - \frac{b^2}{2}(1+z)^6. \tag{24}$$

Taking the derivative with respect to z of Equation (24) along with ρ_m in the continuity equation related to DM, we have

$$(1+z)\frac{d}{dz}\ln(E^2(z)) = 3 + \frac{b^2(1+z)^6}{2H_0^2E^2(z)(1-c^2)} - \frac{Q}{3H_0^3E^3(z)(1-c^2)}. \tag{25}$$

Using Equations (14)–(16), we get the following result

$$\frac{(1+z)}{3}\frac{d}{dz}\ln(E^2(z)) = 1 + \frac{\omega_d}{1+r+\frac{b^2(1+z)^6}{2}} - \frac{b^2(1+z)^6}{2H_0^2E^2(z)}, \tag{26}$$

where r is the coincidence parameter which is defined as $r = \frac{\rho_m}{\rho_d}$ with the help of Equations (17) and (24). Comparing the above equations, it yields

$$\frac{Q}{9H_0^3E^3(z)(1-c^2)} = -\frac{\omega_d}{1+r+\frac{b^2(1+z)^6}{2}} + \frac{b^2(1+z)^6}{2H_0^2E^2(z)}\left(1+\frac{1}{3(1-c^2)}\right). \tag{27}$$

At present time, the above equation becomes

$$Q_0 = -9(1-c^2)\left(\frac{\omega_{d,0}}{1+r_0+\frac{b^2}{2}}\right) + \frac{9b^2(1-c^2)}{2}\left(1+\frac{1}{3(1-c^2)}\right). \tag{28}$$

It is significant to express that the value of Q-term predicts the rate at which the universe expands and coincidence parameter decreases. Using the positivity condition of the Q-term at present time, Equation (28) takes the following form

$$\omega_{d,0} < (1+r+\frac{b^2}{2})\left(\frac{3b^2(1-c^2)+b^2}{b(1-c^2)}\right). \tag{29}$$

The normalized Hubble parameter in terms of coincidence parameter is obtained by the ratio of Equations (17) and (24) such that

$$E^2(z) = -\frac{b^2(1+z)^6}{6H_0^2c^2(r(z)-r_c)}, \tag{30}$$

where $r_c = \frac{(1-c^2)}{(c^2)}$ is a constant quantity. This parameter shows the singular behavior at $r(z) = r_c$. At present time, $r_c = \frac{r_0+\frac{b^2}{6}}{1-\frac{b^2}{6}}$, where $c^2 = \frac{1-\frac{b^2}{6}}{1+r_0}$. For the coincidence parameter, we can consider a CPL-type parametrization form [42] $r(z) = r_o + \epsilon_o\frac{z}{1+z}$, where $\epsilon_o = r'_o$. We can notice that above parametrization becomes singular at $z = -1$ and it has a linear behavior and bounded nature for low and high value of redshift respectively. Taking into account the above parametrization, we get the value of redshift z_s, such that $z_s = -\frac{r_o-r_c}{\epsilon_o(1+\frac{(r_o-r_c)}{\epsilon_o})}$. For the singular behavior, we have the condition $-1 < z_s < 0$. After some manipulation, we obtain

$$r(z) - r_c = \epsilon_o\left(\frac{(z-z_s)}{(1+z_s)(1+z)}\right) \geqslant 0 \Longrightarrow z \geqslant z_s, \tag{31}$$

which yields $\frac{(-\epsilon_o z_s)}{(1+z_s)} \geq 0$ at present time. Substituting these results in Equation (31), it can be written as

$$E^2(z) = -\eta b^2 \frac{(1+z)^7}{6H_0^2(z-z_s)}, \tag{32}$$

here $\eta := \frac{(1+z_s)}{c^2\epsilon_o} > 0$ since $\epsilon_o > 0$. Moreover, we define the function $\theta(z) := \frac{(1+z)}{(z-z_s)}$ and substitute Equation (32) in (26), we get the following result

$$1 + \frac{\omega_d}{1 + r + \frac{b^2(1+z)^6}{2}} = \frac{9 + \eta\theta(7+\theta)}{3\eta\theta}. \tag{33}$$

Using above result the expression (27) for the Q-term can be written as

$$\frac{Q}{9H_0^3E^3(z)} = (1-c^2)\left(\frac{9-\eta\theta(4+\theta)}{3\eta\theta}\right) + \frac{b^2(1+z)^6(1-c^2)}{2H_0^2E^2(z)} + \frac{b^2(1+z)^6}{6H_0^2E^2(z)}. \tag{34}$$

For ω_{1d} the expression for the Q-term takes the following form

$$\begin{aligned} Q_1 &= 9H_0^3\left(\frac{b^2(1+z)^6}{2H_0^2(3-3c^2\omega_0)} + A(1+z)^{3+3c^2\omega_0}\right)^{\frac{3}{2}}(1-c^2) \\ &\times \left(-\frac{\omega_0}{1+r_o+\epsilon_o\frac{z}{1+z}+\frac{b^2(1+z)^6}{2}} + \frac{b^2(1+z)^6}{2H_0^2\left(\frac{b^2(1+z)^6}{2H_0^2(3-3c^2\omega_0)} + A(1+z)^{3+3c^2\omega_0}\right)}\right. \\ &\times \left.\left(1+\frac{1}{3(1-c^2)}\right)\right). \end{aligned} \tag{35}$$

Similarly for ω_{2d}, the Q-term is reduced in the following relation

$$\begin{aligned} Q_2 &= \left(\frac{b^2}{2H_0^2\left(\frac{3-3c^2\omega_0-8c^2\omega_1}{1-\frac{3}{2}(c^2\omega_1)}\right)}(1+z)^{\frac{6-11c^2\omega_1}{1-\frac{3}{2}(c^2\omega_1)}} + B(1+z)^{\frac{3+3c^2\omega_0-3c^2\omega_1}{1-\frac{3}{2}(c^2\omega_1)}}\right)^{\frac{3}{2}} \\ &\times 9H_0^3(1-c^2)\left(-\frac{\omega_0+\omega_1 q}{1+r_o+\epsilon_o\frac{z}{1+z}+\frac{b^2(1+z)^6}{2}}\left(1+\frac{1}{3(1-c^2)}\right)\right. \\ &+ \left.\frac{b^2(1+z)^6}{2H_0^2\left(\frac{b^2}{2H_0^2\left(\frac{3-3c^2\omega_0-8c^2\omega_1}{1-\frac{3}{2}(c^2\omega_1)}\right)}(1+z)^{\frac{6-11c^2\omega_1}{1-\frac{3}{2}(c^2\omega_1)}} + B(1+z)^{\frac{3+3c^2\omega_0-3c^2\omega_1}{1-\frac{3}{2}(c^2\omega_1)}}\right)}\right). \end{aligned} \tag{36}$$

In Figure 1, the plot of Q_1 as a function of z is expressed for three different values of $\omega_0 = -0.8, -0.9, -1$. The specific values for the other constants are $b = 3$, $H_0 = 67$, $c = 0.8$, $A = -0.002$, $\epsilon_o = 0.1$ and $r_o = 0.43$. We can observe that Q_1 inclines the positive trajectory. It is mentioned [77] that the interaction term must not change its sign during cosmic evolution and is observationally verified. The plot of Q_2 versus z for ω_{2d} as shown in Figure 2. The particular values of other constants are $\omega_1 = -0.2, -0.5, -0.8$, $B = 0.002$ and remaining are same as in the above case. It can be seen that Q_2 gives the positive behavior for all epochs related to z.

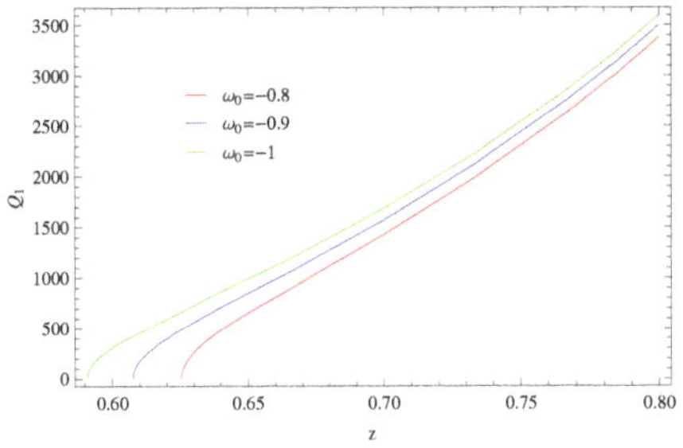

Figure 1. Plot of Q_1 corresponding to z for ω_{1d}.

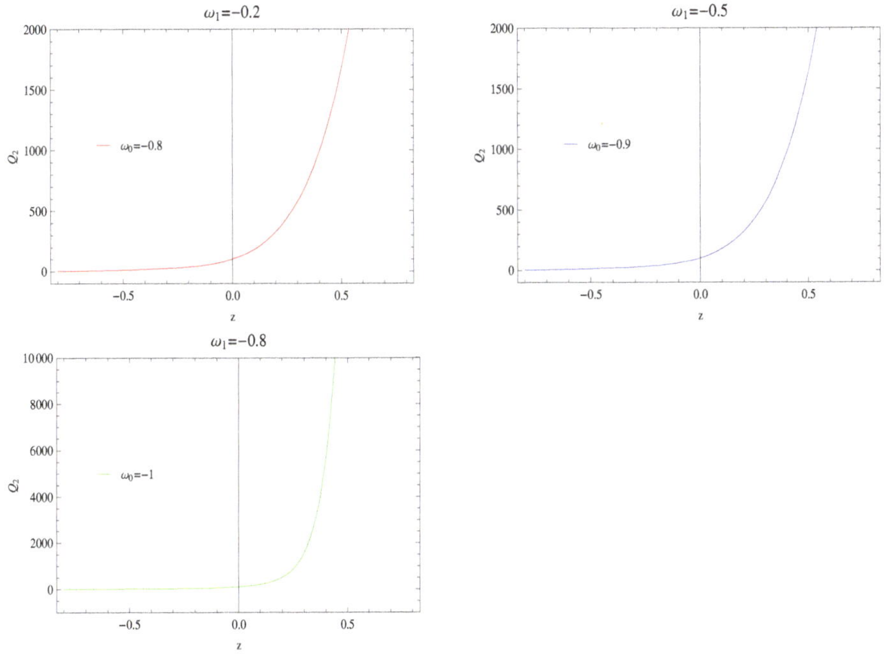

Figure 2. Plot of Q_2 corresponding to z for ω_{2d}.

4. Cosmological Parameters

In this section, we construct some cosmological parameters such as the deceleration parameter, stability analysis, statefinder and Om-diagnostic corresponding to parametrizations of EoS parameter in the presence of dynamical Chern–Simons modified gravity.

4.1. Deceleration Parameter

The deceleration parameter can be described as follows

$$q = -1 - \frac{\dot{H}}{H^2}. \tag{37}$$

This parameter characterizes the accelerated as well as decelerated phases of the universe. For $q \in [-1, 0)$, it shows the accelerated phase of the universe and $q \geq 0$ exhibits the decelerated phase of the universe. The time derivative of Hubble parameter gives the following relation in terms of redshift function

$$\frac{\dot{H}}{H^2} = -\frac{(1+z)}{E}\frac{dE}{dz}. \tag{38}$$

Inserting Equation (38) into (37), we have

$$q = -1 + \frac{(1+z)}{E}\frac{dE}{dz}. \tag{39}$$

The deceleration parameter for ω_{1d} can be evaluated by using Equations (22) and (39) such that

$$\begin{aligned} q_1 &= -1 + \frac{(1+z)}{2\left(\frac{b^2(1+z)^6}{2H_0^2(3-3c^2\omega_0)} + A(1+z)^{3+3c^2\omega_0}\right)} \\ &\times \left(\frac{3b^2(1+z)^5}{H_0^2(3-3c^2\omega_0)} + A(1+z)^{2+3c^2\omega_0}(3+3c^2\omega_0)\right). \end{aligned} \tag{40}$$

The plot of this equation in shown in Figure 3 (left) versus z for three different values of ω_0. The particular values of other constants are same as in above case. For $z > 0$, the deceleration parameter transits towards the range for accelerated phase. For present and future epochs, this parameter represents the accelerated phase of the evolving universe. Substituting the Equation (23) into (39), the expression of deceleration parameter for ω_{2d} takes the following form

$$\begin{aligned} q_2 &= -1 + (1+z)\left(\frac{b^2(1+z)^{\frac{6-11c^2\omega_1}{1-\frac{3}{2}(c^2\omega_1)}}}{H_0^2\left(\frac{3-3c^2\omega_0-8c^2\omega_1}{1-\frac{3}{2}(c^2\omega_1)}\right)} + 2B(1+z)^{\frac{3+3c^2\omega_0-3c^2\omega_1}{1-\frac{3}{2}(c^2\omega_1)}}\right)^{-1} \\ &\times \left[\frac{(3-3c^2\omega_1+3c^2\omega_0)B(1+z)^{-1+\frac{3-3c^2\omega_1+3c^2\omega_0}{1-\frac{3}{2}(c^2\omega_1)}}}{1-\frac{3}{2}(c^2\omega_1)} + \frac{6-11c^2\omega_1}{1-\frac{3}{2}c^2\omega_1}\right. \\ &\times \left.\frac{b^2(1+z)^{-1+\frac{6-11c^2\omega_1}{1-\frac{3}{2}(c^2\omega_1)}}}{2H_0^2\left(3-\frac{8c^2\omega 1}{1-\frac{3}{2}(c^2\omega_1)}-3c^2\omega_0\right)}\right]. \end{aligned} \tag{41}$$

For ω_{2d}, we plot the deceleration parameter q_2 as shown in Figure 3 (right) for same parametric values. In this scenario, the deceleration parameter exhibits accelerated phase of the universe since it remains between -1 and 0 for all values of (ω_0, ω_1).

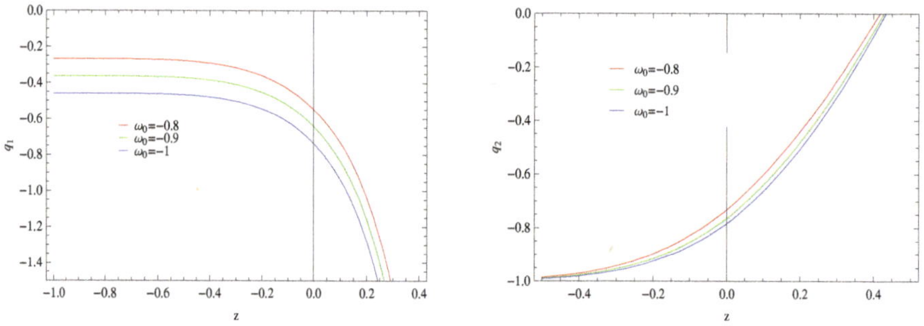

Figure 3. Plot of q_1 for ω_{1d} and q_2 for ω_{2d} corresponding to z taking $(\omega_0, \omega_1) = (-0.8, -0.2)$, $(-0.9, -0.5), (-1, -0.8)$.

4.2. Stability Analysis

The squared speed of sound can be described as follows

$$v_s^2 = \frac{dp}{d\rho}. \tag{42}$$

The squared speed of sound parameter is used to discuss the stability of model. That is, $v_s^2 < 0$ leads to the unstable behavior of the model while $v_s^2 \geq 0$ corresponds to the stable behavior. Inserting Equations (12), (13) and (22) in (42), we obtain the squared speed of sound for ω_{1d} as follows

$$\begin{aligned} v_{s1}^2 &= \Bigg(- 3b^2(1+z)^5 + H_0^2(1+z)\Big(\frac{15b^2(1+z)^4}{H_0^2(3-3c^2\omega_0)} + A(1+z)^{1+3c^2\omega_0} \\ &\times (2+3c^2\omega_0)(3+3c^2\omega_0)\Big) - 2H_0^2\Big(\frac{3b^2(1+z)^5}{H_0^2(3-3c^2\omega_0)} + A(1+z)^{2+3c^2\omega_0} \\ &+ (3+3c^2\omega_0)\Big)\Bigg)\Bigg(3AH_0^2(1+z)^{2+3c^2\omega_0}(3+3c^2\omega_0) + \frac{3b^2(1+z)^5c^2\omega_0}{1-c^2\omega_0}\Bigg)^{-1}. \end{aligned} \tag{43}$$

Taking into account Equations (12), (13) and (23) in (42), the relation of squared speed of sound for ω_{2d} takes the following form

$$\begin{aligned} v_{s2}^2 &= \Bigg[- 3b^2(1+z)^5 + \frac{3b^2(1+z)^{5-\frac{11c^2\omega_1}{1-\frac{3}{2}c^2\omega_1}}(6-\frac{11c^2\omega_1}{1-\frac{3}{2}c^2\omega_1})}{2(3-3c^2\omega_0-\frac{8c^2\omega_1}{1-\frac{3}{2}c^2\omega_1})} + 3BH_0^2 \\ &\times (1+z)^{2+3c^2\omega_1-\frac{3c^2\omega_1}{1-\frac{3}{2}c^2\omega_1}}(3+3c^2\omega_1-\frac{3c^2\omega_1}{1-\frac{3}{2}c^2\omega_1})\Bigg]^{-1}\Bigg[-\frac{b^2}{2}(1+z)^6 \\ &- 3BH_0^2(1+z)^{3+3c^2\omega_1-\frac{3c^2\omega_1}{1-\frac{3}{2}c^2\omega_1}} - \frac{3b^2(1+z)^{6-\frac{11c^2\omega_1}{1-\frac{3}{2}c^2\omega_1}}}{2(3-3c^2\omega_0-\frac{8c^2\omega_1}{1-\frac{3}{2}c^2\omega_1})} \\ &+ \frac{b^2(1+z)^{6-\frac{11c^2\omega_1}{1-\frac{3}{2}c^2\omega_1}}(6-\frac{11c^2\omega_1}{1-\frac{3}{2}c^2\omega_1})}{(3-3c^2\omega_0-\frac{8c^2\omega_1}{1-\frac{3}{2}c^2\omega_1})} + 2BH_0^2(1+z)^{3+3c^2\omega_1-\frac{3c^2\omega_1}{1-\frac{3}{2}c^2\omega_1}} \\ &\times (3+3c^2\omega_1\frac{3c^2\omega_1}{1-\frac{3}{2}c^2\omega_1}) - \frac{b^2(1+z)^{5-\frac{11c^2\omega_1}{1-\frac{3}{2}c^2\omega_1}}(6-\frac{11c^2\omega_1}{1-\frac{3}{2}c^2\omega_1})}{2(3-3c^2\omega_0-\frac{8c^2\omega_1}{1-\frac{3}{2}c^2\omega_1})} \\ &- H_0^2B(1+z)^{2+3c^2\omega_1-\frac{3c^2\omega_1}{1-\frac{3}{2}c^2\omega_1}}(3+3c^2\omega_1-\frac{3c^2\omega_1}{1-\frac{3}{2}c^2\omega_1}) - 3b^2(1+z)^5 \\ &+ \frac{b^2(1+z)^{5-\frac{11c^2\omega_1}{1-\frac{3}{2}c^2\omega_1}}(5-\frac{11c^2\omega_1}{1-\frac{3}{2}c^2\omega_1})(6-\frac{11c^2\omega_1}{1-\frac{3}{2}c^2\omega_1})}{(3-3c^2\omega_0-\frac{8c^2\omega_1}{1-\frac{3}{2}c^2\omega_1})} + 2BH_0^2 \\ &\times (1+z)^{2+3c^2\omega_1-\frac{3c^2\omega_1}{1-\frac{3}{2}c^2\omega_1}}(2+3c^2\omega_1-\frac{3c^2\omega_1}{1-\frac{3}{2}c^2\omega_1}) \\ &\times (3+3c^2\omega_1-\frac{3c^2\omega_1}{1-\frac{3}{2}c^2\omega_1})\Bigg]. \end{aligned} \tag{44}$$

The plot of v_{s1}^2 is expressed in Figure 4 (left). It can be seen that the trajectories of squared speed of sound show positive behavior for a positive range of z (except some values) which gives the stability of the model. However, for a small range of positive values of z, $z = 0$ and $z < 0$, the model expresses unstable behavior. In Figure 4 (right), the graph of squared speed of sound versus redshift parameter is given. The trajectories for $\omega_0 = -0.8, -0.9$ give the positive behavior for all values of z while for $\omega_0 = -1$, the squared speed of sound represents negative behavior for all values. This shows the stable behavior in first case while in latter case, the model is unstable.

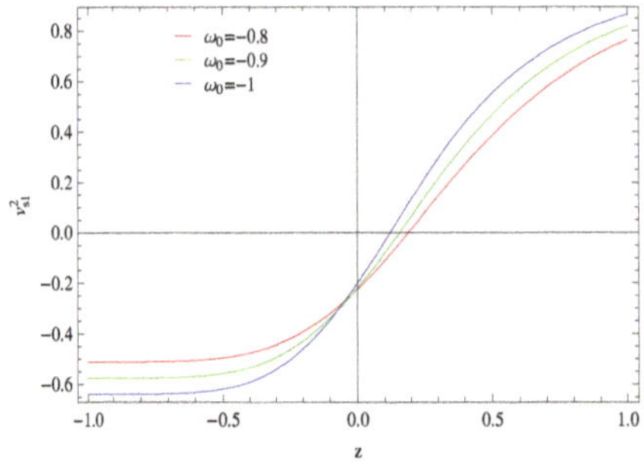

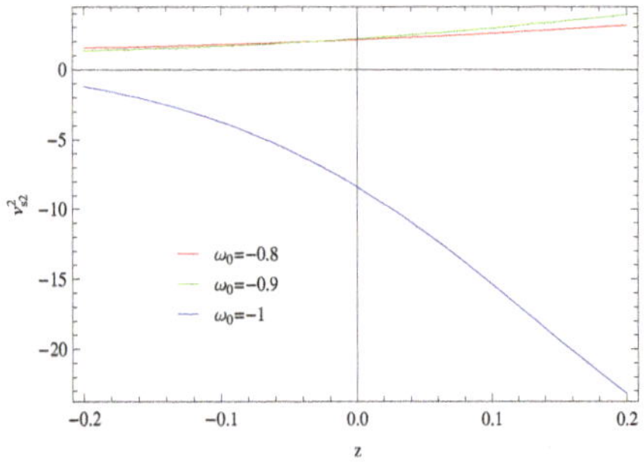

Figure 4. Plot of v_{s1}^2 for ω_{1d} and v_{s2}^2 for ω_{2d} corresponding to z taking $(\omega_0, \omega_1) = (-0.8, -0.2)$, $(-0.9, -0.5)$, $(-1, -0.8)$.

4.3. *Statefinder Parameters*

The statefinder parameters (r, s) are two new cosmological parameters introduced by Sahni [78] which are defined for flat universe model as

$$s = \frac{r-1}{3(q-\frac{1}{2})}, \quad r = 1 + \frac{3\dot{H}}{H^2} + \frac{\ddot{H}}{H^3}, \tag{45}$$

which help in differentiating the DE models. That is, for $(r, s) = (1, 0)$ then it shows the ΛCDM limit, $(r, s) = (1, 1)$ represents the CDM limit. Also, $s > 0$ and $r < 1$ represent the DE regions such that phantom and quintessence and $r > 1$, $s < 0$ give the Chaplygin gas behavior. We can obtain statefinder parameters (r, s) for ω_{1d} by using Equations (22) and (38) in (45), such that

$$\begin{aligned} r_1 &= 1 + \frac{3(1+z)}{2\left(\frac{b^2(1+z)^6}{2H_0^2(3-3c^2)\omega_0} + A(1+z)^{3+3c^2\omega_0}\right)} \times \left(\frac{9b^2(1+z)^5}{H_0^2(3-3c^2\omega_0)} + 3A\right. \\ &\times \left.(1+z)^{2+3c^2\omega_0}(3+3c^2\omega_0)\right) - \left(\frac{1}{2(\frac{b^2(1+z)^6}{2H_0^2(3-3c^2)\omega_0} + A(1+z)^{3+3c^2\omega_0})}\right) \\ &\times \left(\frac{3b^2(1+z)^6}{H_0^2(3-3c^2\omega_0)} + A(1+z)^{3+3c^2\omega_0}(3+3c^2\omega_0)\right)\left(\frac{15b^2(1+z)^6}{H_0^2(3-3c^2\omega_0)}\right. \\ &+ \left. A(1+z)^{3+3c^2\omega_0}(3+3c^2\omega_0)(2+3c^2\omega_0)\right), \end{aligned} \tag{46}$$

$$\begin{aligned} s_1 &= \left[\frac{3(1+z)}{\frac{b^2(1+z)^6}{H_0^2(3-3c^2)\omega_0} + 2A(1+z)^{3+3c^2\omega_0}} \times \left(\frac{9b^2(1+z)^5}{H_0^2(3-3c^2\omega_0)} + 3A\right.\right. \\ &\times \left.(1+z)^{2+3c^2\omega_0}(3+3c^2\omega_0)\right) - \left(\frac{1}{\frac{b^2(1+z)^6}{H_0^2(3-3c^2)\omega_0} + 2A(1+z)^{3+3c^2\omega_0}}\right) \end{aligned}$$

$$
\begin{aligned}
&\times \quad \left(\frac{3b^2(1+z)^6}{H_0^2(3-3c^2\omega_0)} + A(1+z)^{3+3c^2\omega_0}(3+3c^2\omega_0)\right)\left(\frac{15b^2(1+z)^6}{H_0^2(3-3c^2\omega_0)}\right.\\
&+ \quad A(1+z)^{3+3c^2\omega_0}(3+3c^2\omega_0)(2+3c^2\omega_0)\Bigg)\Bigg]\Bigg[3\Bigg(-\frac{3}{2}-(1+z)\\
&\times \quad \left(\frac{b^2(1+z)^6}{2H_0^2(3-3c^2)\omega_0} + A(1+z)^{3+3c^2\omega_0}\right)^{-1} \times \left(\frac{3b^2(1+z)^5}{H_0^2(3-3c^2\omega_0)}\right.\\
&+ \quad A(1+z)^{2+3c^2\omega_0}(3+3c^2\omega_0)\Bigg)\Bigg)\Bigg]^{-1}. \qquad (47)
\end{aligned}
$$

Inserting Equations (23) and (38) in (45), the statefinder parameters for ω_{2d} take the following form

$$
\begin{aligned}
r_2 \;=\;& 1+3(1+z)\Bigg(\frac{b^2}{H_0^2}\Big(\frac{3-3c^2\omega_0-8c^2\omega_1}{1-\frac{3}{2}c^2\omega_1}\Big)(1+z)^{6-\frac{11c^2\omega_1}{1-\frac{3}{2}c^2\omega_1}} + 2B\\
&\times \quad (1+z)^{\frac{3+3c^2\omega_0-3c^2\omega_1}{1-\frac{3}{2}c^2\omega_1}}\Bigg)^{-1}\Bigg[\frac{3b^2(1+z)^{6-\frac{11c^2\omega_1}{1-\frac{3}{2}c^2\omega_1}}(1-\frac{3}{2}c^2\omega_1)(6-\frac{11c^2\omega_1}{1-\frac{3}{2}c^2\omega_1})}{2H_0^2(3-3c^2\omega_0-8c^2\omega_1)}\\
&+ \quad 3B(1+z)^{3+3c^2\omega_0-\frac{3c^2\omega_1}{1-\frac{3}{2}c^2\omega_1}}\Bigg] - \Bigg[\frac{b^2}{H_0^2}\Big(\frac{3-3c^2\omega_0-8c^2\omega_1}{1-\frac{3}{2}c^2\omega_1}\Big)(1+z)^{\frac{6-11c^2\omega_1}{1-\frac{3}{2}c^2\omega_1}}\\
&+ \quad 2B(1+z)^{\frac{3+3c^2\omega_0+3c^2\omega_1}{1-\frac{3}{2}c^2\omega_1}}\Bigg]^{-1}\Bigg[\Bigg(\frac{b^2(1+z)^{6-\frac{11c^2\omega_1}{1-\frac{3}{2}c^2\omega_1}}(1-\frac{3}{2}c^2\omega_1)(6-\frac{11c^2\omega_1}{1-\frac{3}{2}c^2\omega_1})}{2H_0^2(3-3c^2\omega_0-8c^2\omega_1)}\\
&+ \quad B(1+z)^{3+3c^2\omega_0-\frac{3c^2\omega_1}{1-\frac{3}{2}c^2\omega_1}}\Bigg) + \Bigg(B(1+z)^{3+3c^2\omega_0-\frac{3c^2\omega_1}{1-\frac{3}{2}c^2\omega_1}}(3+3c^2\omega_0\\
&- \quad \frac{3c^2\omega_1}{1-\frac{3}{2}c^2\omega_1})(2+3c^2\omega_0-\frac{3c^2\omega_1}{1-\frac{3}{2}c^2\omega_1}) + \frac{b^2(1-\frac{3}{2}c^2\omega_1)(6-\frac{11c^2\omega_1}{1-\frac{3}{2}c^2\omega_1})}{2H_0^2(3-3c^2\omega_0-8c^2\omega_1)}\\
&\times \quad (5-\frac{11c^2\omega_1}{1-\frac{3}{2}c^2\omega_1})(1+z)^{6-\frac{11c^2\omega_1}{1-\frac{3}{2}c^2\omega_1}}\Bigg)\Bigg], \qquad (48)\\
s_2 \;=\;& 3\Bigg(\frac{(1+z)}{\frac{b^2}{H_0^2}(\frac{3-3c^2\omega_0-8c^2\omega_1}{1-\frac{3}{2}c^2\omega_1})(1+z)^{6-\frac{11c^2\omega_1}{1-\frac{3}{2}c^2\omega_1}} + 2B(1+z)^{\frac{3+3c^2\omega_0-3c^2\omega_1}{1-\frac{3}{2}c^2\omega_1}}}\Bigg)\\
&\times \quad \Bigg[\frac{3b^2(1+z)^{6-\frac{11c^2\omega_1}{1-\frac{3}{2}c^2\omega_1}}(1-\frac{3}{2}c^2\omega_1)(6-\frac{11c^2\omega_1}{1-\frac{3}{2}c^2\omega_1})}{2H_0^2(3-3c^2\omega_0-8c^2\omega_1)} + 3B(1+z)^{3+3c^2\omega_0}\\
&\times \quad (1+z)^{-\frac{3c^2\omega_1}{1-\frac{3}{2}c^2\omega_1}}\Bigg] - \Bigg[\frac{b^2}{H_0^2}\Big(\frac{3-3c^2\omega_0-8c^2\omega_1}{1-\frac{3}{2}c^2\omega_1}\Big)(1+z)^{\frac{6-11c^2\omega_1}{1-\frac{3}{2}c^2\omega_1}} + 2B\\
&\times \quad (1+z)^{\frac{3+3c^2\omega_0+3c^2\omega_1}{1-\frac{3}{2}c^2\omega_1}}\Bigg]^{-1}\Bigg[\Bigg(\frac{b^2(1+z)^{6-\frac{11c^2\omega_1}{1-\frac{3}{2}c^2\omega_1}}(1-\frac{3}{2}c^2\omega_1)(6-\frac{11c^2\omega_1}{1-\frac{3}{2}c^2\omega_1})}{2H_0^2(3-3c^2\omega_0-8c^2\omega_1)}\\
&+ \quad B(1+z)^{3+3c^2\omega_0-\frac{3c^2\omega_1}{1-\frac{3}{2}c^2\omega_1}}\Bigg) + \Bigg(B(1+z)^{3+3c^2\omega_0-\frac{3c^2\omega_1}{1-\frac{3}{2}c^2\omega_1}}(3+3c^2\omega_0\\
&- \quad \frac{3c^2\omega_1}{1-\frac{3}{2}c^2\omega_1})(2+3c^2\omega_0-\frac{3c^2\omega_1}{1-\frac{3}{2}c^2\omega_1}) + \frac{b^2(1-\frac{3}{2}c^2\omega_1)(6-\frac{11c^2\omega_1}{1-\frac{3}{2}c^2\omega_1})}{2H_0^2(3-3c^2\omega_0-8c^2\omega_1)}
\end{aligned}
$$

$$
\begin{aligned}
&\times \quad (5-\frac{11c^2\omega_1}{1-\frac{3}{2}c^2\omega_1})(1+z)^{6-\frac{11c^2\omega_1}{1-\frac{3}{2}c^2\omega_1}}\Big)\Big]\Big[3\Big(-\frac{3}{2}-(1+z)\\
&\times \quad \left(\frac{b^2}{2H_0^2}(\frac{3-3c^2\omega_0-8c^2\omega_1}{1-\frac{3}{2}c^2\omega_1})(1+z)^{\frac{6-11c^2\omega_1}{1-\frac{3}{2}c^2\omega_1}}+B(1+z)^{\frac{3+3c^2\omega_0+3c^2\omega_1}{1-\frac{3}{2}c^2\omega_1}}\right)^{-1}\\
&\times \quad \left(\frac{3B(1+z)^{-1+\frac{3}{1-\frac{3}{2}c^2\omega_1}}}{1-\frac{3}{2}c^2\omega_1}+\frac{b^2(6-11c^2\omega_1)(1+z)^{-1+\frac{6-11c^2\omega_1}{1-\frac{3}{2}c^2\omega_1}}}{(1-\frac{3}{2}c^2\omega_1)H_0^2(3-\frac{8c^2\omega 1}{1-\frac{3}{2}c^2\omega_1}-3c^2\omega_0)}\right)\Big)\Big]^{-1}. \qquad (49)
\end{aligned}
$$

In Figure 5 (left), the graph of s_1 displayed against r_1 for three different values of ω_0. We can observe that the (r,s) parameters corresponds to Chaplygin gas behavior for the underlying scenario. However, the trajectory for $\omega_0=-1$ does not yield any result for some region which is related to $r>1$, $s>0$. The model constitutes the ΛCDM limit $(r,s)=(1,0)$ for the trajectories of $\omega_0=-0.8,-0.9$. In the right side plot, we draw s_2 versus r_2 for ω_{2d} which gives the $r<1$ and $s>0$ for the trajectory $\omega_0=-0.9, \omega_1=-0.5$. This shows the DE eras, phantom and quintessence. The remaining two trajectories do not give any fruitful results.

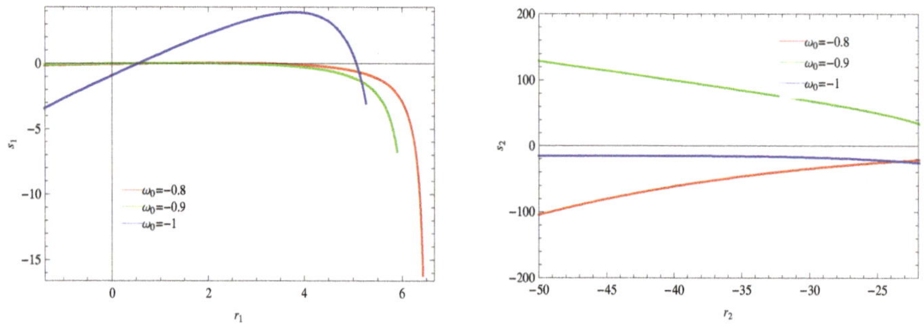

Figure 5. Plot of r_1 corresponding to s_1 for ω_{1d} and r_2 corresponding to s_2 for ω_{2d}.

4.4. Om-Diagnostic

The Om-diagnostic is another tool to differentiate different phases of the universe. The positive trajectory of Om-diagnostic represents the DE era like phantom while quintessence era is obtained from negative behavior. This parameter is given by

$$
Om = \frac{(\frac{H}{H_0})^2-1}{(1+z)^3-1}. \qquad (50)
$$

The Om-diagnostic for ω_{1d} and ω_{2d} can be obtained by substitution of Equations (22) and (23) in above relation, such that

$$
Om_1 = \frac{\frac{b^2(1+z)^6}{2H_0^2(3-3c^2\omega_0)}+A(1+z)^{3+3c^2\omega_0}-1}{(1+z)^3-1}, \qquad (51)
$$

$$
Om_2 = \frac{\frac{b^2(1+z)^{\frac{6-11c^2\omega_1}{1-\frac{3}{2}c^2\omega_1}}}{2H_0^2(\frac{3-3c^2\omega_0-8c^2\omega_1}{1-\frac{3}{2}c^2\omega_1})}+B(1+z)^{\frac{3+3c^2\omega_0-3c^2\omega_1}{1-\frac{3}{2}(c^2\omega_1)}}-1}{(1+z)^3-1}. \qquad (52)
$$

In Figure 6, we draw the Om-diagnostic versus z for ω_{1d} in left plot and for ω_{2d} in the right plot. It can be observed that the trajectories of Om-diagnostic for both cases represent the negative slopes at a past epoch which implies the quintessence era while give positive slopes at future epoch which constitutes the phantom era of the universe.

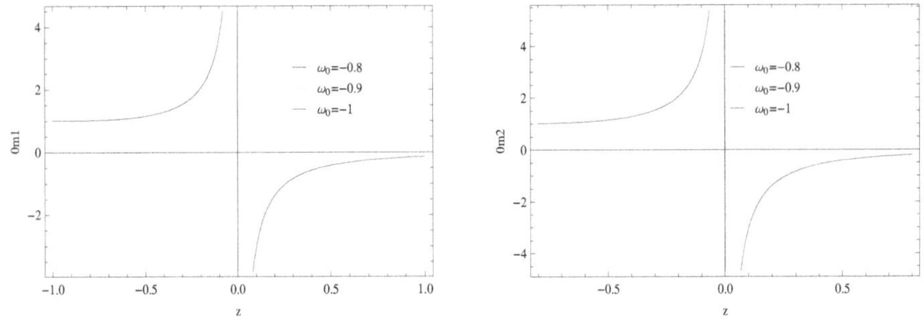

Figure 6. Plot of Om_1 for ω_{1d} and Om_2 for ω_{2d} corresponding to z.

5. Conclusions

In the framework of dynamical Chern–Simons modified gravity, we have assumed the flat FRW spacetime and discussed different DE models by using a collection of observations at low redshift. We have taken the parametrizations of EoS parameter to explore the cosmic evolution of accelerating universe in interacting scenario. The parametrization may be a Taylor series extension in the redshift, a Taylor series extension in the scale factor or any other parametrization of ω. We have evaluated the different cosmological parameters, such as the deceleration parameter, squared speed of sound, Om-diagnostic and statefinder parameters. The deceleration parameter is a cosmological parameter which helps to classify the accelerated as well as decelerated phases of the universe. The squared speed of sound is another cosmological parameter which is used to check the stability of the models. The statefinder parameters differentiate various DE models, their behavior and cosmological evolution at present time. The Om-diagnostic is used to differentiate the phantom and quintessence behavior. The trajectories of the constructed models have been plotted with different constant parametric values.

The interaction term represented the positive behavior and is observationally verified that the interaction term must not change its sign during cosmic evolution. The deceleration parameter indicated the consistent result while squared speed of sound expressed some stable solutions. The statefinder parameters for the first parametrization represented Chaplygin gas model behavior and met the ΛCDM limit for specific choice of parameters and DE era is obtained for second choice of parametrization. The Om diagnostic parameter indicated the phantom and quintessence eras of the universe.

Author Contributions: A.J., S.R. and S.S. completed the manuscript, while K.B. and R.J. done its proof reading.

Funding: This research received no external funding.

Acknowledgments: S.R. and A.J. are thankful to the Higher Education Commission, Islamabad, Pakistan for its financial support under the grant No: 5412/Federal/NRPU/R&D/HEC/2016. The work of K.B. was partially supported by the JSPS KAKENHI Grant Number JP 25800136 and Competitive Research Funds for Fukushima University Faculty (18RI009).

Conflicts of Interest: The authors declare no conflict of interest.

References

1. Tripathi, A.; Sangwan, A.; Jassal, H.K. Dark energy equation of state parameter and its evolution at low redshift. *JCAP* **2017**, *06*, 012. [CrossRef]
2. Perlmutter, S.; Gabi, S.; Goldhaber, G.; Goobar, A.; Groom, D.E.; Hook, I.M.; Kim, A.G.; Kim, M.Y.; Lee, J.C.; Pain, R.; et al. Measurements* of the Cosmological Parameters Ω and Λ from the First Seven Supernovae at $z \geq 0.35$. *Astrophys. J.* **1997**, *483*, 565. [CrossRef]
3. Perlmutter, S.; Aldering, G.; Goldhaber, G.; Knop, R.A.; Nugent, P.; Castro, P.G.; Deustua, S.; Fabbro, S.; Goobar, A.; Groom, D.E.; et al. Measurements of Ω and Λ from 42 high-redshift supernovae. *Astrophys. J.* **1999**, *517*, 565. [CrossRef]
4. Riess, A.G.; Filippenko, A.V.; Challis, P.; Clocchiatti, A.; Diercks, A.; Garnavich, P.M.; Gillil, R.L.; Hogan, C.J.; Jha, S.; Kirshner, R.P.; et al. Observational evidence from supernovae for an accelerating universe and a cosmological constant. *Astron. J.* **1998**, *116*, 1009. [CrossRef]
5. Astier, P.; Guy, J.; Regnault, N.; Pain, R.; Aubourg, E.; Balam, D.; Basa, S.; Carlberg, R.G.; Fabbro, S.; Fouchez, D.; et al. The Supernova Legacy Survey: Measurement of, and w from the first year data set. *Astron. Astrophys. J.* **2006**, *447*, 31–48. [CrossRef]
6. Garnavich, P.M.; Jha, S.; Challis, P.; Clocchiatti, A.; Diercks, A.; Filippenko, A.V.; Gillil, R.L.; Hogan, C.J.; Kirshner, R.P.; Leibundgut, B.; et al. Supernova limits on the cosmic equation of state. *Astrophys. J.* **1998**, *509*, 74. [CrossRef]
7. Tonry, J.L.; Schmidt, B.P.; Barris, B.; Candia, P.; Challis, P.; Clocchiatti, A.; Coil, A.L.; Filippenko, A.V.; Garnavich, P.; Hogan, C.; et al. Cosmological results from high-z supernovae. *Astrophys. J.* **2003**, *594*, 1. [CrossRef]
8. Barris, B.J.; Tonry, J.L.; Blondin, S.; Challis, P.; Chornock, R.; Clocchiatti, A.; Filippenko, A.V.; Garnavich, P.; Holl, S.T.; Jha, S.; et al. Twenty-three high-redshift supernovae from the Institute for Astronomy Deep Survey: Doubling the supernova sample at $z > 0.7$. *Astrophys. J.* **2004**, *602*, 571. [CrossRef]
9. Goobar, A.; Perlmutter, S.; Goldhaber, G.; Knop, R.A.; Nugent, P.; Castro, P.G.; Deustua, S.; Fabbro, S.; Groom, D.E.; Hook, I.M.; et al. The acceleration of the Universe: Measurements of cosmological parameters from type Ia supernovae. *Phys. Scr. J.* **2000**, *85*, 47. [CrossRef]
10. González-Gaitán, S.; Conley, A.; Bianco, F.B.; Howell, D.A.; Sullivan, M.; Perrett, K.; Carlberg, R.; Astier, P.; Balam, D.; Ball, C.; et al. The Rise Time of Normal and Subluminous Type Ia Supernovae. *Astrophys. J.* **2012**, *745*, 44. [CrossRef]
11. Buckley, J.; Byrum, K.; Dingus, B.; Falcone, A.; Kaaret, P.; Krawzcynski, H.; Pohl, M.; Vassiliev, V.; Williams, D.A. The Status and future of ground-based TeV gamma-ray astronomy. A White Paper prepared for the Division of Astrophysics of the American Physical Society. *arXiv* **2008**, arXiv:0810.0444.
12. Nojiri, S.; Odintsov, S.D. Unified cosmic history in modified gravity: From F(R) theory to Lorentz non-invariant models. *Phys. Rep.* **2011**, *505*, 59–144, doi:10.1016/j.physrep.2011.04.001. [CrossRef]
13. Capozziello, S.; Laurentis, M.D. Extended Theories of Gravity. *Phys. Rep.* **2011**, *509*, 167–321, doi:10.1016/j.physrep.2011.09.003. [CrossRef]
14. Nojiri, S.; Odintsov, S.D.; Oikonomou, V.K. Modified Gravity Theories on a Nutshell: Inflation, Bounce and Late-time Evolution. *Phys. Rep.* **2017**, *692*, 1–104, doi:10.1016/j.physrep.2017.06.001. [CrossRef]
15. Faraoni, V.; Capozziello, S. Beyond Einstein Gravity: A Survey of Gravitational Theories for Cosmology and Astrophysics. *Fundam. Theor. Phys.* **2010**, *170*, doi:10.1007/978-94-007-0165-6. [CrossRef]
16. Bamba, K.; Odintsov, S.D. Inflationary cosmology in modified gravity theories. *Symmetry* **2015**, *7*, 220, doi:10.3390/sym7010220. [CrossRef]
17. Cai, Y.F.; Capozziello, S.; Laurentis, M.D.; Saridakis, E.N. f(T) teleparallel gravity and cosmology. *Rep. Prog. Phys.* **2016**, *79*, 106901, doi:10.1088/0034-4885/79/10/106901. [CrossRef]
18. Pfenniger, D.; Combes, F.; Martinet, L. Is dark matter in spiral galaxies cold gas? I. Observational constraints and dynamical clues about galaxy evolution. *Astron. Astrophys.* **1994**, *285*, 79–83.
19. Clowe, D.; Bradač, M.; Gonzalez, A.H.; Markevitch, M.; Rall, S.W.; Jones, C.; Zaritsky, D. A Direct Empirical Proof of the Existence of Dark Matter. *Astrophys. J. Lett.* **2006**, *648*, L109–L113. [CrossRef]
20. Alcock, C.; Allsman, R.A.; Axelrod, T.S.; Bennett, D.P.; Cook, K.H.; Freeman, K.C.; Griest, K.; Guern, J.A.; Lehner, M.J.; Marshall, S.L.; et al. The MACHO Project First-Year Large Magellanic Cloud Results: The Microlensing Rate and the Nature of the Galactic Dark Halo. *Astrophys. J.* **1996**, *461*, 84. [CrossRef]

21. Zlatev, I.; Wang, L.; Steinhardt, P.J. Quintessence, cosmic coincidence, and the cosmological constant. *Phys. Rev. Lett.* **1999**, *82*, 896. [CrossRef]
22. Elizalde, E.; Nojiri, S.; Odinstov, S.D. Late-time cosmology in a (phantom) scalar-tensor theory: Dark energy and the cosmic speed-up. *Phys. Rev. D* **2004**, *70*, 043539. [CrossRef]
23. Nojiri, S.; Odintsov, S.D.; Tsujikawa, S. Properties of singularities in the (phantom) dark energy universe. *Phys. Rev. D* **2005**, *71*, 063004. [CrossRef]
24. Anisimov, A.; Babichev, E.; Vikman, A. B-Inflation. *J. Cosmol. Astropart. Phys.* **2005**, *6*, 006. [CrossRef]
25. Setare, M.R. Interacting generalized Chaplygin gas model in non-flat universe. *Eur. Phys. J. C* **2007**, *52*, 689–692. [CrossRef]
26. Bento, M.C.; Bertolami, O.; Sen, A.A. Generalized Chaplygin gas, accelerated expansion, and dark-energy-matter unification. *Phys. Rev. D* **2002**, *66*, 043507. [CrossRef]
27. Kamenshchik, A.; Moschella, U.; Pasquier, V. An alternative to quintessence. *Phys. Lett. B* **2001**, *511*, 265–268. [CrossRef]
28. Chiba, T.; Okabe, T.; Yamaguchi, M. Kinetically driven quintessence. *Phys. Rev. D* **2000**, *62*, 023511. [CrossRef]
29. Armendariz-Picon, C.; Mukhanov, V.; Steinhardt, P. Dynamical Solution to the Problem of a Small Cosmological Constant and Late-Time Cosmic Acceleration. *J. Phys. Rev. Lett.* **2000**, *85*, 4438. [CrossRef]
30. Armendariz-Picon, C.; Damour, T.; Mukhanov, V. k-Inflation. *Phys. Lett. B* **1999**, *458*, 209–218. [CrossRef]
31. Wei, H.; Cai, R.G. A new model of agegraphic dark energy. *Phys. Lett. B* **2008**, *660*, 113–117. [CrossRef]
32. Cai, R.G. A dark energy model characterized by the age of the Universe. *Phys. Lett. B* **2007**, *657*, 228–231. [CrossRef]
33. Hsu, S.D.H. Entropy bounds and dark energy. *Phys. Lett. B* **2004**, *594*, 13–16. [CrossRef]
34. Li, M. A model of holographic dark energy. *Phys. Lett. B* **2004**, *603*, 1–5. [CrossRef]
35. Setare, M.R. The holographic dark energy in non-flat Brans–Dicke cosmology. *Phys. Lett. B* **2007**, *644*, 99–103. [CrossRef]
36. Caldwell, R.R. A phantom menace? Cosmological consequences of a dark energy component with super-negative equation of state. *Phys. Lett. B* **2002**, *545*, 23–29. [CrossRef]
37. Nojiri, S.; de Odintsov, S.D. Sitter brane universe induced by phantom and quantum effects. *Phys. Lett. B* **2003**, *565*, 1–9. [CrossRef]
38. Nojiri, S.; Odintsov, S.D. Quantum de Sitter cosmology and phantom matter. *Phys. Lett. B* **2003**, *562*, 147–152. [CrossRef]
39. Tavayef, M.; Sheykhi, A.; Bamba, K.; Moradpour, H. Tsallis holographic dark energy. *Phys. Lett. B* **2018**, *781*, 195–200. [CrossRef]
40. Wang, B.; Abdalla, E.; Atrio-Barandela, F.; Pavon, D. Dark matter and dark energy interactions: Theoretical challenges, cosmological implications and observational signatures. *Rep. Prog. Phys.* **2016**, *79*, 096901. [CrossRef]
41. Joan, S. Dark energy: A quantum fossil from the inflationary universe? *J. Phys. A* **2008**, *41*, 164066.
42. Chevallier, M.; Polarski, D. Accelerating universes with scaling dark matter. *Int. J. Mod. Phys. D* **2001**, *10*, 213. [CrossRef]
43. Linder, E.V. Exploring the Expansion History of the Universe. *Phys. Rev. Lett.* **2003**, *90*, 091301. [CrossRef]
44. Jassal, H.K.; Bagla, J.S.; Padmanabhan, T. Observational constraints on low redshift evolution of dark energy: How consistent are different observations? *Phys. Rev. D* **2005**, *72*, 103503. [CrossRef]
45. Feng, L.; Lu, T. A new equation of state for dark energy model. *JCAP* **2011**, *11*, 034. [CrossRef]
46. Efstathiou, G. Constraining the equation of state of the Universe from distant Type Ia supernovae and cosmic microwave background anisotropies. *Mon. Not. R. Astron. Soc.* **1999**, *310*, 842. [CrossRef]
47. Lee, S. Constraints on the dark energy equation of state from the separation of CMB peaks and the evolution of α. *Phys. Rev. D* **2005**, *71*, 123528. [CrossRef]
48. Hannestad, S.; Mörtsell, E. Cosmological constraints on the dark energy equation of state and its evolution. *JCAP* **2004**, *9*, 001. [CrossRef]
49. Wang, Y.; Kostov, V.; Freese, K.; Frieman, J.A.; Gondolo, P. Probing the evolution of the dark energy density with future supernova surveys. *JCAP* **2004**, *12*, 003. [CrossRef]
50. Wang, Y.; Tegmark, M. New dark energy constraints from supernovae, microwave background, and galaxy clustering. *Phys. Rev. Lett.* **2004**, *92*, 241302. [CrossRef]

51. Bassett, B.A.; Corasaniti, P.S.; Kunz, M. The Essence of Quintessence and the Cost of Compression. *Astrophys. J.* **2004**, *617*, L1. [CrossRef]
52. Huterer, D.; Turner, M.S. Prospects for probing the dark energy via supernova distance measurements. *Phys. Rev. D* **1999**, *60*, 081301. [CrossRef]
53. Huterer, D.; Turner, M.S. Probing dark energy: Methods and strategies. *Phys. Rev. D* **2001**, *64*, 123527. [CrossRef]
54. Weller, J.; Albrecht, A. Opportunities for Future Supernova Studies of Cosmic Acceleration. *Phys. Rev. Lett.* **2001**, *86*, 1939. [CrossRef]
55. Pantazis, G.; Nesseris, S.; Perivolaropoulosv, L. Comparison of thawing and freezing dark energy parametrizations. *Phys. Rev. D* **2016**, *93*, 103503. [CrossRef]
56. Rani, N.; Jain, D.; Mahajan, S. Transition redshift: new constraints from parametric and nonparametric methods. *JCAP* **2015**, *12*, 045. [CrossRef]
57. Copel, E.J.; Sami, M.; Tsujikawa, S. Dynamics of dark energy. *Int. J. Mod. Phys. D* **2006**, *15*, 1753.
58. Bamba, K.; Capozziello, S.; Nojiri, S.; Odintsov, S.D. Dark energy cosmology: The equivalent description via different theoretical models and cosmography tests. *Astrophys. Space Sci.* **2012**, *342*, 155. [CrossRef]
59. Capozziello, S.; Cardone, V.F.; Elizalde, E.; Nojiri, S.I.; Odintsov, S.D. Observational constraints on dark energy with generalized equations of state. *Phys. Rev. D* **2006**, *73*, 043512. [CrossRef]
60. Jackiw, R.; Pi, S.Y. Chern-Simons modification of general relativity. *Phys. Rev. D* **2003**, *68*, 104012. [CrossRef]
61. Ashtekar, A.; Balachran, A.P.; Jo, S. The CP problem in quantum gravity. *Int. J. Mod. Phys. A* **1989**, *4*, 1493–1514. [CrossRef]
62. Jawad, A.; Sohail, A. Cosmological evolution of modified QCD ghost dark energy in dynamical Chern-Simons gravity. *Astrophys. Space Sci.* **2015**, *55*, 359. [CrossRef]
63. Jawad, A.; Rani, S. Cosmological Evolution of Pilgrim Dark Energy in f(G) Gravity. *Adv. High Energy Phys.* **2015**, *10*, 952156. [CrossRef]
64. Jawad, A. Reconstruction of $f(\tilde{R})$ models via well-known scale factors. *Eur. Phys. J. Plus* **2014**, *129*, 207. [CrossRef]
65. Jawad, A.; Majeed, A. Correspondence of pilgrim dark energy with scalar field models. *Astrophy. Space Sci.* **2015**, *356*, 375. [CrossRef]
66. Jawad, A. Cosmological analysis of pilgrim dark energy in loop quantum cosmology. *Eur. Phys. J. C* **2015**, *75*, 206. [CrossRef]
67. Jawad, A.; Chattopadhyay, S.; Pasqua, A. A holographic reconstruction of the modifiedf(R) Horava-Lifshitz gravity with scale factor in power-law form. *Astrophy. Space Sci.* **2013**, *346*, 273. [CrossRef]
68. Jawad, A.; Chattopadhyay, S.; Pasqua, A. Reconstruction of f(G) gravity with the new agegraphic dark-energy model. *Eur. Phys. J. Plus* **2013**, *128*, 88. [CrossRef]
69. Jawad, A.; Pasqua, A.; Chattopadhyay, S. Correspondence between f(G) gravity and holographic dark energy via power-law solution. *Astrophy. Space Sci.* **2013**, *344*, 489. [CrossRef]
70. Jawad, A.; Pasqua, A.; Chattopadhyay, S. Holographic reconstruction of f (G) gravity for scale factors pertaining to emergent, logamediate and intermediate scenarios. *Eur. Phys. J. Plus* **2013**, *128*, 156. [CrossRef]
71. Jawad, A. Analysis of QCD ghost gravity. *Astrophys. Space Sci.* **2014**, *353*, 691. [CrossRef]
72. Younus, M.; Jawad, A.; Qummer, S.; Moradpour, H.; Rani, S. Cosmological Implications of the Generalized Entropy Based Holographic Dark Energy Models in Dynamical Chern-Simons Modified Gravity. *Adv. High Energy Phys.* **2019**, *2019*, 1287932. [CrossRef]
73. Cohen, A.; Kaplan, D.; Nelson, A. Effective field theory, black holes, and the cosmological constant. *Phys. Rev. Lett.* **1999**, *82*, 4971. [CrossRef]
74. Chattopadhyay, S.; Jawad, A.; Rani, S. Holographic Polytropic Gravity Models. *Adv. High Energy Phys.* **2015**, *2015*, 798902. [CrossRef]
75. Jawad, A.; Rani, S.; Salako, I.G.; Gulshan, F. Aspects of some new versions of pilgrim dark energy in DGP braneworld. *Eur. Phys. J. Plus* **2016**, *131*, 236. [CrossRef]
76. Elizaldel, E.; Khurshudyan, M.; Nojiri, S. Cosmological singularities in interacting dark energy models with an $w(q)$ parametrization. *arxiv* **2018**, arxiv:1809.01961v1.

77. Cruz, M.; Lepe, S. Holographic approach for dark energy–dark matter interaction in curved FLRW spacetime. *Class. Quantum Gravity* **2018**, *35*, 155013. [CrossRef]
78. Sahni, V.; Saini, T.D.; Starobinsky, A.A.; Alam, U. Statefinder—A new geometrical diagnostic of dark energy. *JEPT Lett.* **2003**, *77*, 201. [CrossRef]

symmetry

MDPI

Article

Topological Gravity Motivated by Renormalization Group

Taisaku Mori [1] and Shin'ichi Nojiri [1,2,*]

1 Department of Physics, Nagoya University, Nagoya 464-8602, Japan; mori.taisaku@k.mbox.nagoya-u.ac.jp
2 Kobayashi-Maskawa Institute for the Origin of Particles and the Universe, Nagoya University, Nagoya 464-8602, Japan
* Correspondence: nojiri@gravity.phys.nagoya-u.ac.jp

Received: 15 August 2018; Accepted: 7 September 2018; Published: 11 September 2018

Abstract: Recently, we have proposed models of topological field theory including gravity in *Mod. Phys. Lett. A* **2016**, *31*, 1650213 and *Phys. Rev. D* **2017**, *96*, 024009, in order to solve the problem of the cosmological constant. The Lagrangian densities of the models are BRS (Becchi-Rouet-Stora) exact and therefore the models can be regarded as topological theories. In the models, the coupling constants, including the cosmological constant, look as if they run with the scale of the universe and its behavior is very similar to the renormalization group. Motivated by these models, we propose new models with an the infrared fixed point, which may correspond to the late time universe, and an ultraviolet fixed point, which may correspond to the early universe. In particular, we construct a model with the solutions corresponding to the de Sitter space-time both in the ultraviolet and the infrared fixed points.

1. Introduction

In *Mod. Phys. Lett. A* **2016**, *31*, 1650213 [1] and *Phys. Rev. D* **2017**, *96*, 024009 [2], models of topological field theory including gravity have been proposed in order to solve the cosmological constant problem. The accelerating expansion of the present universe may be generated by the small cosmological constant. Although the cosmological constant could be identified with a vacuum energy, the vacuum energy receives very large quantum corrections from matters and therefore in order to obtain a realistic very small vacuum energy, very fine-tuning of the counter term for the vacuum energy is necessary (The discussion about the small but non-vanishing vacuum energy is given in [3], for example.) Motivated by this problem of large quantum corrections to the vacuum energy, models of unimodular gravity [4–30] have been proposed. There have been also proposed many scenarios, such as the sequestering mechanism [31–38]. Among of the possible scenarios, we have proposed the models of the topological field theory including gravity in [1] and the cosmology described by these models has been discussed in [2].

The large quantum corrections from matter appear not only in the cosmological constant but other coupling constants. Even if we include the quantum corrections only from matter, the following coupling constants α, β, γ, and δ include large quantum corrections,

$$\mathcal{L}_{\rm qc} = \alpha R + \beta R^2 + \gamma R_{\mu\nu} R^{\mu\nu} + \delta R_{\mu\nu\rho\sigma} R^{\mu\nu\rho\sigma} \, . \tag{1}$$

The coefficient α diverges quadratically and β, γ, and δ diverge logarithmically. We should note that if we include the quantum corrections from the graviton, there appear infinite numbers of divergent quantum corrections, which is one of the reasons why the general relativity is not renormalizable. By using the formulation for the divergence in the cosmological constant proposed in [1,2], these divergences can be tuned to be finite [2,39]. In this formulations, the coupling constants, α, β, γ, δ, and other coupling constants including the cosmological constant are replaced by the scalar fields. Then the divergences coming from the quantum corrections can be absorbed into a redefinition of the scalar fields. The fields depend on the cosmological time, or the scale of the universe. In this sense, the scalar fields, which corresponds to the coupling constants, run with a scale as in the renormalization group. Motivated by the above observation, in this paper, we propose new models where there appear an infrared fixed point, which may correspond to the late time universe, and an ultraviolet fixed point, which may correspond to the early universe. Especially we construct a model with solutions connecting two asymptotic de Sitter space-times, which correspond to the ultraviolet and the infrared fixed points.

In the next section, we review the models of topological gravity presented in [1,2,39]. In Section 3, we propose new models where there appear an infrared fixed point, which may correspond to the late time universe, and an ultraviolet fixed point, which may correspond to the early universe. Especially we construct a model, where the solutions expresses the flow from the de Sitter space-time corresponding to the ultraviolet fixed point to the de Sitter space-time both in the infrared fixed point. The last section is devoted to the summary, where we mention on the problems which have not been solved in this paper and some possibilities to solve them are shown in Appendix A.

2. Review of the Models of Topological Field Theory Including Gravity

We start to review the model proposed in [1]. The action of the model is given by

$$S' = \int d^4x \sqrt{-g} \left\{ \mathcal{L}_{\mathrm{gravity}} + \mathcal{L}_{\mathrm{TP}} \right\} + S_{\mathrm{matter}} , \quad \mathcal{L}_{\mathrm{TP}} \equiv -\lambda + \partial_\mu \lambda \partial^\mu \varphi - \partial_\mu b \partial^\mu c . \tag{2}$$

Here $\mathcal{L}_{\mathrm{gravity}}$ is the Lagrangian density of gravity, which may be arbitrary. The Lagrangian density $\mathcal{L}_{\mathrm{gravity}}$ may include the cosmological constant. In the action (2), S_{matter} is the action of matters, λ and φ are ordinary scalar fields while b is the anti-ghost field and c is the ghost field. The (anti-)ghost fields b and c are fermionic (Grassmann odd) scalar. (The action without c and b has been proposed in [40] in order to solve the problem of time. The cosmological perturbation in the model motivated in the model (2) has been investigated in [41]). Please note that no parameter or coupling constant appear in the action (2) except in the parts of S_{matter} and $\mathcal{L}_{\mathrm{gravity}}$.

We separate the gravity Lagrangian density $\mathcal{L}_{\mathrm{gravity}}$ into the sum of some constant Λ, which corresponds to the cosmological constant and may include the large quantum corrections from matter, and the remaining part $\mathcal{L}^{(0)}_{\mathrm{gravity}}$ as $\mathcal{L}_{\mathrm{gravity}} = \mathcal{L}^{(0)}_{\mathrm{gravity}} - \Lambda$. By shifting the scalar field λ by a constant Λ as $\lambda \to \lambda - \Lambda$, the action (2) can be rewritten as

$$S' = \int d^4x \sqrt{-g} \left\{ \mathcal{L}^{(0)}_{\mathrm{gravity}} - \lambda + \partial_\mu \lambda \partial^\mu \varphi - \partial_\mu b \partial^\mu c \right\} - \Lambda \int d^4x \sqrt{-g} \nabla_\mu \partial^\mu \varphi + S_{\mathrm{matter}} . \tag{3}$$

Since the cosmological constant Λ appears as a coefficient of total derivative in the action (3), there is no contribution from the constant Λ to any dynamics in the model. Thus we have succeeded to tune the large quantum corrections from matter to vanish.

As a quantum field theory, the action (2) generates negative norm states [1], The negative norm states can be, however, removed by defining the physical states which are annihilated by the BRS (Becchi-Rouet-Stora) charge [42]. Please note that the action (2) is invariant under the following infinite number of BRS transformations,

$$\delta\lambda = \delta c = 0 , \quad \delta\varphi = \epsilon c , \quad \delta b = \epsilon \left(\lambda - \lambda_0 \right) . \tag{4}$$

Here ϵ is a Grassmann odd fermionic parameter and λ_0 should satisfy,

$$0 = \nabla^\mu \partial_\mu \lambda_0 , \tag{5}$$

which is just equation for λ: $\left(0 = \nabla^\mu \partial_\mu \lambda \right)$ obtained by the variation of the action (2) with respect to φ. (The existence of the BRS transformation where λ_0 satisfies Equation (5) was pointed out by R. Saitou.) In the BRS formalism, the physical states are BRS invariant and the unphysical states including the negative norm states are removed by the quartet mechanism proposed by Kugo and Ojima in the context of the gauge theory [43,44]. (We can assign the ghost number, which is conserved, 1 for c and -1 for b and ϵ. The four scalar fields λ, φ, b, and c are called a quartet [43,44].) Because $\lambda - \lambda_0$ in (4) is given by the BRS transformation of the anti-ghost b, however, the BRS invariance breaks down spontaneously when $\lambda - \lambda_0$ does not vanish and therefore it becomes difficult to remove the unphysical states and keep the unitarity of the model. In the real universe, we find $\lambda - \lambda_0 \neq 0$ in general because λ plays the role of the dynamical cosmological constant and therefore BRS symmetry is spontaneously broken in general. We should note, however, that in the real universe, one and only one λ satisfying the equation $0 = \nabla^\mu \partial_\mu \lambda$ is realized. Then if we choose λ_0 to be equal to the λ in the real universe, one and only one BRS symmetry in the infinite number of the BRS symmetries given in (4) remains [2]. The remaining BRS symmery is enough to eliminate the unphysical states. and the unitarity is guaranteed.

We can regard the Lagrangian density $\mathcal{L}_{\rm TP}$ in the action (2) as the Lagrangian density of a topological field theory proposed by Witten [45]. In a topological field theory, the Lagrangian density is given by the BRS transformation of some quantity. We may consider the model where only one scalar field φ is included but the Lagrangian density of the model vanishes identically and therefore the action is trivially invariant under any transformation of φ. Then the transformation of φ can be regarded as a gauge symmetry. We now fix the gauge symmetry by imposing the following gauge condition,

$$1 + \nabla_\mu \partial^\mu \varphi = 0 \,. \tag{6}$$

By following the procedure proposed by Kugo and Uehara [46], we can construct the gauge-fixed Lagrangian with the Fadeev-Popov (FP) ghost c and anti-ghost b by the BRS transformation (4) of $-b\left(1 + \nabla_\mu \partial^\mu \varphi\right)$ by choosing $\lambda_0 = 0$,

$$\delta\left(-b\left(1 + \nabla_\mu \partial^\mu \varphi\right)\right) = \epsilon\left(-\left(\lambda - \lambda_0\right)\left(1 + \nabla_\mu \partial^\mu \varphi\right) + b\nabla_\mu \partial^\mu c\right) = \epsilon\left(\mathcal{L} + \lambda_0 + (\text{total derivative terms})\right) \,. \tag{7}$$

Then we confirm that the Lagrangian density $\mathcal{L}_{\rm TP}$ in (2) is given by the BRS transformation of $-b\left(1 + \nabla_\mu \partial^\mu \varphi\right)$ and the model is surely topological. Because λ does not vanish in the real universe, the BRS invariance is broken. In this sense, the model (2) is not topological in the real universe, which could be the reason why this model gives physical contributions.

The above mechanism can be applied to the divergences in (1) or more general divergences as shown in [2]. When we consider the model in (1), the model in (2) is generalized as follows,

$$\begin{aligned}\mathcal{L} = \; & -\Lambda - \lambda_{(\Lambda)} + \left(\alpha + \lambda_{(\alpha)}\right) R + \left(\beta + \lambda_{(\beta)}\right) R^2 + \left(\gamma + \lambda_{(\gamma)}\right) R_{\mu\nu} R^{\mu\nu} + \left(\delta + \lambda_{(\delta)}\right) R_{\mu\nu\rho\sigma} R^{\mu\nu\rho\sigma} \\ & + \partial_\mu \lambda_{(\Lambda)} \partial^\mu \varphi_{(\Lambda)} - \partial_\mu b_{(\Lambda)} \partial^\mu c_{(\Lambda)} + \partial_\mu \lambda_{(\alpha)} \partial^\mu \varphi_{(\alpha)} - \partial_\mu b_{(\alpha)} \partial^\mu c_{(\alpha)} \\ & + \partial_\mu \lambda_{(\beta)} \partial^\mu \varphi_{(\beta)} - \partial_\mu b_{(\beta)} \partial^\mu c_{(\beta)} + \partial_\mu \lambda_{(\gamma)} \partial^\mu \varphi_{(\gamma)} - \partial_\mu b_{(\gamma)} \partial^\mu c_{(\gamma)} + \partial_\mu \lambda_{(\delta)} \partial^\mu \varphi_{(\delta)} - \partial_\mu b_{(\delta)} \partial^\mu c_{(\delta)} \,.\end{aligned} \tag{8}$$

We now shift the fields $\lambda_{(\Lambda)}$, $\lambda_{(\alpha)}$, $\lambda_{(\beta)}$, $\lambda_{(\gamma)}$, and $\lambda_{(\delta)}$ as follows,

$$\lambda_{(\Lambda)} \to \lambda_{(\lambda)} - \Lambda \,, \quad \lambda_{(\alpha)} \to \lambda_{(\alpha)} - \alpha \,, \quad \lambda_{(\beta)} \to \lambda_{(\beta)} - \beta \,, \quad \lambda_{(\gamma)} \to \lambda_{(\gamma)} - \gamma \,, \quad \lambda_{(\delta)} \to \lambda_{(\delta)} - \delta \,, \tag{9}$$

then the Lagrangian density (8) has the following form,

$$\begin{aligned}\mathcal{L} = & - \lambda_{(\Lambda)} + \lambda_{(\alpha)} R + \lambda_{(\beta)} R^2 + \lambda_{(\gamma)} R_{\mu\nu} R^{\mu\nu} + \lambda_{(\delta)} R_{\mu\nu\rho\sigma} R^{\mu\nu\rho\sigma} \\ & + \partial_\mu \lambda_{(\Lambda)} \partial^\mu \varphi_{(\Lambda)} - \partial_\mu b_{(\Lambda)} \partial^\mu c_{(\Lambda)} + \partial_\mu \lambda_{(\alpha)} \partial^\mu \varphi_{(\alpha)} - \partial_\mu b_{(\alpha)} \partial^\mu c_{(\alpha)} \\ & + \partial_\mu \lambda_{(\beta)} \partial^\mu \varphi_{(\beta)} - \partial_\mu b_{(\beta)} \partial^\mu c_{(\beta)} + \partial_\mu \lambda_{(\gamma)} \partial^\mu \varphi_{(\gamma)} \\ & - \partial_\mu b_{(\gamma)} \partial^\mu c_{(\gamma)} + \partial_\mu \lambda_{(\delta)} \partial^\mu \varphi_{(\delta)} - \partial_\mu b_{(\delta)} \partial^\mu c_{(\delta)} \\ & + (\text{total derivative terms}) \,.\end{aligned} \tag{10}$$

Except the total derivative terms, the obtained Lagrangian density (10) does not include the constants Λ, α, β, γ, and δ, which include the divergences from the quantum corrections. Therefore we can absorb the divergences into the redefinition of the scalar fields $\lambda_{(i)}$, $(i = \Lambda, \alpha, \beta, \gamma, \delta)$ and the divergences becomes irrelevant for the dynamics.

In the initial model (1), the parameters are coupling constants but in the new models, (8) or (10), the parameters are replaced by dynamical scalar fields. This is one of the reasons why the divergence coming from the quantum corrections can be absorbed into the redefinition of the scalar fields. Furthermore because the scalar fields are dynamical, as we will see later, the scalar fields play the role of the running coupling constant.

The Lagrangian density (10) is also invariant under the following BRS transformations

$$\delta\lambda_{(i)} = \delta c_{(i)} = 0\,, \quad \delta\varphi_{(i)} = \epsilon c\,, \quad \delta b_{(i)} = \epsilon\left(\lambda_{(i)} - \lambda_{(i)0}\right)\,, \quad (i = \Lambda, \alpha, \beta, \gamma, \delta)\,, \tag{11}$$

where $\lambda_{(i)0}$'s satisfy the equation,

$$0 = \nabla^\mu \partial_\mu \lambda_{(i)0}\,, \tag{12}$$

as in (5). The Lagrangian density (10) is also given by the BRS transformation (11) with $\lambda_{(i)0} = 0$,

$$\delta\left(\sum_{i=\Lambda,\alpha,\beta,\gamma,\delta}\left(-b_{(i)}\left(\mathcal{O}_{(i)} + \nabla_\mu\partial^\mu\varphi_{(i)}\right)\right)\right) = \epsilon\left(\mathcal{L} + (\text{total derivative terms})\right)\,. \tag{13}$$

As mentioned, due to the quantum correction from the graviton, an infinite number of divergences appear. Let $\mathcal{O}_i$ be possible gravitational operators; then a further generalization of the Lagrangian density (10) is given by

$$\mathcal{L} = \sum_i\left(\lambda_{(i)}\mathcal{O}_{(i)} + \partial_\mu\lambda_{(i)}\partial^\mu\varphi_{(i)} - \partial_\mu b_{(i)}\partial^\mu c_{(i)}\right)\,. \tag{14}$$

Then all the divergences are absorbed into the redefinition of λ_i. The Lagrangian density (14) is invariant under the BRS transformation and given by the the BRS transformation of some quantity and therefore the model can be regarded as a topological field theory, again.

Well-known higher derivative gravity can be renormalizable, but the ghosts appear and therefore the higher derivative gravity model is not unitary. Although our model may be renormalizable because the divergence does not appear, the problem of the unitarity remains because the Lagrangian density (14) includes the higher derivative terms. In the viewpoint of string theory, for example, we may expect that if we include the infinite number of higher derivative terms, the unitarity could be recovered but this is out of scope in this paper.

Usually the problem of the renormalizability in quantum field theory is the predictability. Even if we consider the quantum theory of gravity starting from the general relativity, if we include an infinite number of the counterterms, the theory becomes finite but due to the infinite number of the counter terms, the model loses predictability. In the model of (14), there could not be the problem of the divergence but because λ_i's become dynamical, we need infinite number of the initial conditions or somethings and therefore even in the model (14), the predictability could be lost. If the λ_i's have infrared fixed points, however, the predictability could be recovered. In the original model (14), however, we have not obtained non-trivial fixed points, which is one of the motivation why we considered the model in next section, where we try to construct the models with the fixed points.

3. Model Motivated by Renormalization Group

We assume that the space-time is given by the FRW (Friedmann-Robertson-Walker) universe with flat spacial part and a scale factor $a(t)$

$$ds^2 = -dt^2 + a(t)^2 \sum_{i=1}^{3} \left(dx^i\right)^2 . \tag{15}$$

Equation (12) tells that the scalar fields $\lambda_{(i)}$ depend on the scale factor $a(t)$ and then become time-dependent. Because $\lambda_{(i)}$ correspond to the coupling with the operator $\mathcal{O}_{(i)}$, Then the scale factor dependence of $\lambda_{(i)}$ is similar to the scale dependence of the renormalized coupling $\lambda_{(i)}$ Motivated by this observation, we consider the models with an infrared fixed point, which may correspond to the late time universe, and an ultraviolet fixed point, which may correspond to the early universe.

We now assume the following BRS transformations instead of (4),

$$\delta\lambda_{(i)} = \delta c_{(i)} = 0 , \quad \delta\varphi_{(i)} = \epsilon c , \quad \delta b_{(i)} = \epsilon\lambda_{(i)} , \tag{16}$$

and consider the Lagrangian density which is given by the BRS transformation (16) of some quantity,

$$\delta\left(\sum_{i=\Lambda,\alpha,\beta,\gamma,\delta} \left(b_{(i)} \left(\mathcal{O}_i + \nabla_\mu \partial^\mu \varphi_{(i)} + f_i\left(\lambda_{(j)}\right) \varphi_{(i)}\right)\right)\right) = \epsilon\left(\mathcal{L} + (\text{total derivative terms})\right) . \tag{17}$$

Here $\mathcal{O}_i$ are possible gravitational operators as in (14). and $f_i\left(\lambda_{(j)}\right)$'s are functions of $\lambda_{(j)}$. Then we obtain

$$\mathcal{L} = \sum_i \left(\lambda_{(i)}\mathcal{O}_{(i)} + \partial_\mu\lambda_{(i)}\partial^\mu\varphi_{(i)} + \lambda_{(i)}f_i\left(\lambda_{(j)}\right)\varphi_{(i)} - \partial_\mu b_{(i)}\partial^\mu c_{(i)} - f_i\left(\lambda_{(j)}\right) b_{(i)}c_{(i)}\right) . \tag{18}$$

The obtained model (18) is different from the original model (1), (8) or (10). We are using a different gauge fixing and the background solution is not BRS invariant. Then, in this background, the model (18) is not topological.

By the variation with respect to $\varphi_{(i)}$, we obtain the following equations,

$$-\nabla_\mu\nabla^\mu\lambda_{(i)} = \lambda_{(i)}f_i\left(\lambda_{(j)}\right) . \tag{19}$$

In the FRW space-time with flat spacial part (15), Equation (19) can be written as follows,

$$\frac{d^2\lambda_{(i)}}{dt^2} + 3H\frac{d\lambda_{(i)}}{dt} = \lambda_{(i)}f_i\left(\lambda_{(j)}\right) . \tag{20}$$

Here H is the Hubble rate defined by using the scale factor in Equation (15) as $H \equiv \dot{a}/a$. By defining τ by $a = \mathrm{e}^\tau$, we find

$$\frac{d}{dt} = H\frac{d}{d\tau} , \frac{d^2}{dt^2} = H^2\frac{d^2}{d\tau^2} + \dot{H}\frac{d}{d\tau} , \tag{21}$$

and therefore we obtain

$$H^2\left\{\frac{d^2\lambda_{(i)}}{d\tau^2} + \left(3 + \frac{\dot{H}}{H^2}\right)\frac{d\lambda_{(i)}}{d\tau}\right\} = \lambda_{(i)}f_i\left(\lambda_{(j)}\right) . \tag{22}$$

Because the change of a can be identified with the scale transformation, we may compare (20) with the renormalization group equation,

$$\frac{d\lambda_{(i)}}{d\tau} = g_i\left(\lambda_{(j)}\right) . \tag{23}$$

In cosmology, the Hubble rate H is usually used as energy scale but an analogy with the renormalization group in the quantum field theory, suggest the possibility to use the scale factor a as the energy. From

$$\frac{d^2\lambda_{(i)}}{d\tau^2} = \sum_k \frac{\partial g_i\left(\lambda_{(j)}\right)}{\partial\lambda_{(k)}} g_k\left(\lambda_{(j)}\right) , \tag{24}$$

we find

$$f_i\left(\lambda_{(j)}\right) = \frac{H^2}{\lambda_{(i)}}\left\{\sum_k \frac{\partial g_i\left(\lambda_{(j)}\right)}{\partial\lambda_{(k)}} g_k\left(\lambda_{(j)}\right) + \left(3+\frac{\dot{H}}{H^2}\right) g_i\left(\lambda_{(j)}\right)\right\} . \tag{25}$$

The interpretation of Equation (20) as a renormalization group equation requires $f_i\left(\lambda_{(j)}\right)$ to be time independent. Therefore the above identification (25) can have any meaning only if H is a constant at least near the fixed points, that is, the space-time should be, at least asymptotically, the de Sitter space-time. Later we consider the model where two fixed points are connected by the renormalization group. The two fixed points correspond to the ultraviolet (UV) and infrared (IR) limits. Between the two fixed points, H cannot be a constant because H takes different values in the two fixed points. As we will see later, the scale dependence of H can be absorbed into the redefinition of $f_i\left(\lambda_{(j)}\right)$ or $g_i\left(\lambda_{(j)}\right)$. We may assume that the renormalization equations (23) has a ultraviolet or infrared fixed point. If the universe asymptotically goes to the de Sitter universe in the early time or late time. Then if we choose $f_i\left(\lambda_{(j)}\right)$ by (25), the early universe corresponds to the ultraviolet (UV) fixed point and the late time universe to the infrared (IR) fixed point. Because the shift of τ corresponds to the change of the scale and τ is defined by using scale factor as $a = \mathrm{e}^\tau$, the UV limit corresponds to $\tau \to -\infty$ and therefore $a \to 0$ and the IR limit to $\tau \to \infty$, that is, $a \to \infty$. In the neighborhood of the UV fixed point λ^*_{UV}, we now assume,

$$\frac{dg_{(i)}\left(\lambda_{(j)}\right)}{d\lambda_{(i)}} > 0 . \tag{26}$$

Then $g_{(i)}\left(\lambda_{(j)}\right)$ can be expressed as,

$$g_{(i)}\left(\lambda_{(j)}\right) \approx k_{(i)\mathrm{UV}}(\lambda_{(j)})\left(\lambda_{(i)} - \lambda_{(i)\mathrm{UV}}\right) , \tag{27}$$

where $k_{(i)\mathrm{UV}}(\lambda_{(j)})$ is a function of $\lambda_{(j)}$ and $k_{(i)\mathrm{UV}}(\lambda_{(j)\mathrm{UV}}) > 0$. By using the approximation that $k_{(i)\mathrm{UV}}(\lambda_{(j)})$ could be regarded as a constant when $\lambda_{(i)} \approx \lambda_{(i)\mathrm{UV}}$, that is, $k_{(i)\mathrm{UV}}(\lambda_{(j)}) \approx k_{(i)\mathrm{UV}}(\lambda_{(j)\mathrm{UV}})$, the solution of (23) with (27) is given by

$$\lambda_{(i)} \approx \lambda_{(i)\mathrm{UV}} + \lambda_{(i)\mathrm{UV}0}\, a(t)^{k_{(i)\mathrm{UV}}(\lambda_{(j)\mathrm{UV}})} . \tag{28}$$

Here $\lambda_{(i)\mathrm{UV}0}$ is a constant of the integration. On the other hand, near the IR fixed point, we replace $k_{(i)\mathrm{UV}} \to -k_{(i)\mathrm{IR}}$ and $\lambda_{(i)\mathrm{UV}} \to \lambda_{(i)\mathrm{IR}}$ in (27) and (28) as follows,

$$g_{(i)}\left(\lambda_{(j)}\right) \approx -\,k_{(i)\mathrm{IR}}(\lambda_{(j)})\left(\lambda_{(i)} - \lambda_{(i)\mathrm{IR}}\right) . \tag{29}$$

Then we find

$$\lambda_{(i)} \approx \lambda_{(i)\mathrm{IR}} + \lambda_{(i)\mathrm{IR}0} \left(\frac{1}{a(t)} \right)^{k_{(i)\mathrm{IR}}\left(\left(\lambda_{(j)\mathrm{IR}}\right)\right)} . \quad (30)$$

Here $\lambda_{(i)\mathrm{IR}0}$ is a constant of the integration. When $a(t) \to 0$ in (28), and $a(t) \to \infty$ in (30), $\lambda_{(i)}$ goes to $\lambda_{(i)\mathrm{UV}}$ and $\lambda_{(i)\mathrm{IR}}$, respectively. Thus, as long as the above condition in the neighborhood of UV (IR) fixed point is satisfied, $\lambda_{(i)} = \lambda_{(i)\mathrm{UV}}$ ($\lambda_{(i)} = \lambda_{(i)\mathrm{IR}}$) is surely the UV (IR) fixed point. When $g_i \left(\lambda_{(j)}\right)$ behaves as (27) near the UV fixed point, Equation (25) tells that $f_i \left(\lambda_{(j)}\right)$ behaves as

$$f_i \left(\lambda_{(j)}\right) = \frac{H^2}{\lambda_{(i)\mathrm{UV}}} \left(k_{(i)\mathrm{UV}}(\lambda_{(j)\mathrm{UV}}) + 3\right) k_{(i)\mathrm{UV}}(\lambda_{(j)\mathrm{UV}}) \left(\lambda_{(i)} - \lambda_{(i)\mathrm{UV}}\right) + \mathcal{O}\left(\left(\lambda_{(i)} - \lambda_{(i)\mathrm{UV}}\right)^2\right) . \quad (31)$$

On the other hand, when $g_i \left(\lambda_{(j)}\right)$ behaves as (29) near the IR fixed point, $f_i \left(\lambda_{(j)}\right)$ behaves as

$$f_i \left(\lambda_{(j)}\right) = \frac{H^2}{\lambda_{(i)\mathrm{IR}}} \left(k_{(i)\mathrm{IR}}(\lambda_{(j)\mathrm{IR}}) - 3\right) k_{(i)\mathrm{IR}}(\lambda_{(j)\mathrm{IR}}) \left(\lambda_{(i)} - \lambda_{(i)\mathrm{IR}}\right) + \mathcal{O}\left(\left(\lambda_{(i)} - \lambda_{(i)\mathrm{IR}}\right)^2\right) . \quad (32)$$

When we consider the Einstein gravity with cosmological constant, the action is given by,

$$S = \int d^4x\sqrt{-g}\left[\lambda_{(\alpha)}R - \lambda_{(\Lambda)} + \sum_{i=\Lambda,\alpha}\left(\partial_\mu\lambda_{(i)}\partial^\mu\varphi_{(i)} - \partial_\mu b_{(i)}\partial^\mu c_{(i)} + \lambda_{(i)}f_{(i)}(\lambda_{(j)})\varphi_{(i)}\right)\right] + S_{\mathrm{matter}} . \quad (33)$$

Here S_{matter} is the action of matters. Varying the action (33) with respect to the metric $g^{\mu\nu}$, we obtain the following equation,

$$\lambda_{(\alpha)}G_{\mu\nu} + \frac{1}{2}\lambda_{(\Lambda)}g_{\mu\nu} - \left(\nabla_\mu\nabla_\nu - \nabla^2\right)\lambda_{(\alpha)} + \sum_{i=\Lambda,\alpha}\left[\frac{1}{2}g_{\mu\nu}\left(\partial_\rho\lambda_{(i)}\partial^\rho\varphi_{(i)} + \lambda_{(i)}f_{(i)}(\lambda_{(j)})\varphi_{(i)}\right) + \partial_\mu\lambda_{(i)}\partial_\nu\varphi_{(i)}\right] = T_{\mu\nu} . \quad (34)$$

We should note that if the FP ghost and anti-ghost has any classical value, which may correspond to the vacuum expectation value, superselection rule or ghost number conservation is violated and therefore we put them vanish. In (34), $G_{\mu\nu}$ is the Einstein tensor and $T_{\mu\nu}$ is the energy momentum tensor of matters. In the spatially flat FRW background if we assume that $\lambda_{(i)}$ and $\varphi_{(i)}$ depend only on the cosmological time t, the $(0,0)$-component of Equation (34) has the following form,

$$H^2 = \frac{1}{6\lambda_{(\alpha)}}\left\{\lambda_{(\Lambda)} - 3H\dot{\lambda}_{(\alpha)} - \sum_{i=\Lambda,\alpha}\left(\dot{\lambda}_{(i)}\dot{\varphi}_{(i)} - \lambda_{(i)}f_i(\lambda_j)\varphi_{(i)}\right)\right\} \quad (35)$$

In the the neighborhood of the UV fixed point, substituting (25) and (28) into the above expression, we obtain,

$$\begin{aligned} H^2 \approx & \frac{1}{6\lambda_{(\alpha)}}\Bigg(\lambda_{(\Lambda)\mathrm{UV}} + \lambda_{(\Lambda)\mathrm{UV}0}a(t)^{k_{(\Lambda)\mathrm{UV}}(\lambda_{(j)\mathrm{UV}})} - 3H\dot{\lambda}_{(\alpha)} \\ & + \sum_{i=\Lambda,\alpha} k_{(i)}(\lambda_{(j)})Ha(t)^{k_{(i)}(\lambda_{(j)})}\lambda_{(i)\mathrm{UV}0}\dot{\varphi}_{(i)}\Bigg) \\ & + \frac{H^2}{6\lambda_{(\alpha)}}\sum_{i=\Lambda,\alpha}\left\{\sum_k \frac{\partial g_i\left(\lambda_{(j)}\right)}{\partial\lambda_{(k)}}g_i\left(\lambda_{(j)}\right) + \left(3 + \frac{\dot{H}}{H^2}\right)g_i\left(\lambda_{(j)}\right)\right\}\varphi_{(i)} , \end{aligned} \quad (36)$$

Then in the UV limit

$$a(t) \to 0 , \quad g_{(i)} \to 0 , \quad \lambda_{(i)} \to \lambda_{(i)\mathrm{UV}} , \quad (37)$$

we obtain the de-Sitter solution, where H is a constant,

$$H = H_{\mathrm{UV}} = \sqrt{\frac{\lambda_{(\Lambda)\mathrm{UV}}}{6\lambda_{(\alpha)\mathrm{UV}}}} = \mathrm{const.} \tag{38}$$

On the other hand, near the IR fixed point, instead of (36), we obtain

$$\begin{aligned} H^2 \approx & \frac{1}{6\lambda_{(\alpha)}} \left(\lambda_{(\Lambda)\mathrm{IR}} + \lambda_{(\Lambda)\mathrm{IR}0} a(t)^{-k_{(\Lambda)\mathrm{IR}}\left(\lambda_{(j)\mathrm{IR}}\right)} \right. \\ & \left. - 3H\dot{\lambda}_{(\alpha)} - \sum_{i=\Lambda,\alpha} k_{(i)} \left(\lambda_{(j)}\right) H a(t)^{-k_{(i)}\left(\lambda_{(j)}\right)} \lambda_{(i)\mathrm{IR}0} \dot{\varphi}_{(i)} \right) \\ & + \frac{H^2}{6\lambda_{(\alpha)}} \sum_{i=\Lambda,\alpha} \left\{ \sum_k \frac{\partial g_i\left(\lambda_{(j)}\right)}{\partial \lambda_{(k)}} g_i\left(\lambda_{(j)}\right) + \left(3 + \frac{\dot{H}}{H^2}\right) g_i\left(\lambda_{(j)}\right) \right\} \varphi_{(i)} , \end{aligned} \tag{39}$$

Then in the IR limit

$$a(t) \to \infty , \quad g_{(i)} \to 0 , \quad \lambda_{(i)} \to \lambda_{(i)\mathrm{IR}} , \tag{40}$$

we obtain the de-Sitter solution, where

$$H = H_{\mathrm{IR}} = \sqrt{\frac{\lambda_{(\Lambda)\mathrm{IR}}}{6\lambda_{(\alpha)\mathrm{IR}}}} = \mathrm{const.} \tag{41}$$

We now try to construct a model, where the IR fixed point is connected with the UV fixed point by the renormalization flow. As an example, we may consider the following model

$$f_{(i)}\left(\lambda_{(j)}\right) = C_{(i)}\left(\lambda_{(j)}\right)\left(\lambda_{(i)} - \lambda_{(i)\mathrm{UV}}\right)\left(\lambda_{(i)} - \lambda_{(i)\mathrm{IR}}\right) , \tag{42}$$

Here $C_{(i)}\left(\lambda_{(j)}\right)$ is a positive function. By using (38) and comparing (31) and (42), we find

$$\frac{\lambda_{(\Lambda)\mathrm{UV}}}{6\lambda_{(\alpha)\mathrm{UV}}\lambda_{(i)\mathrm{UV}}} \left(k_{(i)\mathrm{UV}}(\lambda_{(j)\mathrm{UV}}) + 3\right) k_{(i)\mathrm{UV}}(\lambda_{(j)\mathrm{UV}}) = C_{(i)}\left(\lambda_{(j)\mathrm{UV}}\right)\left(\lambda_{(i)\mathrm{UV}} - \lambda_{(i)\mathrm{IR}}\right) , \tag{43}$$

which can be solved with respect to $k_{(i)\mathrm{UV}} > 0$, as follows,

$$k_{(i)\mathrm{UV}} = -\frac{3}{2} + \frac{1}{2}\sqrt{9 + \frac{24\lambda_{(\alpha)\mathrm{UV}}\lambda_{(i)\mathrm{UV}}C_{(i)}\left(\lambda_{(j)\mathrm{UV}}\right)}{\lambda_{(\Lambda)\mathrm{UV}}}\left(\lambda_{(i)\mathrm{UV}} - \lambda_{(i)\mathrm{IR}}\right)} . \tag{44}$$

On the other hand, by using (41) and comparing (32) and (42), we find

$$\frac{\lambda_{(\Lambda)\mathrm{IR}}}{6\lambda_{(\alpha)\mathrm{IR}}\lambda_{(i)\mathrm{IR}}} \left(k_{(i)\mathrm{IR}}(\lambda_{(j)\mathrm{IR}}) - 3\right) k_{(i)\mathrm{IR}}(\lambda_{(j)\mathrm{IR}}) = -C_{(i)}\left(\lambda_{(j)\mathrm{IR}}\right)\left(\lambda_{(i)\mathrm{UV}} - \lambda_{(i)\mathrm{IR}}\right) , \tag{45}$$

which can be solved with respect to $k_{(i)\mathrm{IR}} > 0$, as follows,

$$k_{(i)\mathrm{IR}} = \frac{3}{2} \pm \frac{1}{2}\sqrt{9 - \frac{24\lambda_{(\alpha)\mathrm{IR}}\lambda_{(i)\mathrm{IR}}C_{(i)}\left(\lambda_{(j)\mathrm{IR}}\right)}{\lambda_{(\Lambda)\mathrm{IR}}}\left(\lambda_{(i)\mathrm{UV}} - \lambda_{(i)\mathrm{IR}}\right)} , \tag{46}$$

which requires

$$9 \geq \frac{24\lambda_{(\alpha)\mathrm{IR}}\lambda_{(i)\mathrm{IR}}C_{(i)}\left(\lambda_{(j)\mathrm{IR}}\right)}{\lambda_{(\Lambda)\mathrm{IR}}}\left(\lambda_{(i)\mathrm{UV}} - \lambda_{(i)\mathrm{IR}}\right) . \tag{47}$$

Therefore, as long as we choose $C_{(i)}\left(\lambda_{(j)}\right)$ to satisfy the constraint (47), the model (42) surely connect the IR fixed point with the UV fixed point by the renormalization flow.

4. Summary

Motivated with the model in [1,2,39], we have proposed models of topological field theory including gravity. In those models, the coupling constants are replaced by scalar fields, which run as in the renormalization group following the scale of the universe. As an example, we have constructed a model which connects the inflation in the early universe and the accelerating expansion of the present universe or late time. The de Sitter space-times corresponding to the inflation and the late time accelerating expansion appear as the ultraviolet and infrared fixed points, respectively. There remains, however, several problems, which violate the good properties in the original models in [1,2,39].

1. Because the shift symmetry as in (8) is lost, the models in this paper do not solve the problem of the large quantum correction.
2. Because $\lambda_{(i)}$ in (11) has a non-trivial value, the BRS symmetry in (11) should be broken.
3. Although the original model in [1,2,39] has no parameters, the models proposed in this paper should have several parameters.

Therefore it could be interesting if we constructed any model which solves some of the above problems by keeping the structure similar to the renormalization group. Some ideas that to try to solve these problems are given in Appendix A.

In summary, we have not succeeded to solved all the problems but we may have shown that there might be possibilities to solve them. In this paper, we have considered models where the scalar fields $\lambda_{(i)}$'s play the role of the running coupling constants as in the renormalization group. We have treated the scalar fields classically although the renormalization group, of course, comes from the quantum corrections. Therefore the models proposed in this paper might be realized by an effective field theory connecting the low energy region with the high energy regions. If the models are really given as effective theories, the models need not always to satisfy all the unitarity conditions.

We have anyway succeeded in constructing such models and we have shown that we can construct the model with fixed point. The models have, however, arbitrariness, which could be removed by the constraints from the observations and/or the consistencies of the models. We like to pursue the problem in the future work.

Author Contributions: S.N. gave a basic idea of this work and T.M. elaborated the idea and proceed this work with S.N.

Funding: This research was funded by (in part) by MEXT KAKENHI Grant-in-Aid for Scientific Research on Innovative Areas "Cosmic Acceleration" No. 15H05890 (S.N.) and the JSPS Grant-in-Aid for Scientific Research (C) No. 18K03615 (S.N.).

Acknowledgments: This work is supported (in part) by MEXT KAKENHI Grant-in-Aid for Scientific Research on Innovative Areas "Cosmic Acceleration" No. 15H05890 (S.N.) and the JSPS Grant-in-Aid for Scientific Research (C) No. 18K03615 (S.N.).

Conflicts of Interest: The authors declare no conflict of interest.

Appendix A. Some Propositions to Improve the Models

In this appendix, we consider models, which may solve the problem given in summary section. We believe the models in this section may give some clues to solve the problems.

An example of the model, which may solve the second problem, could be

$$\begin{aligned} &\delta\left(\sum_{i=\Lambda,\alpha,\beta,\gamma,\delta} \left(b_{(i)}\left(\mathcal{O}_i + \nabla_\mu \partial^\mu \varphi_{(i)} \pm k_{(0)i}\varphi_{(i)}\right)\right)\right) = \epsilon\left(\mathcal{L} + (\text{total derivative terms})\right), \\ &\mathcal{L} = \sum_i \left(\lambda_{(i)}\mathcal{O}_{(i)} + \partial_\mu\lambda_{(i)}\partial^\mu\varphi_{(i)} \pm k_{(0)i}\lambda_{(i)}\varphi_{(i)} - \partial_\mu b_{(i)}\partial^\mu c_{(i)} \mp k_{(0)i}b_{(i)}c_{(i)}\right). \end{aligned} \tag{A1}$$

Then $\lambda_{(i)} = 0$ is a ultraviolet (infrared) fixed point for $+k_{(0)i}$ $\left(-k_{(0)i}\right)$. By the variation of $\varphi_{(i)}$, we obtain

$$0 = -\nabla^{\mu}\partial_{\mu}\lambda_{(i)} \pm k_{(0)i}\lambda_{(i)} \,. \tag{A2}$$

Let a solution of (A2) be $\lambda_{(i)} = \lambda^{\rm cl}_{(i)}$. Then the action given by the Lagrangian density $\mathcal{L}$ in (A1) is invariant under the following BRS transformation instead of (16),

$$\delta\lambda_{(i)} = \delta c_{(i)} = 0\,, \quad \delta\varphi_{(i)} = \epsilon c\,, \quad \delta b_{(i)} = \epsilon\left(\lambda_{(i)} - \lambda^{\rm cl}_{(i)}\right)\,, \tag{A3}$$

Then because one of the solutions in $\lambda^{\rm cl}_{(i)}$ is realized in the real world, the BRS symmetry corresponding to the solution $\lambda^{\rm cl}_{(i)}$ is not broken and the unitarity can be preserved.

Another kind of the solution may be given by the following kind of model,

$$S = \int d^4x\sqrt{-g}\left\{\frac{R}{2\kappa^2} - \lambda + \mathcal{L}\left(g_{\mu\nu}, X, Y_{\mu\nu}\right)\right\}\,, \quad X \equiv -\partial_\mu\lambda\partial^\mu\lambda\,, \quad Y_{\mu\nu} \equiv \nabla_\mu\partial_\nu\lambda\,. \tag{A4}$$

Here $\mathcal{L}$ could be the Lagrangian density of the k-essence or the Galileon model. Because $\mathcal{L}$ is invariant under the shift of λ by a constant λ_0: $\lambda \to \lambda + \lambda_0$, the vacuum energy can be absorbed into the definition of λ and the first problem could be solved. Then if we choose $\mathcal{L}$ to give a unitary model, we need not to consider the second problem. When we consider $\mathcal{L}$ of the k-essence, $\mathcal{L} = \mathcal{L}(X)$, for simplicity, by the variation of λ, we obtain

$$0 = 1 - 2\nabla^{\mu}\left(\partial_\mu\lambda\mathcal{L}'\left(X\right)\right)\,. \tag{A5}$$

In the FRW universe with the flat spacial part (15), Equation (A5) has the following form,

$$0 = 1 + 2a(t)^{-3}\frac{d}{dt}\left(a(t)^3\dot{\lambda}\mathcal{L}'\left(\dot{\lambda}^2\right)\right)\,, \tag{A6}$$

which tells that the fixed point, where $\dot{\lambda} = 0$ is not the solution.

References

1. Nojiri, S. Some solutions for one of the cosmological constant problems. *Mod. Phys. Lett. A* **2016**, *31*, 1650213. [CrossRef]
2. Mori, T.; Nitta, D.; Nojiri, S. BRS structure of Simple Model of Cosmological Constant and Cosmology. *Phys. Rev. D* **2017**, *96*, 024009. [CrossRef]
3. Burgess, C.P. *The Cosmological Constant Problem: Why It's Hard to Get Dark Energy from Micro-Physics*; Oxford University Press: Oxford, UK, 2013; pp. 150–188. [CrossRef]
4. Anderson, J.L.; Finkelstein, D. Cosmological constant and fundamental length. *Am. J. Phys.* **1971**, *39*, 901. [CrossRef]
5. Buchmuller, W.; Dragon, N. Einstein Gravity From Restricted Coordinate Invariance. *Phys. Lett. B* **1988**, *207*, 292. [CrossRef]
6. Buchmuller, W.; Dragon, N. Gauge Fixing and the Cosmological Constant. *Phys. Lett. B* **1989**, *223*, 313. [CrossRef]
7. Henneaux, M.; Teitelboim, C. The Cosmological Constant and General Covariance. *Phys. Lett. B* **1989**, *222*, 195. [CrossRef]
8. Unruh, W.G. A Unimodular Theory of Canonical Quantum Gravity. *Phys. Rev. D* **1989**, *40*, 1048. [CrossRef]
9. Ng, Y.J.; van Dam, H. Unimodular Theory of Gravity and the Cosmological Constant. *J. Math. Phys.* **1991**, *32*, 1337. [CrossRef]
10. Finkelstein, D.R.; Galiautdinov, A.A.; Baugh, J.E. Unimodular relativity and cosmological constant. *J. Math. Phys.* **2001**, *42*, 340. [CrossRef]
11. Alvarez, E. Can one tell Einstein's unimodular theory from Einstein's general relativity? *JHEP* **2005**, *503*, 2. [CrossRef]

12. Alvarez, E.; Blas, D.; Garriga, J.; Verdaguer, E. Transverse Fierz-Pauli symmetry. *Nucl. Phys. B* **2006**, *756*, 148. [CrossRef]
13. Abbassi, A.H.; Abbassi, A.M. Density-metric unimodular gravity: Vacuum spherical symmetry. *Class. Quant. Grav.* **2008**, *25*, 175018. [CrossRef]
14. Ellis, G.F.R.; van Elst, H.; Murugan, J.; Uzan, J.P. On the Trace-Free Einstein Equations as a Viable Alternative to General Relativity. *Class. Quant. Grav.* **2011**, *28*, 225007. [CrossRef]
15. Jain, P. A flat space-time model of the Universe. *Mod. Phys. Lett. A* **2012**, *27*, 1250201. [CrossRef]
16. Singh, N.K. Unimodular Constraint on global scale Invariance. *Mod. Phys. Lett. A* **2013**, *28*, 1350130. [CrossRef]
17. Kluson, J. Canonical Analysis of Unimodular Gravity. *Phys. Rev. D* **2015**, *91*, 064058. [CrossRef]
18. Padilla, A.; Saltas, I.D. A note on classical and quantum unimodular gravity. *Eur. Phys. J. C* **2015**, *75*, 561. [CrossRef]
19. Barceló, C.; Carballo-Rubio, R.; Garay, L.J. Unimodular gravity and general relativity from graviton self-interactions. *Phys. Rev. D* **2014**, *89*, 124019. [CrossRef]
20. Barceló, C.; Carballo-Rubio, R.; Garay, L.J. Absence of cosmological constant problem in special relativistic field theory of gravity. *arXiv*, **2014**, arXiv:1406.7713.
21. Burger, D.J.; Ellis, G.F.R.; Murugan, J.; Weltman, A. The KLT relations in unimodular gravity. *arXiv* **2015**, arXiv:1511.08517.
22. Álvarez, E.; González-Martín, S.; Herrero-Valea, M.; Martín, C.P. Quantum Corrections to Unimodular Gravity. *JHEP* **2015**, *1508*, 78. [CrossRef]
23. Jain, P.; Jaiswal, A.; Karmakar, P.; Kashyap, G.; Singh, N.K. Cosmological implications of unimodular gravity. *JCAP* **2012**, *1211*, 3. [CrossRef]
24. Jain, P.; Karmakar, P.; Mitra, S.; Panda, S.; Singh, N.K. Testing Unimodular Gravity. *JCAP* **2012**, *1205*, 20. [CrossRef]
25. Cho, I.; Singh, N.K. Unimodular Theory of Gravity and Inflation. *Class. Quant. Grav.* **2015**, *32*, 135020. [CrossRef]
26. Basak, A.; Fabre, O.; Shankaranarayanan, S. Cosmological perturbation of Unimodular Gravity and General Relativity are identical. *arXiv* **2015**, arXiv:1511.01805.
27. Gao, C.; Brandenberger, R.H.; Cai, Y.; Chen, P. Cosmological Perturbations in Unimodular Gravity. *JCAP* **2014**, *1409*, 21. [CrossRef]
28. Eichhorn, A. The Renormalization Group flow of unimodular f(R) gravity. *JHEP* **2015**, *1504*, 96. [CrossRef]
29. Saltas, I.D. UV structure of quantum unimodular gravity. *Phys. Rev. D* **2014**, *90*. [CrossRef]
30. Nojiri, S.; Odintsov, S.D.; Oikonomou, V.K. Unimodular $F(R)$ Gravity. *arXiv* **2015**, arXiv:1512.07223.
31. Kaloper, N.; Padilla, A. Sequestering the Standard Model Vacuum Energy. *Phys. Rev. Lett.* **2014**, *112*, 091304. [CrossRef] [PubMed]
32. Kaloper, N.; Padilla, A. Vacuum Energy Sequestering: The Framework and Its Cosmological Consequences. *Phys. Rev. D* **2014**, *90*, 084023. [CrossRef]
33. Kaloper, N.; Padilla, A.; Stefanyszyn, D.; Zahariade, G. A Manifestly Local Theory of Vacuum Energy Sequestering. *arXiv* **2015**, arXiv:1505.01492.
34. Batra, P.; Hinterbichler, K.; Hui, L.; Kabat, D.N. Pseudo-redundant vacuum energy. *Phys. Rev. D* **2008**, *78*, 043507. [CrossRef]
35. Shaw, D.J.; Barrow, J.D. A Testable Solution of the Cosmological Constant and Coincidence Problems. *Phys. Rev. D* **2011**, *83*, 043518. [CrossRef]
36. Barrow, J.D.; Shaw, D.J. A New Solution of The Cosmological Constant Problems. *Phys. Rev. Lett.* **2011**, *106*, 101302. [CrossRef] [PubMed]
37. Carballo-Rubio, R. Longitudinal diffeomorphisms obstruct the protection of vacuum energy. *Phys. Rev. D* **2015**, *91*, 124071. [CrossRef]
38. Tsukamoto, T.; Katsuragawa, T.; Nojiri, S. Sequestering mechanism in scalar-tensor gravity. *Phys. Rev. D* **2017**, *96*, 124003. [CrossRef]
39. Nojiri, S. Cosmological constant and renormalization of gravity. *Galaxies* **2018**, *6*, 24. [CrossRef]
40. Shlaer, B. Solution to the problem of time. *arXiv* **2014**, arXiv:1411.8006.
41. Saitou, R.; Gong, Y. de Sitter spacetime with a Becchi-Rouet-Stora quartet. *Int. J. Mod. Phys. D* **2017**, *26*, 1750132. [CrossRef]

42. Becchi, C.; Rouet, A.; Stora, R. Renormalization of Gauge Theories. *Ann. Phys.* **1976**, *98*, 287. [CrossRef]
43. Kugo, T.; Ojima, I. Manifestly Covariant Canonical Formulation of Yang-Mills Field Theories: Physical State Subsidiary Conditions and Physical S Matrix Unitarity. *Phys. Lett. B* **1978**, *73*, 459. [CrossRef]
44. Kugo, T.; Ojima, I. Local Covariant Operator Formalism of Nonabelian Gauge Theories and Quark Confinement Problem. *Prog. Theor. Phys. Suppl.* **1979**, *66*, 1. [CrossRef]
45. Witten, E. Topological Quantum Field Theory. *Commun. Math. Phys.* **1988**, *117*, 353. [CrossRef]
46. Kugo, T.; Uehara, S. General Procedure of Gauge Fixing Based on BRS Invariance Principle. *Nucl. Phys. B* **1982**, *197*, 378. [CrossRef]

symmetry

MDPI

Article

Alternative Uses for Quantum Systems and Devices

Orchidea Maria Lecian

DICEA-Department of Civil, Constructional and Environmental Engineering, Faculty of Civil and Industrial Engineering, Sapienza University of Rome, Via Eudossiana, 18-00184 Rome, Italy; orchideamaria.lecian@uniroma1.it

Received: 31 January 2019; Accepted: 7 March 2019; Published: 2 April 2019

Abstract: Quantum optical systems and devices were analyzed to verify theories both predicting new particles on flat spacetime, and for the verification of Planck-scale physics for cosmological investigation.

Keywords: quantum optical systems; astronomical and space-research instrumentation; instruments, apparatus, and components common to several branches of physics and astronomy; normal galaxies, extragalactic objects and systems; field theory; comparative planetology; properties of specific particles; quantum optics; fundamental astronomy

---

1. Introduction

The origin of galaxies can be testified [1] through semianalytic calculations for the growth of supermassive black holes to discriminate between seeding models and pertinent accretion modes according to the abundances and maximum masses of the formers.

Information about interstellar gas in high-redshift galaxies [2] is relevant in providing the tools for the numerical simulation of pressure and energy and verifying the behavior of their celestial bodies. The evolution of galactic structures and their compact objects in the vicinity of galaxy centers, as well as the refinements of General Relativity (GR)tests [3], can be achieved by accurately measuring the frequencies of the periapsis and Lense–Thirring precessions. By means of a single, linear interferometer [4], the evolution of a binary system can be followed as described by their leading-order quadrupole gravitational radiation by calculating the source rate (of gravitational radiation) and the observation range. For coalescing neutron-star binary systems, optimization of recycling frequency values can be calculated in the cases of the maximization of the detections rate or of the measurement precision. The signal-to-noise ratio is improved in longer-time observations. Signal and noise superpositions constitute probability volumes iff the likelihood ratio (i.e., the noise-to-signal ratio) has a single extremum that either is close to the physical state of the signal or its series converges close to the latter. Correlation coefficients and standard deviation should be accurate enough to compensate for the neglect of the first Post-Newtonian (PN) order in source modelization. The properties of galaxies and celestial bodies can also be accounted for by hypothesizing the existence of new particles and/or novel features of existing particles. Such analyses allow one to exploit the specificities of particles and of their aggregation states to further modelize the known characteristics of such bodies.

The studiedoptical quantum systems that rise by the quantum properties of particles, as well as those caused by the quantum-gravitational nature of spacetime and their interaction, allow one to also exploit such quantum properties for phenomena taking place at scales larger than the Planckian. Lab experiments aimed at verifying these descriptions, as well as observational surveys for astrophysical phenomena, are therefore affected by the quantum description of particles and spacetime in the resulting optical systems [5–8]. The role of possible modifications of the dispersion relations (such as those porposed in [9]) are therefore to be restricted to the analyzed length scales.

The quantum description of spacetime at scales larger than Planck length can give rise to phenomena ascribed to the semiclassical nature of the spacetime as well as its quantum properties, which affect the description of physical phenomena above the Planckian scale. The evidence for new physical phenomena at any scale larger than the Planckian can be described after the phenomenological description of effective quantum properties of spacetime in the solution to the gravitational-field equations.

The macroscopic appearance of the present universe as well as its matter content can therefore be completed with the quantum corrections to the associated optical systems, to which quantum-gravitational resulting corrections can also be applied.

For rapidly rotating stars, high angular resolution observations at near-IR regions and temperature-difference detection between the poles and the equator are possible through these techniques [10]. It is therefore possible to verify rotational instabilities by analysing the nature of emissions and their asymmetry, and the mass-loss rate, in order to measure the critical speed of the stellar wind(s)—by the brightness of the blobs, their size, and their morphology—at the poles (that characterizes and accompany these events).

A laser-based system was designed [11] that, with the calibration of interferometers, is able to compensate for dispersive technologies. The resulting device is suited for exoplanet detection by stellar spectroscopy and velocimetry.

Active Galatic Nuclei (AGN's) milliarcsecond-emission regions were shown to provide data for measuring the mass function of quasar black holes [12]. This is achieved by combining the phases observed in two identical telescopes to control the anisotropy of the UV emission (i.e., from interstellar medium). At these frequencies, an interferometer-based baseline can individuate the shape of the line-emitting region. The ratio of the total emission flux to that measured through a single instrument sets a lower bound to the size of the region. The continuum spectrum does differently correspond to the star background in the considered galaxy and/or continuum light scattered from the nucleus.

By combining three telescopes, it is possible to calculate the phase off-set for a baseline interferometer [13]. This techniques allows with the precision to resolve a pointlike source position of a celestial body such as an Earth-like planet within the Solar System.

The devices of very-long-baseline interferometry are equipped [14] to routine disk-based recording systems for Gbps data rates by both cm and mm networks. For this, receiver systems and coherence time improve both baseline and image/noise sensitivity for fiber-based communication networks and real-time networks in order to access the radio detection of microJy sky pixels to analyze needed sources, such that phase-referencing preparation should not be especially required before each self-calibration of the chosen target.

The Sagnac-Fabry-Pérot interferometer [15] is a device eligible both for the detection of gravitational waves and as an instrument for particle physics experiments.

The paper is organized as follows.

In Section 2, several models predicting new features for particles, fundamental-symmetry violations, and new particles are reviewed, and alternative verification experiments are proposed, for which at least some features of these models can be tested.

In Section 3, cosmological theories for Solar System planet and exoplanet formation are revised; the results of astrophysical experiments that are useful for the verification of such theories are outlined.

Concluding remarks about perspectives for the continuation of investigations are developed.

2. New-Particle Detection

The Berry geometrical phase is due to a nonrelativistic system, whose Hamiltonian $\mathcal{H}_\lambda \equiv \mathcal{H}_\lambda(\lambda_1, \lambda_2, ..., \lambda_n)$ depends continuously on a family of slowly changing parameters λ_i, $i = 1, ..., n$ [16]. In £ space dimensions, it determines a broken $O(3)$ symmetry, as the Hamiltonian $\mathcal{H}_\lambda$ does not commute with generators of an $O(3)$ symmetry [17–19].

Without following here the analyses of References [20,21], we report that, in curved spacetime, the wavefunction can be factorized as

$$\Phi(x) = e^{-i\Phi_G(x)}\varphi(x), \tag{1}$$

with $\Phi_G(x)$ being the gravitational part of the wavefunction,

$$\Phi_G(x) = -\frac{1}{4}\int_P^x dz^\sigma(\gamma_{\alpha\sigma,\beta}(z) - \gamma_{\beta\sigma,\alpha}(z))[(x^\alpha - z\alpha)k^\beta - (x^\beta - z\beta)k^\alpha] + \frac{1}{2}\int_P^x dz^\sigma\gamma_{\alpha\sigma}k^\sigma, \tag{2}$$

with plane-wave momentum k^α satisfying $k^\alpha k_\alpha = m_k^2$. This way, unless strong fine tunings are imposed on $\varphi(x)$ in Equation (1), there results broken $O(3)$ 3-dimensional space symmetry.

2.1. Weak Gravitational Fields

In the case of broken $O(3)$ space symmetry, the velocity-distribution function for velocities characterizing the wavepackets is constructed originating quantum states describing asymptotical $(-\infty)$ KLSZ states. In the case of a weak gravitational field, the velocity distribution for particles [22] in the laboratory frame departs from that calculated on Minkowski flat spacetime as

$$f_v = \frac{1}{8\pi^3 det[(\varsigma_v)^2]} exp\left[-\frac{1}{2}(\vec{v} - \vec{v}_\odot)^T \varsigma_v^{-2}(\vec{v} - \vec{v}_\odot)\right] \tag{3}$$

with $\varsigma_v \equiv diag[\varsigma_x, \varsigma_y, \varsigma_z]$ as the velocity dispersion tensor, which encodes the solution to the Einstein field equations (EFE')speculiarity through its metric tensor components. This situation gives rise to a velocity anisotropy:

$$\beta(\mathbf{r}) \equiv 1 - \frac{\varsigma_y^2 + \varsigma_z^2}{2\varsigma_x^2} \tag{4}$$

which can be detected by a ionization chamber able to recover the track parameters $(X, Y, Z, \theta, \phi, S)$.

For the detection of dark matter, given a weakly-interacting massive particle (WIMP)χ of mass m_χ, from parameter space (m_χ, ς_i) it is possible to evaluate the WIMP-nucleon cross section $\sigma_{W-nucleon}$.

In Reference [22], a model-independent cross section of dark matter on protons $\varsigma_{i,p}$ is found as $\varsigma_i \simeq 10^{-3}pb$ for scintillators targeted of CsI(Tl), ^{19} F [23,24], respectively.

F targets were studied in Reference [25] for Earth-based experiments analyzing atmospheric-origin particles.

Detectors for anisotropic ultraenergetic cosmic rays of galactic origin are schematized in References [26–28]; the dark-matter-induced symmetry violations are examined in [29].

2.2. Fractional Charge

An instrument aimed at detecting fractional-charge particles is the rotor electrometer. It was designed as a Faraday container with an arbitrary high-impedance amplifier, endowed with copper pads, for which different charges reach the container walls at different velocities, such that the time of flight can be calculated, i.e., after a tuning the impedance suited for the charge to be detected [30,31].

For fractional quantum numbers, see also Reference [32].

2.3. Further Particles

Differently, the findings of References [33,34] can be compared. In Reference [34], the electric dipole moment of the electron and that of the neutron are evaluated as a constraint to CP violation arising from a broken $SU(3)$-symmetry, which can lead to theories characterized either by baryonic number N_b or leptonic number N_l violation.

Proton–proton collision outcomes are interests of study at the LHC facility, see, e.g., References [35,36], respectively.

This phenomenon can be compared with broken spacial $O(3)$ symmetry originating from a geometrical Berry phase Equation (1) on curved spacetime, and the remaining degrees of freedom can be used for further purposes. The difference can be confirmed by the individuation of sparticles whose mass dispersion relation Δm_{ij} for masses m_{ij} is given by

$$\frac{\Delta m_{jk}^2}{m_0^2} = \frac{\lambda_j \lambda_k^*}{\pi^2} \ln \frac{M_{Pl}}{M_G}, \tag{5}$$

where λ_i factorizes the (requested) coupling constant, m_0 is the mass of the common (standard-model) scalar (normalized to Planck mass M_{Pl}), and M_G is the mass for a (massive) gravitational mode.

Interacting Further Particles

After the breach of higher-dimensional structures [37–40], nonperturbative degrees of freedom give rise to Compton-length waves (particles) whose masses M_C are comparable with Planck mass M_{Pl}; they interact very weakly and gravitationally. In particular, masses M_C are of order $M_C \simeq R/M_{Pl}$, with R the lower bound on the compactification (energy) scale, and their gravitational interaction can modify ordinary Newtonian gravity.

A possibility was envisaged to verify the existence of particles such as those described by Equation (5) by cantilever detectors and/or silicon-based microelectromechanical systems [38–40]. In the following, alternative procedures are proposed for the sake of comparison with other theories and models.

2.4. *Verifying New Particles by Alternative Experiments*

Detectors for Earth-based experiments looking for WIMPs of mass m_W, $m_W 80 Gev$ scattering on smaller particles were proposed in Reference [41]; nevertheless, interaction signals happening in the Sun are considered as well.

The main differences between generic light scalars and axions were discussed in Reference [42] on the basis of P and T violations.The regions of the parameter space available for axions exclude, by electric dipole moment bounds, those for a Fifth-Force recognition as spin-dependent and mediated by an axion-like particle; nonetheless, for a generic scalar unaffected by CP violation, a Fifth-Force description is still possible.

The signal containing a spin-flipping effect calculated after the cross section of the absorption by a scanning Fabry-Pérot interferometer as a function of a 'relaxation time' can be 'cleaned' [43] in order to obtain the true description of the emission rate and the absorption one.

For a beam of electrons prepared for a Fabry-Pérot interferometer according to a required velocity distribution precision and (three-dimensional space) radial resolution, for Thomson scattering of laser electrons from an electron beam, Doppler-shifted wavelength of photons backscattered under 180 degrees, velocity distribution radially resolved in space, absolute electron energy, and the degree of space-charge compensation can be measured [44]. Measurement of longitudinal and transverse electron temperature is determined up to a lower bound for the ratio, respectively, and it has an upper bound (of 10/2) for velocity distribution. It further reveals fractional space-charge compensation; moreover, it is suited for higher laser intensity, i.e., by appropriate placement and use of the cavity mirrors of a confocal resonator.

This technique provides, for the first time, nondestructive measurement of velocity distribution in an electron beam radially resolved in space. The results presented here comprise the direct measurement of the absolute electron energy and the degree of space-charge compensation in the electron beam. The determination of an upper bound of 10/2 for the ratio of longitudinal-to-transverse electron temperature implies the first direct measurement of flattened velocity distribution.

Differently, it is also possible to look for new predicted particles by adapting previously proposed experiments and apparati for the required tasks.

Noise-minimization techniques involving changing mirror disposition for Michelson interferometers were reviewed in Reference [45].

For nonlinear interferometers [46], optical switching (for example, but not only, of mirrors) can be obtained via cross-phase modulation of a lossy (particle-beam) line, i.e., for a Sagnac interferometer.

The Sagnac effect can be explored by studying the role of spin-rotation coupling for circularly polarized light in order to testify on the photon-helicity coupling to rotation: for this, an analogous experiment of neutron interferometry can be performed [47]. The frequency shift and a constant optical phase shift for the prepared beam of neutrons can be tested by multiplying angular velocity Ω by the time of flight of a photon between two interferometers ends $\Delta t = l/c$ to obtain helicity-rotation phase shift $\Delta\Phi = 2\Omega l/c$ as the same phase shift predicted in the rotating frame at the detector.

The presence of different particles in the (Earth-based) lab system, such as those described for Equation (5), can be revealed by a different helicity-rotation phase shift $\tilde{\Delta}\Phi$.

Their gravitational and other kinds of interaction with neutrons in the prepared beam would modify neutron kinetic energy K_n. Indeed, velocities $\vec{v}$ correspond to the average of the wavepacket of the prepared neutron beam; should neutrons undergo interactions, their velocities after interaction(s) $\vec{v}' = \vec{v}_{n\ interact} \neq \vec{v}$ might be changed, i.e., in any case of inelastic scattering interaction(s). The helicity-rotation phase shift(s) can be measured by evaluating the requested time for end-to-end interferometer path covering, $\Delta\tilde{t}$, their velocities $\vec{v}_{n\ interact} \neq \vec{v}_{n\ i} \equiv c^2\vec{k}_i/\omega_i$ and their velocity distribution, being $\Phi \equiv \Phi(\vec{k})$ and $\Phi' \equiv \Phi(\vec{k}_{interact})$.

Differently, in the case of a weak gravitational field, the velocities of the new interacting particles (not prepared in the neutron beam) in the experiment environment would be further modified, e.g., such as established in Equation (3), for which different helicity-rotation phase shift(s) $\tilde{\Delta}\Phi\ ''$ would be detected.

The presence of different kinds of particles would be predictable in the case of different values for $\tilde{\Delta}\Phi\ ''$.

The effectiveness of a gravitational (but not necessarily only Berry) phase, such as the one in Equation (2), multiplying wavefunction Equation (1) (from which the neutron wavepacket is prepared), would lead to two different results, $\tilde{\Delta}\Phi_G\ '''$ and $\tilde{\Delta}\Phi\ '''$ for the measures of the helicity-rotation phase shift(s) according to whether the new particles interact gravitationally or not.

2.5. Semiclassical Descriptions

Analysis of a semiclassical regime for the quantum nature of gravity based on the notion of precausality was developed in Reference [48]. Among precausality requirements, the necessity to imply quantum modifications to matter fields rather than on the geometrical description of spacetime was also established for lab experiments in Minkowski spacetime. In curved spacetime, EFE nonlinearity plays a crucial role in determining the viability of a geometrical gravitational theory, and as far as quantum-gravity corrections to nongravitationally interacting high-energy matter fields are concerned [49].

Quantum optical corrections for Maxwell equations [5] are predicted for short-distance experiments, for which a Fock occupation space can be defined for the quantum optical system. Such corrections [50] can be framed within models interpreting the statistical correlations as the outcome of theories with local hidden variables.

An experiment with correlated light beams in coupled interferometers allows for semiclassical-limit analysis [51].

Among quantum (nonsemiclassical) effects, the production of Planck-sized black holes can be discerned in this way [52].

The modification of the thermodynamical properties of macroscopic materials [53] can, after these controls, be exploited to study the possibility of modified energy–momentum relations.

For low-energy matter fields, the phenomenon of gravitational decoherence can be investigated by studying a quantum system interacting with the external environment. This way, a modification to the quantum fields brought by the (Minkowski) lab gravitational field can be modeled as spacetime fluctuations of quantum-gravitational origin acting on the matter fields [54], as well as for experiments taking place in larger scales than Planck length.

Quantum-gravitational decoherence can, in this way, be differentiated from quantum decoherence by experiment settings including cold-ion traps [55].

The interpretation possibilities trapped-ion crystals and the generating functionals for self-interacting scalar fields [55,56] further be done for Reference [57] determining the corresponding two-point correlation function.

The correlation between quantum signals and interferometers was schematized in Reference [58] by setting the different theoretical interpretation of two-mode squeezed vacuum states and two independent squeezed states by considering the output states as influenced by the transmission coefficient of the beam splitter.

Quantum fluctuations of electrons in a storage ring can further be modeled as a Markov process for lattice-gauge theories [59].

Helium properties in several different aggregation states, by taking advantage of their features as macroscopic quantum states [60], have been exploited and proposed to be exploited for the detection of gravitational waves [61]. From a theoretical point of view, such experiments [62] allow to obtain hints about the topology of the Riemannian manifold generated as an EFE solution, not only in the weak-field limit. Applications for the determination of the mass of white dwarfs [63] and about the evolution of galaxy formation [63] have also been performed.

The experimental device, consisting of fiberoptic gyroscopes, allow establish a reliable offset with regard to the Earth-rotation effects [64]; the remaining noise can be studied as a quantum property of the aggregation state of the material [65].

Semiclassical Experiments

The modification of energy–momentum dispersion relations (as from analysis of its spectral decomposition) was proposed in the literature with the aim to propose properties of quantum spacetime foam and some of its semiclassical limits features.

In particular, it is possible to study the phenomenological implication of the foamy structure by investigating the properties of macroscopic materials with regard to their reflection and refraction specificities by comparing the atoms and molecules constituting the solid-state structure, either crystalline or amorphous. This is done by approximating the corresponding potential (wells) as black-hole-like potentials. In this case, the chosen interacting particle (photon) is small enough with regard to the potential wells and the Planck scale, but the experiment is conducted at length sizes larger than the Planck scale [7,8]. The overall gravitational regime of the lab system, however, is still Minkowskian, and there exists a valid paradigm to discriminate and calibrate interaction(s) between the system and the external environment.

Photon transit in a (macroscopic) block of diaelectric material is supposed to cause a (photon) momentum transfer; there exist appropriate temperatures at which a momentum change caused by the diffractive diaelectric index, for which the momentum transferred to the block can produce appreciable (position) reaction shift of the block as the photon exits the block. The diffractive diaelectric properties, caused by its solid-state structure, can be approximated to the effects of a lattice of (small-size) black holes, which can account for quantum-gravitational properties of the spacetime inside the block and, in particular, its foamy features. The described experiment [6] consists of letting a photon cross a block of diaelectric material of mass $\tilde{M}$ and volume $V_{\tilde{M}} = L_1 L_2 L_3$, whose refraction index n_{ref} can be evaluated after the absolute value of the Poynting vector, and whose center of mass should have moved of displacement ΔX_k at the exit of the photon after k double reflections, i.e.,

$$\Delta X_k = L_1 \frac{h\omega}{2\pi \tilde{M} c^2}(n - 1 + 2k) \tag{6}$$

with ω being the frequency of the photon. Any modification to the measured displacement has to be ascribed to quantum-gravitational phenomena, which can manifest in the modification of the photon energy, in the modification of the diffraction index of the diaelectric block, and/or after the spacetime foamy structure modifies the potential of the solid-state structure, photon energy, and their interaction.

3. Sky Investigations

3.1. 'Post-Keplerian' Objects

The values of spin and of the orientation of the massive black hole at the galactic center can be constrained by analyzing the motion of pulsars around it [66]. To do so, considering pulsar precession or any other quantities averaged on it have proven less efficient than considering the pulsar as a test particle moving in a Kerr metric. In the latter strategy at first PN order, no counterfilter has been theorized that is able to remove the Keplerian [67] 'noise' due to the other considered Newtonian 'material' in the galactic region. In Reference [68], S stars are considered for their almost Keplerian behavior from which perturbation due to a background Schwarzschild metric can be isolated; this way, the black-hole-spin-induced quadrupole moment can be measured under the description of redshift measurements distributed along the orbital path and more intensive at the pericenter. Under the assumption of a different background metric, the orbit of the star is influenced by the spin of the galactic black hole (BH) at PN order, as after analysis of photon-propagation delays, for which the geodesics path is governed by a Hamiltonian

$$H_{geod} = H_{Mink} + H_{Schw} + H_{FD} + H_q, \tag{7}$$

where flat spacetime Hamiltonian, the Schwarzschild Hamiltonian, the frame-dragging Hamiltonian, and the BH quadrupole moment have been defined, respectively: the Schwarzschild Hamiltonian is of the v^4 order, and both the frame-dragging Hamiltonian and the BH quadrupole moment one are of the v^6 order.

Further items of information can be obtained at the PN order for a star orbiting the galactic BH by a Keplerian nonprecessing orbit by simplifying the stellar Doppler shift as described by PN parameter β as $\beta \equiv \beta(r; a)$

$$\beta^2 = \frac{r_s}{r} - \frac{r_s}{2_a}, \tag{8}$$

with a being the orbital major semiaxis, and r_s the Schwarzschild radius of the BH. The periapse shift due to any kind of dark matter is negligible at this order.

The precession of a star orbiting the galactic BH can be expressed [69] as a function of astrometric deviation δ_x, as a function of galactic BH spin χ and a, $\delta_x \equiv \delta_x(\chi, a)$

$$\delta_x = \chi \frac{1}{a^{1/2}(1-e^2)^{3/2}} \tag{9}$$

For double-neutron-star binaries, orbital-period derivatives of orbital period $\dot{p}$ can acquire improvements by wide-bandwidth coherent-dispersion devices, as pointed out in References [70–72].

Appropriate controls from binary pulsar systems can [73] set upper and/or lower bounds on the parameters and/or the parameter space of theories whose low-energy limit admits a strong-field limit, different from GR, a different value for universal and/or a running Newton constant G, and on the energy density of the low-frequency (limit of) gravitational waves.

By studying the rate equation for the derivative of mass M growth of a large body, given as Ω, the Keplerian orbital frequency of the large body orbiting a star of mass M_* at orbital distance α and Σ_P the surface density of the field planetesimals, R the radius of the large body, [74,75]

$$\frac{dM}{dt} = \pi R^2 \Omega \Sigma_P F_G, \tag{10}$$

(References [76,77]), with F_G as the gravitational focusing factor.

The magnitude of Neptune's orbital expansion [78] has imposed a lower limit of about 5 AU; numerical results indicate the inclination distribution as sensitive to the rate of orbital evolution for giant planets, for which longer timescales of orbit evolution are correlated to higher inclinations.

For this, optical interferometers, infrared long-baseline and long-baseline (sub)millimeter interferometers, and high-sensitivity infrared observatories, are compared.

3.2. Verifying New Celestial Bodies by Alternative Experiments

3.2.1. Optical Interferometers

Optical interferometers are useful in the study of galaxies, celestial bodies in galaxies, and Newtonian and Keplerian material in and around them. Protoplanetary disks (i.e., around a star) can be analyzed for the information they carry for structure evolution [79] for the role of grains, dust, polycyclic aromatic hydrocarbons, and minerals. Extinction cross-section C_{exti} of a radiation field with solid (macroscopic) particles equals that for absorption C_{abs}, summed to that for scattering C_{sca}, and also equals the imaginary part of total electric polarizability α expressed as the sum of the latter, α_{j_k}, in the three direction of the semiaxes, i.e., $j = j_{\hat{x}}, j_{\hat{y}}, j_{\hat{z}}$, of the ellipsoid-shaped orbiting particle describing the grain, such as

$$C_{exti} = C_{abs} + C_{abs} = Im(\alpha) = Im\left[V(\epsilon_1 + i\epsilon_2 - \epsilon_m)\sum_{\varrho=1}^{\varrho=3}\frac{1}{(\epsilon_m + L_\varrho(\epsilon_1 + i\epsilon_2 - \epsilon_m))}\right] \tag{11}$$

with ϵ as the dielectric function, L_ϱ geometrical factors such that $\sum_{\varrho=1}^{\varrho=3} L_\varrho = 1$, and m the complex refractive index individuating (also) a mass, as solid materials are described by their own optical constants. The presence of different kinds of dust individuates different sizes for the formation of planets, as well as for their size (mass). The composition of dust and dust grains reveals structure age. Spectral analysis reveals the composition in brain minerals, dust, dust grains, different kinds of dust, and crystalline and amorphous material. Grain growth [80] can be individuated both by spectroscopy and by mm observation according to grain size.

Dust-temperature determination can be achieved by analysis of different vibrations of the lattices of heavy ions, and/or groups of ions having low bond energies, and/or when the signal-to-noise ratio is high.

3.2.2. Transition Lines

Studying CO transition lines CO(3-2) or CO(2-1) at submillimetric (submm) scales allows one to infer the interaction properties between a BH and a spheroidal celestial body (of comparable features with regard to the former) [81], and helps shed light on the role of quasars and quasarlike objects in the evolution of galaxies. Galaxy emission lines were partially surveyed in Reference [82]. The same emission lines also provide information on star-disk size. An increased ratio among the lines might indicate [83] an increase of the temperature of the gas corresponding to the upper layers of the disks; higher angular resolution for scanning the dust region might indicate the presence of a warp; nevertheless, the variation of disk thinness is unlikely to be due to photoevaporation, grain growth, and binarity. Differently, the presence of a planet could be considered as responsible for warp shaping, the creation of an inner whole, and different angular resolutions for emission lines. Analysis of mm continuum can also detect azimuthal morphology. Emission-line analysis allows to control the spectral-energy distribution model, followed by the studied mechanism. CO (1-0) line observations have proven [84] effective in detecting galaxy-forming areas, and areas up to the optical range. The possibility of gravitational lensing for CO lines was discussed in Reference [85]. The detection of acoustic modes was discussed in Reference [86] as far as mirror suspensions are concerned.

3.2.3. Laser Interferometers

The use of sapphire crystals in laser interferometers was analyzed in Reference [87]. In restrictions due to the availability of the medium, which can be at least partially overcome by applying temperature gradient techniques, the interest in the detection of gravitational waves has been pointed out. Thermoelastic noise can be reduced [88] by changing the beam shape as a non-Gaussian mesa-shaped center-flat circular intensity profile by modifying the mirror shapes. This is necessary [89,90] to improve the techniques with center-flat mirrors and mesa-shaped beam optical cavities. Tests for the angular resolution [91] of a six-antenna millimeter radio interferometer that is able to detect, in a binary system of two stars with their each own disk, two different angular variables. Uniform probability distribution was chosen to impose a lower limit to the spectral index of one of the stars (Star A) as a function of the parameters of the other star (Star B), achieving a 99.7% (3 σ) confidence level, not only to determine different angular positions for two disks for Star A, but also to infer that they were not on the same plane by the nondetection in the scattered light images (within interstellar and galactic media) of the primary disk.

The emission from the target stars has mostly been modelled as thermal; nevertheless, other (and/or further) nonthermal mechanisms, i.e, in this case, free–free and gyrosynchrotron emissions, are eligible candidates to be supposed as a non-negligible fraction of the millimeter flux in the observations. The reached angular resolution enabled upper and lower bounds on disk-grain distribution (of Star A), and allowed for dust-deposit probability.

By the same device, it is possible to analyze [92] gaseous CO emission by the far quasar, its mass, density and temperature, and put lower bounds by comparing the CO line-flux ratios to those of a one-component large velocity gradient [93]. Excitation evaluation allows to evaluate on the quasar's gas and metallic enrichment. Dust emission and gas density at the given redshift allow to infer that not only was star formation possible at the observed time, but also rapid-growth black-hole formation at early cosmological times.

Appropriate continuum observations (at a millimeter scale) and the choice of molecular transitions allow to gain information about the core centers of stars and disks. Mm-continuum observations of two intermediate-mass star-forming regions up to high-mass star-forming regions, while the CN and CS molecular line shows chemical and physical effects [94] that cannot be confused with the opacity properties of celestial bodies. An increase of the dust opacity index and a decrease of the optical depth allows to hypothesize the presence of grains at the core and/or disk centers. The choice of opportune molecular-transition [95] lines allows to classify disk properties in star formation. To analyze continuum observations [96], the increase of the opacity index caused by an insufficient signal-to-noise ratio and UV coverage was proposed to ameliorate observations by increasing the baseline length. For proposals to refine the signal from line contamination (from bolometer data), see, e.g., Reference [97]. For a better opacity index, see Reference [98].

3.2.4. Baseline Interferometers

The detection-rate statistics of compact radio sources were analyzed for particular choices of sky pixelization [99]. For a single-baseline interferometer, they can be detected iff the most flux density coincides with that of a compact structure. Smaller, i.e., thinner, structures could be missed within this investigation pattern.

Arm-cavity-mirror mechanical modes of interferometric detectors might cause parametric instabilities. This instability can be dumped by adding a spring made of piezoelectric material with the task of dumping to the amplifier circuit attached to the detector material, and an extra resistor with the purpose of shunting, then linked to the ground of the circuit by electric wire [100]. The piezoelectric material has the anisotropic structure of Reference [101], such that strain-energy dissipation in the shunted piezoelectric material depends on the material's geometric shape.

Differently, this problem was proposed to be overcome by choosing a cooled silicon mass for the detector material [102].

As far as long-baseline interferometers are concerned, the dust-evaporation boundary region in young-stellar-object disks can be sufficiently resolved [103], such that the physics underlying grain formation can be schematized.

For a single-baseline Earth-based interferometer, differential astrometric observations are affected by stellar aberration in angular resolution [104]; variations of calibration terms among pixels of interest must be introduced to avoid correlations between calibration summands, and both azimuthal derivatives of the position-variable sky and equatorial angle, for which the former implies the lower bound for the accuracy of the velocity absolute value.

3.2.5. Redshift Role

At a redshift of $z = 6.419$, transition lines CO (6-5) and CO (7-6) indicate that the behavior of the interstellar gas allows for quick metal and dust enrichment; from the area of the molecular region and the brightness of the transition lines it is possible to infer [92] star and massive-BH formation can occur at the same cosmological epoch.

The stellar photosphere is suited for optical photometry in order to individuate emission regions by simple geometrical models for sources in the IR. By suitably expressing spectral-energy distribution for dust disks [105], upper and lower bounds for the dust-sublimation temperatures can be imposed after the calculation of the size of the region where the phenomenon takes place, whose radius can be parameterized as a square inverse function of dust temperature.

Calibration of source data and location ones can allow to Fourier-transform the (time) delay to the (event) rate domain, to which appropriate filters can be applied [106] to eliminate radio-frequency interference for early-cosmological investigation. By letting imaging scale as $O(Nlog(N))$, with N being the number of data samples, it is possible to individuate sources at redshifts $\tilde{z}$, $\tilde{z} = 7$ to $\tilde{z} = 11$ [107] to investigate the first epoch of star formation and of reionization. Therefore, weaker sources can also be detected.

Laser-photocathode uses [108] are advantageous in laser interferometry as a coherent transition radiation that can generate radiation is fully characterized by the square modulus of the Fourier transform. The energy spectrum emitted by transition radiation is uniform, such that, according to Reference [109], the frequency spectrum is only a(n exponential) function of the electron-beam form factor. For celestial bodies emitting in the IR spectrum, this is a consistent optimization criterion for system alignment.

4. Outlook

Planet formation can be individuated [110] by the spectral-energy distribution of the observed lines and in the spectrum in the continuum.

The distribution of major exoplanet axes is best accounted for nonlinear model fitting, for which the parameter space can be applied (Bayesian filters, described Markov chains).

Analysis of lines CO(1-0), CO(3-2) and CO(2-1) [111] can reveal the presence of large, massive, cold molecular clouds that exist with kinetic temperatures close to that of CMBtemperature (in the inner disks). Radial velocities and (position) offsets from the center of the star are measured, as well as the CO(3-2) spectrum in mm (wave) array-device observations.

By the same techniques and infrared observations [112,113], for CO (1-0) line observations, we can make numerical simulations [114] of the hydrodynamical properties of dust and gas morphology at the central region of ionizing stars, for which phenomena of star-formation account and compensate for the presence of nuclear gas according to star-formation rate.

Laser-frequency measurements [115] help calibrate frequency absolute and the long-term stability of a fiber Fabry-Pérot interferometer. For small temperatures, i.e., for a spectrum of 1–3 ms^{-1}, it is possible to characterize the Doppler radial velocity shifts at the 1 ms^{-1} of exoplanets.

Laser interferometers have proven efficient [116] in detecting particle interactions linearly in g, such as spin-gravity coupling, and P- and T-violating interactions from an astrophysical point of view. It may also apply to (integral-spin) dark-matter searches, as well as other kinds of investigations.

The existence of continuous spectra within the search for gravitational waves has led to the individuation of planets, for which several techniques have been set.

The needed hypotheses for for adding five parameters describing generic elliptical orbits, i.e., for eccentricities e, such that $e \simeq 0.8$ to computations for the computation of a monochromatic source, were analyzed in Reference [117] for a radio pulsar orbiting a planet, both for Earth-based detectors of gravitational waves as well as for interplanetary spacecrafts.

After pointing out spectral-analysis laser-frequency noise [118], time-delay-based interferometry is effective in comparing different optical paths. By numerically simulating different times of travel, it is possible to extract the spectrum signal of planets and other celestial bodies perturbing the gathered signal in spacecraft interferometry in the Solar System. By numerically simulating parameterized post-Newtonian (PPN) parameters β and γ for Solar System bodies for (geodesics) solutions to the variation of the metric, the opportune time is delayed.

Within gravitational-wave observations, it is possible [119] to discover planets either orbiting compact binaries or passing close to them, with masses of around $\sim 2 \times 10^{30} g$, even at redshift $z \sim 1$. It is possible to resolve an inflation stochastic GW background in frequency range $f_{min} \sim 0.2$ Hz and $f_{max} \sim 1$ Hz. By gravitational-wave-detecting in space, at f_{min} or lower frequency $f \leq f_{min}$, it is possible to resolve extragalactic white-dwarf binaries, and, at higher frequencies, $f \geq f_{min}$, cosmological double neutron-star binaries and double black-hole binaries or black hole–neutron star binaries by assuming a nearly-flat noise spectrum.

Ultrashort-period exoplanets can be discovered [120] as weak sources of gravitational waves close to binary systems, according to the frequencies of emitted gravitational waves $f_{gr} > 10^{-4}$ Hz. By cumulative periastron shift, it is possible to express luminosity as a frequency function, as, usually, the ratio between the apparent luminosities of exoplanets and other celestial bodies to other binary systems reaching Earth-based detectors is widely resolved.

The atmosphere of extrasolar planets orbiting a star is possible by differential-phase measurements near the IR spectrum [121] by the brightness ratio of the planet and star.Indeed, after the possibility of angularly resolving the star, optimization of statistical tests for orbital and spectral parameters is possible. In case the planet's revolution time is not negligible, such optimization could, therefore, be lowered.

In particular, it is possible [122] to evaluate atmosphere cross section as a wavelength function, such as Rayleigh scattering and refraction, i.e., from 115 to 1000 nm, from UV O_2 absorption. As a result, it is possible to infer whether atmosphere for a given planet exists, and to establish the chemical elements or process that determine the planetary radius to near-IR refraction.

Microarcsecond resolution allows for astrometry measurements about the nuclei of active galaxies, and accretion disks of supermassive black holes and their the relativistic jets. Precision allows for the verification of stellar and galactic structure, as well as hypotheses about dark matter and cosmology back to star-formation times, small-scale investigations of quasar and AGN cores, and to investigate binary supermassive black holes.

At microarcsecond precision, the astrometric revelation of quasar parallaxes is rendered accessible [123], which allows to analytically investigate, at the cosmological scale, the parameter space possibly needed to describe dark energy. Indeed, a direct geometric measurement is free of astrophysical systematic effects. The particle-induced effects are summarizzed in [124–134].

By means of far-IR coherent interferometry, even close to quantum noise, it is possible for an interferometer to individuate an Earth-like planet. At high spectral resolution, precise measurements of atmospheric temperature and molecules, pressure, and composition are achievable.

Particle quantum properties and the quantum features of spacetime at Planckian lengths allow to investigate the semiclassical limit of quantum-gravitational expressions. Quantum optical systems resulting in aggregation states of matter allow to account for such quantum features for phenomena taking place at scales larger than the Planckian, for (lab) experiments, and for observational surveys taking place in the background (Minkowski) flat spacetime [5–8].

The paper was organized as follows.

In Section 1, the main motivations for the paper were presented.

In Section 2, theories predicting new features for experimentally known particles and new particles were recalled to specify which experiment systems could be useful for their verification.

In Section 3, experiment devices and techniques were recalled for the verification of fundamental features, such as planet and exoplanet formation and structure, of standard cosmology, were outlined.

In Section 4, brief remarks about perspective investigations were proposed.

Funding: This research received no external funding.

Conflicts of Interest: The author declare no conflict of interest.

References

1. Ricarte, A.; Natarajan, P. The observational signatures of supermassive black hole seeds. *arXiv* **2018**, arXiv:1809.01177.
2. Tamburello, V.; Capelo, P.R.; Mayer, L.; Bellovary, J.M.; Wadsley, J. Supermassive black hole pairs in clumpy galaxies at high redshift: delayed binary formation and concurrent mass growth. *Mon. Not. Roy. Astron. Soc.* **2017**, *464*, 2952–2962. [CrossRef]
3. Bender, P.L. Gravitational wave astronomy, relativity tests, and massive black holes. *IAU Symp.* **2010**, *261*, 240. [CrossRef]
4. Finn, L.S.; Chernoff, D.F. Observing binary inspiral in gravitational radiation: One interferometer. *Phys. Rev. D* **1993**, *47*, 2198. [CrossRef]
5. Faizal, M.; Momeni, D. Universality of short distance corrections to quantum optics. *arXiv* **2018**, arXiv:1811.01934.
6. Frisch, O.R. Take a photon. *Contemp. Phys.* **1965**, *7*, 45. [CrossRef]
7. Bekenstein, J.D. Is a tabletop search for Planck scale signals feasible. *Phys. Rev. D* **2012**, *86*, 124040. [CrossRef]
8. Bekenstein, J.D. Can quantum gravity be exposed in the laboratory? *Found. Phys.* **2014**, *44*, 452. [CrossRef]
9. Anandan, J.; Aharonov, Y. Geometry of quantum evolution. *Phys. Rev. Lett.* **1990**, *65*, 1697–1700. [CrossRef] [PubMed]
10. Van Boekel, R.; Kervella, P.; Scholler, M.; Herbst, T.; Brander, W.; de Koter, A.; Waters, L.B.F.M.; Hillier, D.J.; Paresce, F.; Lenzen, R.; et al. Direct measurement of the size and shape of the present-day stellar wind of eta carinae. *Astron. Astrophys.* **2003**, *410*, L37. [CrossRef]
11. Hajian, A.R.; Behr, B.B.; Cenko, A.T.; Olling, R.P.; Mozurkewich, D.; Armstrong, J.T.; Pohl, B.; Petrossian, S.; Knuth, K.H.; Hindsley, R.B. Initial results from the USNO dispersed Fourier transform spectrograph. *Astrophys. J.* **2007**, *661*, 616. [CrossRef]
12. Voit, G.M. On nulling interferometers and the line-emitting regions of agns. *Astrophys. J.* **1997**, *487*, L109. [CrossRef]
13. Danchi, W.C.; Rajagopal, J.; Kuchner, M.; Richardson, J.; Deming, D. The importance of phase in nulling interferometry and a three telescope closure-phase nulling interferometer concept. *Astrophys. J.* **2006**, *645*, 1554. [CrossRef]
14. Garrett, M.A. When you wish upon a star: Future developments in astronomical VLBI. *ASP Conf. Ser.* **2003**, *306*, 3.
15. Chen, Y.B. Sagnac interferometer as a speed meter type, quantum nondemolition gravitational wave detector. *Phys. Rev. D* **2003**, *67*, 122004. [CrossRef]
16. Thompson, R.; Papini, G. Berry's phase and gravitational wave. In Proceedings of the 5th Canadian Conference on General Relativity and Relativistic Astrophysics, University of Waterloo, Waterloo, ON, Canada, 13–15 May 1993; World Scientific Pub Co Inc.: Hackensack, NJ, USA, 1993.
17. Bruno, A.; Capolupo, A.; Kak, S.; Raimondo, G.; Vitiello, G. Berry-like phase and gauge field in quantum computing. In *Methods, Models, Simulations and Approaches Towards a General Theory of Change*; World Scientific: Singapore, 2012; pp. 83–94.

18. Pachos, J.; Zanardi, P.; Rasseti, M. NonAbelian Berry connections for quantum computation. *Phys. Rev. A* **2000**, *61*, 010305. [CrossRef]
19. Berry, M.V. Quantal phase factors accompanying adiabatic changes. *Proc. R. Soc. Lond.* **1984**, *A392*, 45–57. [CrossRef]
20. Hinterbichler, K. Theoretical aspects of massive gravity. *Rev. Mod. Phys.* **2012**, *84*, 671–710. [CrossRef]
21. Visser, M. Mass for the graviton. *Gen. Rel. Grav.* **1998**, *30*, 1717. [CrossRef]
22. Billard, J.; Mayet, F.; Grignon, C.; Santos, D. Directional detection of dark matter with MIMAC: WIM identification and track reconstruction. *J. Phys. Conf. Ser.* **2001**, *309*, 012015. [CrossRef]
23. Lee, H.S.; Bhang, H.C.; Choi, J.H.; Dao, H.; Hahn, I.S.; Hwang, M.J.; Jung, S.W.; Kang, W.G.; Kim, D.W.; Kim, H.J.; et al. Limits on WIMP-nucleon cross section with CsI(Tl) crystal detectors. *arXiv* **2007**, arXiv:0704.0423.
24. Archambault, S.; Aubin, F.; Auger, M.; Behke, E.; Beltran, B.; Clark, K.; Dai, X.; Davour, A.; Farine, J.; Faust, R.; et al. Dark matter spin-dependent limits for WIMP interactions on F-19 by PICASSO. *Phys. Lett. B* **2009**, *682*, 185. [CrossRef]
25. Goodman, J.A.; Ellsworth, A.S.; Ito, J.R.; MacFall, J.R.; Siohan, F.; Streitmatter, R.E.; Tonwar, S.C.; Vishwanath, R.; Yodh, G.B. Composition of primary cosmic rays above 10^{13}ev from the study of time distributions of energetic hadrons near air shower cores. *AIP Conf. Proc.* **1979**, *49*, 1.
26. Ficthel, C.E.; Linsley, J. High-energy and ultrahigh-energy cosmic rays. *Astrophys. J.* **1986**, *300*, 474. [CrossRef]
27. Alexandrov, A.; Asada, T.; Puonaura, A.; Consiglio, L.; D'Ambrosio, N.; De Lellis, G.; Di Crescenzo, A.; Di Marco, N.; Di Vacri, M.L.; Furuya, S.; et al. Intrinsic neutron background of nuclear emulsions for directional Dark Matter searches. *Astropart. Phys.* **2016**, *80*, 16. [CrossRef]
28. SuperCDMS Collaboration. The SuperCDMS Experiment. *arXiv* **2005**, arXiv:astro-ph/0502435.
29. Carroll, S.M.; Mantry, S.; Ramsey-Musolf, M.J.; Stubbs, C.W. Dark-matter-induced weak equivalence principle violation. *Phys. Rev. Lett.* **2009**, *103*, 011301. [CrossRef]
30. Price, J.C.; Innes, W.R.; Klein, S.; Perl, M.L. The rotor electrometer: A new instrument for bulk matter quark search experiments. *Rev. Sci. Instrum.* **1986**, *57*, 2691. [CrossRef]
31. Innes, W.R.; Perl, M.L.; Price, J.C. A rotor electrometer for fractional charge searches. In Proceedings of the 4th International Conference on Muon Spin Rotation, Relaxation and Resonance, Uppsala, Sweden, 23–27 June 1986; pp. 1–2.
32. Mathai, V.; Wilkin, G. Fractional quantum numbers via complex orbifolds. *arXiv* **2018**, arXiv:1811.11748.
33. Sparnaay, M.J. Measurements of attractive forces between flat plates. *Physica* **1958**, *24*, 751. [CrossRef]
34. Dimopoulos, S.; Hall, L.J. Electric dipole moments as a test of supersymmetric unification. *Phys. Lett. B* **1995**, *344*, 185. [CrossRef]
35. Evans, L.; Bryant, P. LHC Machine. Available online: https://iopscience.iop.org/article/10.1088/1748-0221/3/08/S08001/pdf (accessed on 27 March 2019).
36. Aad, G.; Brad Abbott, B.; Abdallah, J.; Khalek, S.A.; Abdinov, O.; Aben, R.; Abi, B.; Abolins, M.; AbouZeid, O.; Abramowicz, H.; et al. Measurement of the $t\bar{t}$ production cross-section as a function of jet multiplicity and jet transverse momentum in 7 TeV proton-proton collisions with the ATLAS detector. *JHEP* **2015**, *1501*, 020. [CrossRef]
37. Antoniadis, I.; Dimopoulos, S.; Dvali, G.R. Millimeter range forces in superstring theories with weak scale compactification. *Nucl. Phys. B* **1998**, *516*, 70. [CrossRef]
38. Price, J.C. *International Symposium on Experimental Gravitational Physics*; Michelson, P.F., Ed.; World Scientific: Singapore, 1988; pp. 436–439.
39. Kapitulnik, A.; Kenny, T. *NSF Proposal 1997*; National Science Foundation: Alexandria, VA, USA, 1997.
40. Weld, D.M.; Xia, J.; Cabrera, B.; Kapitulnik, A. A new apparatus for detecting micron-scale deviations from newtonian gravity. *Phys. Rev. D* **2008**, *77*, 062006. [CrossRef]
41. Gould, A. Cosmological density of WIMPs from solar and terrestrial annihilations. *Astrophys. J.* **1992**, *388*, 338–344. [CrossRef]
42. Mantry, S.; Pitschmann, M.; Ramsey-Musolf, M.J. Differences between axions and generic light scalars in laboratory experiments. *arXiv* **2014**, arXiv:1411.2162.
43. Gibbs, H.M.; Hull, R.J. Spin-exchange cross sections for Rb-87- Rb-87 and Rb-87- Cs-133 collisions. *Phys. Rev.* **1967**, *153*, 132. [CrossRef]

44. Habfast, C.; Poth, H.; Seligmann, B.; Wolf, A.; Berger, J.; Blatt, P.; Hauck, P.; Meyer, W.; Neumann, R. Measurementof laser light thomson scattered from a cooling electron beam. *Appl. Phys. B* **1987**, *44*, 87. [CrossRef]
45. Biscardi, R.; Ramirez, G.; Williams, G.P.; Zimba, C. Effects of rf sidebands on spectral reproducibility for infrared synchrotron radiation. *Rev. Sci. Instrum.* **1995**, *66*, 1856. [CrossRef]
46. D'Ariano, G.M.; Kumar, P. A quantum mechanical study of optical regenerators based on nonlinear loop mirrors. *IEEE Photonics Tech. Lett.* **1998**, *10*, 699. [CrossRef]
47. Mashhoon, B.; Neutze, R.; Hannam, M.; Stedman, G.E. Observable frequency shifts via spin rotation coupling. *Phys. Lett. A* **1998**, *249*, 161. [CrossRef]
48. Di Casola, E.; Liberati, S.; Sonego, S. Between quantum and classical gravity: Is there a mesoscopic spacetime? *Found. Phys.* **2015**, *45*, 171. [CrossRef]
49. Di Casola, E.; Liberati, S.; Sonego, S. Nonequivalence of equivalence principles. *Am. J. Phys.* **2015**, *83*, 39. [CrossRef]
50. Freedman, S.J.; Clauser, J.F. Experimental test of local hidden-variable theories. *Phys. Rev. Lett.* **1972**, *28*, 938. [CrossRef]
51. Ruo Berchera, I.; Degiovanni, I.P.; Olivares, S.; Genovese, M. Quantum light in coupled interferometers for quantum gravity tests. *Phys. Rev. Lett.* **2013**, *110*, 213601. [CrossRef] [PubMed]
52. Nicolini, P.; Mureika, J.; Spallucci, E.; Winstanley, E.; Bleicher, M. Production and evaporation of Planck scale black holes at the LHC. In Proceedings of the MG13 Meeting on General Relativity Stockholm University, Stockholm, Sweden, 1–7 July 2012.
53. Castellanos, E. Planck scale physics and Bogoliubov spaces in a Bose-Einstein condensate. *EPL* **2013**, *103*, 40004. [CrossRef]
54. Bassi, A.; Grossardt, A.; Ulbricht, H. Gravitational decoherence. *Class. Quant. Grav.* **2017**, *34*, 193002. [CrossRef]
55. Cirac, J.I.; Zoller, P. Quantum computations with cold trapped ions. *Phys. Rev. Lett.* **1995**, *74*, 4091. [CrossRef]
56. Bermudez, A.; Aarts, G.; Mueller, M. Quantum sensors for the generating functional of interacting quantum field theories *Phys. Rev. X* **2017**, *7*, 041012. [CrossRef]
57. De Ramo'n, J.; Garay, L.J.; Marti'n-Marti'nez, E. Direct measurement of the two-point function in quantum fields. *Phys. Rev. D* **2018**, *98*, 105011. [CrossRef]
58. Ruo-Berchera, I.; Degiovanni, I.P.; Olivares, S.; Samantaray, N.; Traina, P.; Genovese, M. One- and two-mode squeezed light in correlated interferometry. *Phys. Rev. A* **2015**, *92*, 053821. [CrossRef]
59. Jowett, J.M. Dynamics of electrons in storage rings including nonlinear damping and quantum excitation effects. *Conf. Proc. C* **1984**, *830811*, 283.
60. Froewis, F.; Sekatski, P.; Duer, W.; Gisin, N.; Sangouard, N. Macroscopic quantum states: Measures, fragility and implementations. *Rev. Mod. Phys.* **2018**, *90*, 025004. [CrossRef]
61. Singh, S.; De Lorenzo, L.A.; Pikovski, I.; Schwab, K.C. Detecting continuous gravitational waves with superfluid ^{4}He. *New J. Phys.* **2017**, *19*, 073023. [CrossRef]
62. Zloshchastiev, K.G. Acoustic phase lenses in superfluid He as models of composite space-times in general relativity: Classical and quantum properties with provision for spatial topology. *Acta Phys. Polon. B* **1999**, *30*, 897–905.
63. Yang, Y.; Zabludoff, A.I.; Dave, R.; Eisenstein, D.J.; Pinto, P.A.; Katz, N.; Weinberg, D.H.; Barton, E.J. Probing galaxy formation with he II cooling lines. *Astrophys. J.* **2006**, *640*, 539. [CrossRef]
64. Tajmar, M.; Plesescu, F.; Seifert, B. Anomalousfiber optic gyroscope signals observed above spinning rings at low temperature. *J. Phys. Conf. Ser.* **2009**, *150*, 032101. [CrossRef]
65. Tajmar, M.; Plesescu, F. Fiber-optic-gyroscope measurements close to rotating liquid helium. *AIP Conf. Proc.* **2010**, *1208*, 220. [CrossRef]
66. Zhang, F.; Saha, P. Probing the spinning of the massive black hole in the Galactic Center via pulsar timing: A full relativistic treatment. *Astrophys. J.* **2017**, *849*, 33. [CrossRef]
67. Barausse, E.; Cardoso, V.; Pani, P. Can environmental effects spoil precision gravitational-wave astrophysics? *Phys. Rev. D* **2014**, *89*, 104059. [CrossRef]
68. Angelil, R.; Saha, P.; Merritt, D. Towards relativistic orbit fitting of Galactic center stars and pulsars. *Astrophys. J.* **2010**, *720*, 1303. [CrossRef]

69. Waisberg, I.; Dexter, J.; Gillessen, S.; Pfuhl, O.; Eisenhauer, F.; Plewa, P.M.; Baubock, M.; Jimenez-Rosales, A.; Habibi, M.; Ott, T.; et al. What stellar orbit is needed to measure the spin of the Galactic centre black hole from astrometric data? *Mon. Not. R. Astron. Soc.* **2018**, *476*, 3600. [CrossRef]
70. Stairs, I.H. Testing general relativity with pulsar timing. *arXiv* **2003**, arXiv:astro-ph/0307536.
71. Jodrell Bank Observatory Pulsar Group. *COBRA: Pulsar Documentation*; Jodrell Bank Observatory Pulsar Group: Lower Withington, UK, 2001.
72. Swinburne Pulsar Group, The Caltech, Parkes, Swinburne Recorder Mk II. 2002. Available online: http://astronomy.swin.edu.au/pulsar/ (accessed on 27 November 2002).
73. Taylor, J.H. Pulsar timing and relativistic gravity. *Philos. Trans. R. Soc. Lond. Ser. A* **1992**, *341*, 117–134. [CrossRef]
74. Yagi, K.; Stein, L.C. Black hole based tests of general relativity. *Class. Quant. Grav.* **2016**, *33*, 054001. [CrossRef]
75. Wolf, S.; Malbet, F.; Alexander, R.; Berger, J.-P.; Creech-Eakman, M.; Duchene, G.; Dutrey, A.; Mordasini, C.; Pantin, E.; Pont, F.; et al. Circumstellar disks and planets. Science cases for next-generation optical/infrared long-baseline interferometers. *Astron. Astrophys. Rev.* **2012**, *20*, 52. [CrossRef]
76. Safronov, V.S. *Evolution of the Protoplanetary Cloud and Formation of the Earth and the Planets*; Serie: NASA technical translation, F-677; Program for Scientific Translations: Jerusalem, Israel, 1972.
77. Armitage, P.A. *Astrophysics of Planet Formation*; Cambridge University Press: Cambridge, UK; New York, NY, USA, 2010.
78. Malhotra, R. The origin of pluto's orbit: Implications for the solar system beyond neptune. *Astron. J.* **1995**, *110*, 420. [CrossRef]
79. Henning, T.; Meeus, G. Dust processing and mineralogy in protoplanetary accretion disks. *arXiv* **2009**, arXiv:0911.1010.
80. Brauer, F.; Dullemond, C.P.; Henning, T. Coagulation, fragmentation and radial motion of solid particles in protoplanetary disks. *Astron. Astrophys.* **2008**, *480*, 859. [CrossRef]
81. Coppin, K.; Swinbank, A.M.; Neri, R.; Cox, P.; Alexander, D.M.; Smail, I.; Page, M.J.; Stevens, J.A.; Knudsen, K.K.; Ivson, R.J.; et al. Testing the evolutionary link between submillimetre galaxies and quasars: CO observations of QSOs at z 2. *Mon. Not. R. Astron. Soc.* **2008**, *389*, 45. [CrossRef]
82. Hippelein, H.; Maier, C.; Meisenheimer, K.; Wolf, C.; Fried, J.W.; von Kuhlmann, B.; Kummel, M.; Phelps, S.; Roser, H.-J. Star forming rates between z = 0.25 and z = 1.2 from the CADIS emission line survey. *Astron. Astrophys.* **2003**, *402*, 65. [CrossRef]
83. Hughes, A.M.; Andrews, S.A.; Espaillat, C.; Wilner, D.J.; Calvet, N.; D'Alessio, P.; Qi, C.; Williams, J.P.; Hogerheijde, M.R. A spatially resolved inner hole in the disk around GM aurigae. *Astrophys. J.* **2008**, *698*, 131. [CrossRef]
84. Lisenfeld, U.; Braine, J.; Duc, P.A.; Brinks, E.; Charmandaris, V.; Leon, S. Molecular and ionized gas in the tidal tail in Stephan's Quintet. *Astron. Astrophys.* **2004**, *426*, 471. [CrossRef]
85. Downes, D.; Solomon, P.M. Molecular gas and dust at Z = 2.6 in smm j14011+0252: a strongly lensed, ultraluminous galaxy, not a huge, massive disk. *Astrophys. J.* **2003**, *582*, 37. [CrossRef]
86. Braccini, S.; Casciano, C.; Coredo, F.; Frasconi, F.; Gregori, G.P.; Majorana, E.; Paparo, G.; Passaquieti, R.; Puppo, P.; Rapagnani, P.; et al. Monitoring the acoustic emission of the blades of the mirror suspension for a gravitational wave interferometer. *Phys. Lett. A* **2002**, **301**, 389. [CrossRef]
87. Barish, B.C.; Camp, J.; Kells, W.P.; Sanders, G.H.; Whitcomb, S.E.; Zhang, L.; Zhu, R.-Y.; Deng, P.; Xu, J.; Zhou, G.; et al. Development of large size sapphire crystals for laserinterferometer gravitational-wave observatory. *IEEE Trans. Nucl. Sci.* **2002**, *49*, 1233. [CrossRef]
88. D'Ambrosio, E.; O'Shaughnessy, R.W.; Strigin, S.; Thorne, K.S.; Vyatchanin, S. Reducing thermoelastic noise in gravitational-wave interferometers by flattening the light beams. *arXiv* **2004**, arXiv:gr-qc/0409075.
89. D'Ambrosio, E.; O'Shaughnessy, R.; Thorne, K. LIGO Report Number G000223-00-D. Available online: http://admdbsrv.ligo.caltech.edu/dcc/ (accessed on 16 August 2000).
90. Braginsky, V.; D'Ambrosio, E.; O'Shaughnessy, R.; Strigin, S.; Thorne, K.; Vyatchanin, S. LIGO Report Number T030009-00-R. Available online: https://dcc.ligo.org/public/0027/T030009/000/T030009-00.pdf (accessed on 23 January 2003).
91. Duchene, G.; Menard, F.; Stapelfeldt, K.; Duvert, G. A layered edge-on circumstellar disk around HK Tau B. *Astron. Astrophys.* **2003**, *400*, 559–565. [CrossRef]

92. Bertoldi, F.; Cox, P.; Neri, R.; Carilli, C.L.; Walter, F.; Omont, A.; Beelen, A.; Henkel, C.; Fan, X.; Strauss, M.A.; et al. High-excitation CO in a quasar host galaxy at z = 6.42. *Astron. Astrophys.* **2003**, *409*, L47. [CrossRef]
93. Mao, R.Q.; Henkel, C.; Schulz, A.; Zielinsky, M.; Mauersberger, R.; Stoerzer, H.; Wilson, T.L.; Gensheimer, P. Dense gas in nearby galaxies. XIII. CO submillimeter line emission from the starburst galaxy M 82. *Astron. Astrophys.* **2000**, *358*, 433.
94. Beuther, H.; Schilke, P.; Wyrowski, F. High-spatial-resolution CN and CS observation of two regions of massive star formation. *Astrophys. J.* **2004**, *615*, 832. [CrossRef]
95. Yorke, H.W.; Sonnhalter, C. On the Formation of Massive Stars. *ApJ* **2002**, *569*, 846. [CrossRef]
96. Kumar, M.S.N.; Fernandes, A.J.L.; Hunter, T.R.; Davis, C.J.; Kurtz, S. A massive disk/envelope in shocked H_2 emission around an UCHII region. *Astron. Astrophys.* **2003**, *412*, 175. [CrossRef]
97. Gueth, F.; Bachiller, R.; Tafalla, M. Dust emission from young outflows: The case of L 1157. *Astron. Astrophys.* **2003**, *401*, L5. [CrossRef]
98. Hogerheijde, M.R.; Sandell, G. Testing Envelope Models of Young Stellar Objects with Submillimeter Continuum and Molecular-Line Observations. *ApJ* **2000**, *534*, 880. [CrossRef]
99. Porcas, R.W.; Alef, W.; Ghosh, T.; Salter, C.J.; Garrington, S.T. Compact structure in first survey sources. In Proceedings of the 7th European VLBI Network Symposium on New Developments in VLBI Science and Technology and EVN Users Meeting, Toledo, Spain, 12–15 October 2004.
100. Gras, S.; Fritschel, P.; Barsotti, L.; Evans, M. Resonant dampers for parametric instabilities in gravitational wave detectors. *Phys. Rev. D* **2015**, *92*, 082001. [CrossRef]
101. Hagood, N.; von Flotow, A. Damping of structural vibrations with piezoelectric materials and passive electrical networks. *J. Sound Vib.* **1991**, *146*, 243. [CrossRef]
102. Zhang, J.; Zhao, C.; Ju, L.; Blair, D. Study of parametric instability in gravitational wave detectors with silicon test masses. *Class. Quant. Grav.* **2017**, *34*, 055006. [CrossRef]
103. Tannirkulam, A.K.; Harries, T.J.; Monnier, J.D. The inner rim of YSO disks: Effects of dust grain evolution. *Astrophys. J.* **2007**, *661*, 374. [CrossRef]
104. Turyshev, S.G. Relativistic stellar aberration for the space interferometry mission (2). *arXiv* **2002**, arXiv:gr-qc/0205062.
105. Akeson, R.L.; Boden, A.F.; Monnier, J.D.; Millan-Gabet, R.; Beichman, C.; Beletic, J.; Hartmann, L.; Hillenbrand, L.; Koresko, C.; Sargent, A.; et al. Keck interferometer observations of classical and weak line T tauri stars. *Astrophys. J.* **2005**, *635*, 1173. [CrossRef]
106. Parsons, A.R.; Backer, D.C. Calibration of low-frequency, wide-field radio interferometers using delay/delay-rate filtering. *Astron. J.* **2009**, *138*, 219. [CrossRef]
107. Bradley, R.; Backer, D.; Parsons, A.; Parashare, C.; Gugliucci, N.E. *A Precision Array to Probe the Epoch of Reionization*; Bulletin of the American Astronomical Society: New York, NY, USA, 2005; p. 1216.
108. Nozawa, I.; Gohdo, M.; Kan, K.; Kondoh, T.; Ogata, A.; Yang, J.; Yoshida, Y. *Bunch Length Measurement of Femtosecond Electron Beam by Monitoring Coherent Transition Radiation*; JACoW: Geneva, Switzerland, 2015. [CrossRef]
109. Frank, I.M.; Ginzburg, V.L. Radiation of a Uniformly Moving Electron Due to Its Transition from One Medium to Another. *J. Phys.* **1945**, *9*, 353.
110. Orellana, M.; Cieza, L.A.; Schreiber, M.R.; Merin, B.; Brown, J.M.; Pellizza, L.J.; Romero, G.A. Transition disks: 4 candidates for ongoing giant planet formation in Ophiuchus (Research Note). *Astron. Astrophys.* **2012**, *539*, A41. [CrossRef]
111. Loinard, L.; Allen, R.J. Cold massive molecular clouds in the inner disk of m31. *arXiv* **1998**, arXiv:astro-ph/9801164.
112. Unwin, S.C.; Shao, M.; Tanner, A.M.; Allen, R.J.; Beichman, C.A.; Boboltz, D.; Catanzarite, J.H.; Chaboyer, B.C.; Ciardi, D.R.; Edberg, S.J.; et al. Taking the measure of the universe: Precision astrometry with SIM PlanetQuest. *Publ. Astron. Soc. Pac.* **2008**, *120*, 38. [CrossRef]
113. Lloyd, J.P. Habitable Planet Detection and Characterization with Far Infrared Coherent Interferometry. *arXiv* **2011**, arXiv:1104.4112.
114. Sheth, K.; Regan, M.W.; Vogel, S.N.; Teuben, P.J. Molecular gas, dust and star formation in the barred spiral ngc 5383. *Astrophys. J.* **2000**, *532*, 221. [CrossRef]

115. Jennings, J.; Halverson, S.; Terrien, R.; Mahadevan, S.; Ycas, G.; Diddams, S.A. Frequency stability characterization of a broadband fiber Fabry-Pérot interferometer. *Opt. Express* **2017**, *25*, 15599. [CrossRef]
116. Stadnik, Y. *Manifestations of Dark Matter and Variations of the Fundamental Constants of Nature in Atoms and Astrophysical Phenomena*; Springer: Cham, Switzerland, 2017.
117. Dhurandhar, S.V.; Vecchio, A. Searching for continuous gravitational wave sources in binary systems. *Phys. Rev. D* **2001**, *63*, 122001. [CrossRef]
118. Wang, G.; Ni, W.T. Orbit optimization for ASTROD-GW and its time delay interferometry with two arms using CGC ephemeris. *Chin. Phys. B* **2013**, *22*, 049501. [CrossRef]
119. Seto, N. Detecting planets around compact binaries with gravitational wave detectors in space. *Astrophys. J.* **2008**, *677*, L55. [CrossRef]
120. Cunha, J.V.; Silva, F.E.; Lima, J.A.S. Gravitational waves from ultra short period exoplanets. *Mon. Not. R. Astron. Soc.* **2018**, *480*, L28. [CrossRef]
121. Joergens, V.; Quirrenbach, A. Modeling of closure phase measurements with amber/vlti—towards characterization of exoplanetary atmospheres. *Proc. SPIE Int. Soc. Opt. Eng.* **2004**, *5491*, 551. [CrossRef]
122. Betremieux, Y.; Kaltenegger, L. Transmission spectrum of earth as a transiting exoplanet from the ultraviolet to the near-infrared. *Astrophys. J.* **2013**, *772*, L31. [CrossRef]
123. Ding, F.; Croft, R.A.C. Future dark energy constraints from measurements of quasar parallax: Gaia, SIM and beyond. *Mon. Not. R. Astron. Soc.* **2009**, *397*, 1739. [CrossRef]
124. Papini, G. Zitterbewegung and gravitational Berry phase. *Phys. Lett. A* **2012**, *376*, 1287. [CrossRef]
125. Winterflood, J.; Blair, D.G.; Notcutt, M.; Schilling, R. Position control system for suspended masses in laser interferometer gravitational wave detectors. *Rev. Sci. Instrum.* **1995**, *66*, 2763. [CrossRef]
126. Fujimoto, R. X-Ray Spectroscopic Observations of Intermediate Polars and Mass Determination of White Dwarfs. Ph.D. Thesis, Tokyo University, Tokyo, Japan, 1998.
127. Hees, A.; Do, T.; Ghez, A.M.; Martinez, G.D.; Naoz, S.; Becklin, E.E.; Boehle, A.; Chappel, S.; Chu, D.; Dehghanfar, A.; et al. Testing General Relativity with stellar orbits around the supermassive black hole in our Galactic center. *Phys. Rev. Lett.* **2017**, *118*, 211101. [CrossRef] [PubMed]
128. Weinberg, N.N.; Milosavljevic, M.; Ghez, A.M. Stellar dynamics at the galactic center with a thirty meter telescope. *Astrophys. J.* **2005**, *622*, 878. [CrossRef]
129. Psaltis, D. Testing general relativity with the event horizon telescope. *arXiv* **2018**, arXiv:1806.09740.
130. Zucker, S.; Alexander, T.; Gillessen, S.; Eisenhauer, F.; Genzel, R. Probing post-newtonian gravity near the galactic black hole with stellar doppler measurements. *Astrophys. J.* **2006**, *639*, L21. [CrossRef]
131. Barnes, P.D., Jr.; Caldwell, D.; DaSilva, A. Low background underground facilities for the direct detection of dark matter. In Proceedings of the 1990 Summer Study on High Energy Physics, Snowmass, CO, USA, 25 June–13 July 1990.
132. Giacomelli, G. High-energy astrophysics: Status of observations at large underground detectors. In Proceedings of the 2nd International Workshop on Theoretical and Phenomenological Aspects of Underground Physics, Toledo, Spain, 9–13 September 1991.
133. Giacomelli, G. High-energy underground physics and astrophysics. *Nucl. Phys. (Proc. Suppl.)* **1993**, *33*, 57–76. [CrossRef]
134. Beier, E.W.; Frank, E.D.; Frati, W.; Kim, S.B.; Mann, A.K.; Newcomer, F.M.; Van Berg, R.; Zhang, W.; Hirata, K.S.; Inoue, K.; et al. Survey of atmospheric neutrino data and implications for neutrino mass and mixing. *Phys. Lett. B* **1992**, *283*, 446. [CrossRef]

symmetry

MDPI

Article

The Lanczos Equation on Light-Like Hypersurfaces in a Cosmologically Viable Class of Kinetic Gravity Braiding Theories

Bence Racskó [1] and László Á. Gergely [2,*]

[1] Department of Theoretical Physics, University of Szeged, Tisza L. krt. 84-86, H-6720 Szeged, Hungary; daeron806@gmail.com

[2] Institute of Physics, University of Szeged, Dóm tér 9, H-6720 Szeged, Hungary

* Correspondence: laszlo.a.gergely@gmail.com or gergely@physx.u-szeged.hu

Received: 11 April 2019; Accepted: 22 April 2019; Published: 2 May 2019

Abstract: We discuss junction conditions across null hypersurfaces in a class of scalar–tensor gravity theories (i) with second-order dynamics, (ii) obeying the recent constraints imposed by gravitational wave propagation, and (iii) allowing for a cosmologically viable evolution. These requirements select kinetic gravity braiding models with linear kinetic term dependence and scalar field-dependent coupling to curvature. We explore a pseudo-orthonormal tetrad and its allowed gauge fixing with one null vector standing as the normal and the other being transversal to the hypersurface. We derive a generalization of the Lanczos equation in a 2 + 1 decomposed form, relating the energy density, current, and isotropic pressure of a distributional source to the jumps in the transverse curvature and transverse derivative of the scalar. Additionally, we discuss a scalar junction condition and its implications for the distributional source.

Keywords: scalar–tensor gravity; junction conditions; null hypersurfaces

1. Introduction

Scalar–tensor gravity theories give viable modifications of general relativity in which accelerated expansion could be recovered without dark energy at late times; well-tested solar system constraints could be obeyed (for example through the Vainshtein mechanism); and the recent constraint from gravitational wave detection [1–7] on the propagation speed of the tensorial modes could be successfully implemented. Indeed, from the class of Hordeski theories ensuring second-order dynamics for both the scalar field and the metric tensor [8,9], a subclass has been identified [10–13] in which gravitational waves propagate with the speed of light (as verified both from the almost coincident detection with accompanying γ-rays in the case of the neutron star binary merger and from a stringent test of the dispersion relations disruling massive modes for the 10 black hole mergers). This subclass contains cubic derivative couplings of the scalar field in the Lagrangian, known as kinetic gravity braiding [14,15]. In the Jordan frame, the curvature couples with the scalar through an unspecified function of the scalar field.

This class of scalar–tensor gravity models could be further restricted by the requirement to ensure a viable cosmological evolution. In Ref. [16], it has been proven that for a kinetic gravity braiding model with Lagrangian only linearly and quadratically depending on the kinetic term $X = -(\nabla\phi)^2/2$, an autonomous system of equations governs the dynamics, leading to a number of fixed points for the background dynamics, with three of them representing consecutive radiation-, matter-, and dark energy-dominated regimes (see for example Figure 1 of Ref. [16]). The same model was further analyzed from the string theory-motivated point of view of avoiding de Sitter regimes, which are not embeddable in string theory [17]. Cross-correlating this model class with the requirement of the propagation of tensorial modes with the speed of light, the quadratic dependence has to be dropped. In this paper, we consider this class of kinetic gravity braiding models with only linear dependence on the kinetic terms and analyze the junctions across null hypersurfaces.

Junction conditions in general relativity are known either for spatial or temporal hypersurfaces [18] or for null hypersurfaces [19,20]. The latter are more sophisticated, as the normal to the hypersurface is not suitable for a 3 + 1 space-time decomposition, being also tangent at the same time. The decomposition can be done with respect to a transverse vector, with the gauge arising from its non-unique choice dropping out from the final results [19], or by employing a pseudo-orthonormal basis with two null vectors, one of them playing the role of the normal, the other being transversal [20]. The distributional contribution arising in the curvature from the possible discontinuity of the metric derivative across the hypersurface is related to singular sources on the hypersurface through the Lanczos equation. The same technique led to the derivation of the dynamics on a brane embedded in a 5-dimensional bulk [21–23].

In the full Horndeski class of scalar–tensor gravity theories, junction conditions across spatial or temporal hypersurfaces have been derived [24,25], but the null case stays uncovered, despite its importance being undoubted as all electromagnetic and gravitational shock-waves propagate along such hypersurfaces.

Here we propose to derive such junction conditions for the class of kinetic gravity braiding theories with a linear kinetic terms, which, as discussed above, are both cosmologically viable and obey the gravitational wave constraints. This generalizes our earlier work on null junctions in Brans–Dicke theories [26].

The notations are as follows: space-time indices are Greek, 2-dimensional spatial indices are Latin capital letters. The soldering of any quantity A, with values A^+ and A^- on the two sides of the hypersurface, is $\tilde{A} = A^+\Theta(f) + A^-\Theta(-f)$, where Θ is the step function. The average on the hypersurface is denoted as $\langle A\rangle = (A^+ + A^-)/2$ and the jump over the hypersurface as $[A] = A^+ - A^-$.

2. Equations of Motion

The assumed Lagrangian

$$L_{GKGB} = \underbrace{B(\phi)X + V(\phi)}_{L_2} \underbrace{-2\xi(\phi)\Box\phi X}_{L_3} \underbrace{+\frac{1}{2}F(\phi)R}_{L_4} \tag{1}$$

with B, ξ, F arbitrary functions of the scalar field yields the following expressions through the variation of metric

$$E^{(2)}_{\mu\nu} = -\frac{1}{2}B(\phi)\left(Xg_{\mu\nu} - \phi_\mu\phi_\nu\right) - \frac{1}{2}V(\phi)g_{\mu\nu}, \tag{2}$$

$$E^{(3)}_{\mu\nu} = \xi(\phi)\Box\phi\phi_\mu\phi_\nu + 2\xi'(\phi)X\left(\phi_\mu\phi_\nu + Xg_{\mu\nu}\right) + 2\xi(\phi)X_{(\mu}\phi_{\nu)} - \xi(\phi)X_\kappa\phi^\kappa g_{\mu\nu}, \tag{3}$$

$$E^{(4)}_{\mu\nu} = \frac{1}{2}\left\{F(\phi)G_{\mu\nu} + \left(F'(\phi)\Box\phi - 2F''(\phi)X\right)g_{\mu\nu} - F'(\phi)\phi_{\mu\nu} - F''(\phi)\phi_\mu\phi_\nu\right\}, \tag{4}$$

and through the variation of the scalar field

$$E^{(2)}_{\phi} = B(\phi)\Box\phi - B'(\phi)X + V'(\phi), \tag{5}$$

$$E^{(3)}_{\phi} = \xi(\phi)\left\{(\Box\phi)^2 - \phi_{\mu\nu}\phi^{\mu\nu} - R^{\mu\nu}\phi_\mu\phi_\nu\right\} - 2\xi''(\phi)X^2, \tag{6}$$

$$E^{(4)}_{\phi} = \frac{1}{2}F'(\phi)R, \tag{7}$$

where $\phi_\mu \equiv \nabla_\mu \phi$ and $\phi_{\mu\nu} \equiv \nabla_\nu \nabla_\mu \phi$. The Ricci curvature tensor appears in the expression $E_\phi^{(3)}$ through the Ricci identity $[\nabla_\mu, \nabla_\nu] V^\kappa = R^\kappa{}_{\lambda\mu\nu} V^\lambda$, which has been used to get rid of third derivatives of ϕ.

These are the left-hand sides of the equations of motion (EoMs). The right-hand sides are half of the energy–momentum tensor for the metric variation of the matter action and zero for the scalar field variation, as in the Jordan frame the matter does not couple to the scalar field.

3. Junction Conditions

3.1. The Extrinsic Formulation

We employ a pseudo-orthonormal basis with two null vectors N^μ and L^μ, the first being the normal (surface gradient, which is also tangent) to the hypersurface Σ and the other playing the role of the transverse vector, with respect to which we perform a (2 + 1) + 1 decomposition [20]. The normalization is $L^\mu N_\mu = -1$. The continuity of both the metric tensor $g_{\mu\nu}$ and scalar ϕ are imposed over the hypersurface: $[\phi] = [g_{\mu\nu}] = 0$. Their first derivatives in the null transverse direction $\phi_L \equiv L^\mu \partial_\mu \phi$ and $L^\rho \partial_\rho g_{\mu\nu}$ may have a jump

$$\zeta = [\phi_L], \quad c_{\mu\nu} = \left[L^\rho \partial_\rho g_{\mu\nu}\right], \tag{8}$$

and since all tangential derivatives are assumed to be continuous, we have

$$[\phi_\mu] = -N_\mu \zeta, \quad [\partial_\kappa g_{\mu\nu}] = -N_\kappa c_{\mu\nu}. \tag{9}$$

The second-order derivatives appearing in the equations of motion

$$E_{\mu\nu} \equiv \tilde{E}_{\mu\nu} + \mathscr{E}_{\mu\nu}\delta(f) = \frac{1}{2}\left(\tilde{T}_{\mu\nu} + \mathscr{T}_{\mu\nu}\delta(f)\right), \tag{10}$$

$$E_\phi \equiv \tilde{E}_\phi + \mathscr{E}_\phi \delta(f) = 0 \tag{11}$$

lead to the distributional contributions $\mathscr{E}_{\mu\nu}$ and $\mathscr{E}_\phi$ along the thin shell, arising from the derivative of the step function. All quantities with a tilde are the regular contributions to the respective quantities. For consistency, we also include a distributional energy–momentum tensor $\mathscr{T}_{\mu\nu}$ together with the regular one $\tilde{T}_{\mu\nu}$. In the argument of the delta distribution, f denotes a function which generates the hypersurface as its zero set. For convenience, we also assume that $N_\mu = \nabla_\mu f$.

We introduce the notations

$$c_\mu = c_{\mu\nu} N^\nu, \quad c^\dagger = c_\mu N^\mu, \quad c = c^\mu_\mu \tag{12}$$

and explicitly give the jump of the connection as

$$[\Gamma^\kappa_{\mu\nu}] = -\frac{1}{2}\left(N_\mu c^\kappa_\nu + N_\nu c^\kappa_\mu - N^\kappa c_{\mu\nu}\right), \tag{13}$$

hence the singular parts of the curvature tensor and its traces become

$$\mathscr{R}^\kappa{}_{\lambda\mu\nu} = -\frac{1}{2}\left(N_\mu c^\kappa_\nu N_\lambda - N_\nu c^\kappa_\mu N_\lambda + N_\nu c_{\mu\lambda} N^\kappa - N_\mu c_{\nu\lambda} N^\kappa\right), \tag{14}$$

$$\mathscr{R}_{\mu\nu} = -\frac{1}{2}\left(N_\mu c_\nu + N_\nu c_\mu - N_\mu N_\nu c\right), \tag{15}$$

$$\mathscr{R} = -c^\dagger. \tag{16}$$

In particular, the singular part of the Einstein tensor is

$$\mathscr{G}_{\mu\nu} = -\frac{1}{2}\left(N_\mu c_\nu + N_\nu c_\mu - N_\mu N_\nu c - c^\dagger g_{\mu\nu}\right). \tag{17}$$

We also give the jumps and singular parts of the quantities constructed from the scalar field. As a calligraphic version of ϕ is not catchy, in the decomposition $A = \tilde{A} + \mathscr{A}\delta(f)$ we introduce the alternative notation $\mathscr{A} \equiv \mathrm{Sing}(A)$, denoting the singular part of the arbitrary quantity A.

For the scalar field, we have

$$\mathrm{Sing}(\phi_{\mu\nu}) = -\zeta N_\mu N_\nu, \quad \mathrm{Sing}(\Box\phi) = -\zeta N_\mu N^\mu = 0, \quad [X] = \phi_N \zeta, \tag{18}$$

where $\phi_N = N^\mu \phi_\mu$ is the normal derivative. We note that the value of ϕ_N on the hypersurface is unambigous, being a tangential derivative, which is continuous.

Explicit calculation gives the hypersurface contributions to the left-hand side of the tensorial EoMs:

$$\mathscr{E}^{(2)}_{\mu\nu} = 0, \tag{19}$$

$$\mathscr{E}^{(3)}_{\mu\nu} = \xi(\phi)\zeta\left(2\phi_N N_{(\mu}\langle\phi_{\nu)}\rangle - \phi_N^2 g_{\mu\nu}\right), \tag{20}$$

$$\mathscr{E}^{(4)}_{\mu\nu} = \frac{1}{2}\left(F(\phi)\mathscr{G}_{\mu\nu} + F'(\phi)\zeta N_\mu N_\nu\right) \tag{21}$$

and the hypersurface contributions to the left-hand side of the scalar EoMs:

$$\mathscr{E}^{(2)}_{\phi} = 0, \tag{22}$$

$$\mathscr{E}^{(3)}_{\phi} = \xi(\phi)\left(2\zeta N^\mu N^\nu\langle\phi_{\mu\nu}\rangle + \phi_N c^\mu\langle\phi_\mu\rangle - \frac{1}{2}\phi_N^2 c\right), \tag{23}$$

$$\mathscr{E}^{(4)}_{\phi} = -\frac{1}{2}F'(\phi)c^{+}. \tag{24}$$

3.2. The Intrinsic Formulation

The above equations are expressed in a four-dimensional coordinate system smooth across the hypersurface. Such coordinate systems may be difficult to construct, hence it would be more practical to use coordinantes intrinsic to the junction hypersurface.

The hypersurface contributions to the left-hand side of the tensor EoMs (20) and (21) are tangential in the sense that

$$\mathscr{E}^{(3)}_{\mu\nu}N^\nu = \mathscr{E}^{(4)}_{\mu\nu}N^\nu = 0. \tag{25}$$

Hence, we may expand them in a basis adapted to the junction hypersurface Σ. We choose this basis as $\left(L^\mu, N^\mu, e^\mu_2, e^\mu_3\right)$, where the e^μ_A are two spacelike tangent vector fields to Σ, satisfying

$$N_\mu e^\mu_A = L_\mu e^\mu_A = 0. \tag{26}$$

For a fixed choice of N^μ, we may always choose e^μ_A such that the vector fields $(N^\mu, e^\mu_2, e^\mu_3)$ form a holonomic set, but this is not imperative (we may also choose them to form a pseudo-orthonormal system). The following statements are also valid in the anholonomic case. The inner products of the spacelike vectors generate a spacelike induced metric

$$q_{AB} = g_{\mu\nu}e^\mu_A e^\nu_B \tag{27}$$

on the two-dimensional subspaces spanned by the vectors e^μ_A. Its inverse is denoted q^{AB} (capital Latin indices are raised and lowered by either the metric or its inverse). The completeness relation of the adapted basis is

$$g^{\mu\nu} = -L^\mu N^\nu - N^\mu L^\nu + q^{AB}e^\mu_A e^\nu_B. \tag{28}$$

We further denote $e_1^\mu = N^\mu$, with the Latin indices $a, b, ...$ taking the values $1, 2, ...$ The extrinsic curvature $K_{ab} = e_a^\mu e_b^\nu \frac{1}{2}\mathcal{L}_N g_{\mu\nu}$ is unsuitable to describe the transversal change in the metric as N^μ is also tangential. For this reason, we introduce the transverse curvature [20]:

$$\mathcal{K}_{ab} = \frac{1}{2}e_a^\mu e_b^\nu \mathcal{L}_L g_{\mu\nu}, \tag{29}$$

with its jump related to $c_{\mu\nu}$ as

$$[\mathcal{K}_{ab}] = \frac{1}{2}e_a^\mu e_b^\nu c_{\mu\nu}. \tag{30}$$

The singular part (hypersurface contribution) of the Einstein equation is but the generalized Lanczos equation

$$\mathscr{E}^{\mu\nu} = \frac{1}{2}\mathscr{T}^{\mu\nu}, \tag{31}$$

where $\mathscr{E}$ is the sum of the terms (20) and (21). As the left-hand side is purely tangential, the distributional stress–energy–momentum tensor admits the decomposition

$$\mathscr{T}^{\mu\nu} = \rho N^\mu N^\nu + j^A\left(N^\mu e_A^\nu + e_A^\mu N^\nu\right) + p^{AB}e_A^\mu e_B^\nu, \tag{32}$$

where ρ, j^A, and p^{AB} are the energy density, current vector, and stress tensor of the distributional source, respectively. These quantities, defined as the components emerging with respect to the intrinsic triad of vectors, can be evaluated even when the bulk coordinates do not match smoothly along Σ. They are defined as

$$\rho = 2\mathscr{E}_{\mu\nu}L^\mu L^\nu, \quad j_A = -2\mathscr{E}_{\mu\nu}L^\mu e_A^\nu, \quad p^{AB} = 2\mathscr{E}_{\mu\nu}e_A^\mu e_B^\nu. \tag{33}$$

The 2 + 1 decomposition of Equation (31) yields an isotropic pressure $p^{AB} = pq^{AB}$ and

$$\rho = F(\phi)[\mathcal{K}_{AB}]q^{AB} + F'(\phi)[\phi_L] - 2\xi(\phi)\phi_N[\phi_L^2], \tag{34}$$

$$j_A = -F(\phi)[\mathcal{K}_{NA}] + 2\xi(\phi)[\phi_L]\phi_N\phi_A, \tag{35}$$

$$p = F(\phi)[\mathcal{K}_{NN}] - 2\xi(\phi)[\phi_L]\phi_N^2, \tag{36}$$

where $\phi_A = e_A^\mu\phi_\mu$, $\mathcal{K}_{NA} \equiv \mathcal{K}_{1A}$, and $\mathcal{K}_{NN} \equiv \mathcal{K}_{11}$.

The scalar equation is

$$\begin{aligned} 0 \quad = \quad & \xi(\phi)\phi_N^2 q^{AB}[\mathcal{K}_{AB}] - 2\xi(\phi)\phi_N\phi^A[\mathcal{K}_{NA}] \\ & + \left(F'(\phi) + 2\xi(\phi)\phi_N\langle\phi_L\rangle\right)[\mathcal{K}_{NN}] - 2\xi(\phi)[\phi_L]\left(\phi_{NN} - \langle\mathcal{K}_{NN}\rangle\phi_N\right), \end{aligned} \tag{37}$$

which contains jumps and averages. However, by exploring the relation $[A]\langle B\rangle + \langle A\rangle[B] = [AB]$, the averages can be transformed away to obtain

$$\begin{aligned} 0 = & \xi(\phi)\phi_N^2 q^{AB}[\mathcal{K}_{AB}] - 2\xi(\phi)\phi_N\phi^A[\mathcal{K}_{NA}] + F'(\phi)[\mathcal{K}_{NN}] \\ & - 2\xi(\phi)[\phi_L]\phi_{NN} + 2\xi(\phi)\phi_N[\phi_L\mathcal{K}_{NN}]. \end{aligned} \tag{38}$$

Equations (34)–(36) provide generalizations of the Lanczos equation, and Equation (38) a constraint on the distributional sources.

3.3. Gauge Fixing

At this point, it is worthwhile to remember that there is still gauge freedom in the tetrad choice. The normal vector field is autoparallel [20]

$$N^\nu\nabla_\nu N^\mu = \kappa N^\mu \tag{39}$$

with the non-affinity parameter $\kappa = \mathcal{K}_{NN}$. If the null fields are rescaled as $\bar{N}^{\mu} = e^{\alpha} N^{\mu}$ and $\bar{L}^{\mu} = e^{-\alpha} L^{\mu}$, with some function α defined on the hypersurface, then the non-affinity parameter changes as

$$\bar{\kappa} = e^{\alpha} \left(N^{\nu} \nabla_{\nu} \alpha + \kappa \right), \tag{40}$$

while

$$\phi_{\bar{L}} = e^{-\alpha} \phi_{L}.$$

Hence,

$$\phi_{\bar{L}} \bar{\mathcal{K}}_{\bar{N}\bar{N}} = \phi_{L} \left(N^{\nu} \nabla_{\nu} \alpha + \mathcal{K}_{NN} \right). \tag{41}$$

It is possible to achieve

$$[\phi_{\bar{L}} \bar{\mathcal{K}}_{\bar{N}\bar{N}}] = 0 \tag{42}$$

through any solution of the differential equation

$$\frac{\partial \alpha}{\partial \lambda} = -\langle \kappa \rangle - \frac{\langle \phi_L \rangle}{[\phi_L]} [\kappa], \tag{43}$$

where λ is a coordinate adapted to N^{μ}, and the ratio $\langle \phi_L \rangle / [\phi_L]$ is a function on the hypersurface, being evaluated there. Hence, in this gauge, the last term of Equation (38) drops out.

4. Discussion of the Junction Conditions

From the 2 + 1 decomposed form of the tensorial junction conditions, we may express the jumps in the components of the transverse curvature in terms of the distributional energy density, current, and isotropic pressure, as well as the jump of the transverse derivative of the scalar field and its square, as follows

$$[\mathcal{K}_{AB}] q^{AB} = \frac{\rho}{F} - (\ln F)' [\phi_L] + \frac{2\xi \phi_N}{F} [\phi_L^2], \tag{44}$$

$$[\mathcal{K}_{NA}] = -\frac{j_A}{F} + \frac{2\xi \phi_N}{F} \phi_A [\phi_L], \tag{45}$$

$$[\mathcal{K}_{NN}] = \frac{p}{F} + \frac{2\xi \phi_N^2}{F} [\phi_L]. \tag{46}$$

Then the scalar junction equation (in the gauge where $[\phi_L \mathcal{K}_{NN}] = 0$) becomes

$$\begin{aligned} & F' p + \xi \phi_N \left(\phi_N \rho + 2\phi^A j_A \right) \\ = \; & \xi \left(2F \phi_{NN} - F' \phi_N^2 + 4\xi \phi_N^2 \phi_A \phi^A \right) [\phi_L] - 2\xi^2 \phi_N^3 [\phi_L^2]. \end{aligned} \tag{47}$$

There are two cases when these equations simplify considerably: (A) when there is no cubic derivative coupling $\xi = 0$, and (B) when the normal derivative of the scalar field vanishes $\phi_N = 0$. In both cases, the scalar Equation (47) shows that there is no isotropic pressure $p = 0$, that the third Lanczos Equation (46) implies $[\mathcal{K}_{NN}] = 0$, the second Lanczos Equation (45) gives the current as $j_A = -F(\phi) [\mathcal{K}_{NA}]$, and finally, the first Lanczos Equation (44) constrains the energy density as $\rho = F(\phi) [\mathcal{K}_{AB}] q^{AB} + F'(\phi) [\phi_L]$.

5. Concluding Remarks

By exploring a formalism based on a transverse null vector to the null hypersurface, we derived junction conditions across null shells in the kinetic gravity braiding theories with linear kinetic term dependence, in which the curvature and the scalar couples through a generic scalar field-dependent function. These scalar–tensor theories obey both the gravitational wave constraints and could exhibit a viable cosmological evolution through radiation-, matter-, and dark energy-dominated fixed points.

Our formalism gives the necessary equations to discuss energetic shock waves propagating with the speed of light in these models.

The junction conditions contain the 2 + 1 decomposed form of the tensorial equation, a generalization of the general relativistic Lanczos equation. This relates the jump in the transverse curvature to the distributional energy density, current, and isotropic pressure. In the relations, the jump of the transverse derivative of the scalar and its square are also involved. An additional scalar equation, without counterpart in general relativity, constrains all of these functions.

If either there is no cubic derivative coupling term $\xi = 0$, or the scalar field does not change in the normal direction to the null hypersurface $\phi_N = 0$, the junction conditions simplify considerably, leaving the possibility of a distributional source without pressure

$$\mathscr{T}^{\mu\nu} = \left(F(\phi)\,[\mathcal{K}_{AB}]q^{AB} + F'(\phi)\,[\phi_L]\right) N^{\mu}N^{\nu} - F(\phi)\,[\mathcal{K}_{NA}](N^{\mu}e^{\nu}_{A} + e^{\mu}_{A}N^{\nu}), \tag{48}$$

together with the geometric condition $[\mathcal{K}_{NN}] = 0$. These generalize the corresponding result found for Brans–Dicke theories in the Jordan frame [26].

Author Contributions: Conceptualization, L.Á.G.; methodology, L.Á.G. and B.R.; validation, L.Á.G. and B.R.; formal analysis, B.R.; writing—original draft preparation, B.R. and L.Á.G.; writing—review and editing, B.R. and L.Á.G.; supervision, L.Á.G.; funding acquisition, L.Á.G. and B.R.

Funding: This research was funded by the Hungarian National Research Development and Innovation Office (NKFIH) in the form of grant 123996 and carried out in the framework of European Cooperation in Science and Technology (COST) actions CA15117 (CANTATA) and CA16104 (GWverse) supported by COST. During the preparation of this manuscript, B.R. was supported by the UNKP-18-3 New National Excellence Program of the Hungarian Ministry of Human Capacities.

Conflicts of Interest: The authors declare no conflict of interest.

References

1. LIGO Scientific Collaboration and Virgo Collaboration. Observation of Gravitational Waves from a Binary Black Hole Merger. *Phys. Rev. Lett.* **2016**, *116*, 061102. [CrossRef]
2. LIGO Scientific Collaboration and Virgo Collaboration. GW151226: Observation of Gravitational Waves from a 22-Solar-Mass Binary Black Hole Coalescence. *Phys. Rev. Lett.* **2016**, *116*, 241103. [CrossRef]
3. LIGO Scientific Collaboration and Virgo Collaboration. GW170104: Observation of a 50-Solar-Mass Binary Black Hole Coalescence at Redshift 0.2. *Phys. Rev. Lett.* **2017**, *118*, 221101. [CrossRef] [PubMed]
4. LIGO Scientific Collaboration and Virgo Collaboration. GW170608: Observation of a 19-Solar-Mass Binary Black Hole Coalescence. *Astrophys. J. Lett.* **2017**, *851*, L35. [CrossRef]
5. LIGO Scientific Collaboration and Virgo Collaboration. GW170814: A Three-Detector Observation of Gravitational Waves from a Binary Black Hole Coalescence. *Phys. Rev. Lett.* **2017**, *119*, 141101. [CrossRef]
6. LIGO Scientific Collaboration and Virgo Collaboration. GW170817: Observation of Gravitational Waves from a Binary Neutron Star Inspiral. *Phys. Rev. Lett.* **2017**, *119*, 161101. [CrossRef] [PubMed]
7. LIGO Scientific Collaboration and Virgo Collaboration. GWTC-1: A Gravitational-Wave Transient Catalog of Compact Binary Mergers Observed by LIGO and Virgo during the First and Second Observing Runs. *arXiv* **2018**, arXiv:1811.12907.
8. Horndeski, G.W. Second-order scalar-tensor field equations in a four-dimensional space. *Int. J. Theor. Phys.* **1974**, *10*, 363–384. [CrossRef]
9. Deffayet, C.; Gao, X.; Steer, D.A.; Zahariade, G. From k-essence to generalized Galileons. *Phys. Rev. D* **2011**, *84*, 064039. [CrossRef]
10. Baker, T.; Bellini, E.; Ferreira, P.G.; Lagos, M.; Noller, J.; Sawicki, I. Strong constraints on cosmological gravity from GW170817 and GRB 170817A. *Phys. Rev. Lett.* **2017**, *119*, 251301. [CrossRef]
11. Ezquiaga, J.M.; Zumalacárregui, M. Dark Energy after GW170817: Dead ends and the road ahead. *Phys. Rev. Lett.* **2017**, *119*, 251304. [CrossRef] [PubMed]
12. Creminelli, P.; Vernizzi, F. Dark Energy after GW170817 and GRB170817A. *Phys. Rev. Lett.* **2017**, *119*, 251302. [CrossRef] [PubMed]

13. Sakstein, J.; Jain, B. Implications of the Neutron Star Merger GW170817 for Cosmological Scalar-Tensor Theories. *Phys. Rev. Lett.* **2017**, *119*, 251303. [CrossRef] [PubMed]
14. Kase, R.; Tsujikawa, S. Dark energy in Horndeski theories after GW170817: A review. *arXiv* **2019**, arXiv:1809.08735.
15. Deffayet, C.; Pujolas, O.; Sawicki, I.; Vikman, A. Imperfect Dark Energy from Kinetic Gravity Braiding. *arXiv* **2010**, arXiv:1008.0048.
16. Kase, R.; Tsujikawa, S.; Felice, A.D. Cosmology with a successful Vainshtein screening in theories beyond Horndeski. *Phys. Rev. D* **2016**, *93*, 024007. [CrossRef]
17. Heisenberg, L.; Bartelmann, M.; Brandenberger, R.; Refregier, A. Horndeski in the Swampland. *arXiv* **2019**, arXiv:1902.03939.
18. Israel, W. Singular hypersurfaces and thin shells in general relativity. *Nouvo Cim. B* **1966**, *44*, 1–14. [CrossRef]
19. Barrabés, C.; Israel, W. Thin shells in general relativity and cosmology: The lightlike limit. *Phys. Rev. D* **1991**, *43*, 1129–1142. [CrossRef]
20. Poisson, E. *A Relativist's Toolkit: The Mathematics of Black-Hole Mechanics*; Cambridge University Press: Cambridge, UK, 2004.
21. Shiromizu, T.; Maeda, K.I.; Sasaki, M. The Einstein equations on the 3-brane world. *Phys. Rev. D* **2000**, *62*, 024012. [CrossRef]
22. Gergely, L.Á. Generalized Friedmann branes. *Phys. Rev. D* **2003**, *68*, 124011. [CrossRef]
23. Gergely, L.Á. Friedmann branes with variable tension. *Phys. Rev. D* **2008**, *78*, 084006. [CrossRef]
24. Padilla, A.; Sivanesan, V. Boundary Terms and Junction Conditions for Generalized Scalar-Tensor Theories. *arXiv* **2012**, arXiv:1206.1258.
25. Nishi, S.; Kobayashi, T.; Tanahashi, N.; Yamaguchi, M. Cosmological matching conditions and galilean genesis in Horndeski's theory. *arXiv* **2014**, arXiv:1401.1045.
26. Racskó, B.; Gergely, L.Á. Light-Like Shockwaves in Scalar-Tensor Theories. *Universe* **2018**, *4*, 44. [CrossRef]

symmetry

MDPI

Article

Effect of Quantum Gravity on the Stability of Black Holes

Riasat Ali [1], Kazuharu Bamba [2,*] and Syed Asif Ali Shah [1]

[1] Department of Mathematics, GC University Faisalabad Layyah Campus, Layyah 31200, Pakistan; riasatyasin@gmail.com (R.A.); asifalishah695@gmail.com (S.A.A.S.)

[2] Division of Human Support System, Faculty of Symbiotic Systems Science, Fukushima University, Fukushima 960-1296, Japan

* Correspondence: bamba@sss.fukushima-u.ac.jp

Received: 10 April 2019; Accepted: 26 April 2019; Published: 5 May 2019

Abstract: We investigate the massive vector field equation with the WKB approximation. The tunneling mechanism of charged bosons from the gauged super-gravity black hole is observed. It is shown that the appropriate radiation consistent with black holes can be obtained in general under the condition that back reaction of the emitted charged particle with self-gravitational interaction is neglected. The computed temperatures are dependant on the geometry of black hole and quantum gravity. We also explore the corrections to the charged bosons by analyzing tunneling probability, the emission radiation by taking quantum gravity into consideration and the conservation of charge and energy. Furthermore, we study the quantum gravity effect on radiation and discuss the instability and stability of black hole.

Keywords: higher dimension gauged super-gravity black hole; quantum gravity; quantum tunneling phenomenon; Hawking radiation

1. Introduction

General relativity is associated with the thermodynamics and quantum effect which are strongly supportive of each other. A black hole (BH) is a compact object whose gravitational pull is so intense that can not escape the light. It was proved by Hawking that a BH has an additional property of emitting radiation. Since Hawking's great contribution on BH thermodynamics, the radiation from the BH has attained the attention of many researchers. There are many different process to obtain the Hawking radiation by applying the quantum field equations or the semi-classical phenomena. Different accesses to quantum gravity, as well as BH physics predict a minimum measure length or a maximum evident momentum and associated modifications of the principle of the Heisenberg uncertainty which is called the generalized uncertainty principle (GUP).

The thermal radiation coming from any stationary metric are calculated [1]. The physical image is that the radiation develops in the quasiclassical tunneling of particles from a gravitational barrier. They obtained a thermal spectrum and twice the temperature for Hawking radiation of non-rotating BH. The expression $\exp(-2Im(\int pdr))$ is not invariant under canonical transformation in generally and expressed that this implies half the correct temperature for BH [2]. In the setting of black rings significance, the radiation of the Dirac particles can be calculated by applying the Dirac wave equation in both the charged and uncharged case. The formulate of the field equations of uncharged and charged Dirac particles by using the covariant Dirac wave equation [3]. E. T. Akhmedov et al. [4] calculated Hawking radiation by using the quasi-classical phenomenon. The temporal contribution to gravitational WKB-like calculations are discussed in [5]. The authors analyzed that the quasiclassical method for gravitational backgrounds contains subtleties not found in the usual quantum mechanical tunneling problem.

V. Akhmedova et al. [6] compared the anomaly method and the WKB/tunneling method for finding radiation through non-trivial space-time. They conclude that these both method are not valid for all types of metrics. The discreteness space effect of the GUP are investigated in space [7]. Corda [8] analyzed interferometric detection of gravitational waves: the definitive test for general relativity. He conclude that accurate angular and frequency dependent response functions of interferometers for gravitational waves arising from various theories of gravity will be the definitive test for general relativity. The authors investigated insights and possible resolution to the information loss paradox via the tunneling picture [9]. They observe that the quantum correction give zero temperature for the radiation as the mass of the BH is zero.

From F(R) theory to Lorentz non-invariant models in modified gravity are investigated as [10]. The extended theories of gravity are discussed in [11]. The authors analyzed the problems of gravitational waves and neutrino oscillations through extended gravity theory. The authors [12] examined the rule to all alternative gravities, a particularly significance of scalar-tensor and f(R) theories. Yale [13] analyzed the exact Hawking radiation of scalars, fermions and bosons 1-spin particles applying quantum tunneling phenomena without back reaction. The different dark energy models like Λ cold dark matter, Pseudo-Rip and Little Rip universes, non-singular dark energy universes, the quintessence and phantom cosmologies with different types are analyzed [14].

Sharif and Javed [15] analyzed the Hawking radiation of fermion particles applying quantum tunneling phenomena from traversable wormholes. Corda [16] studied the important issue that the non-strictly continuous character of the Hawking radiation spectrum generates a natural correspondence between Hawking radiation and quasi-normal modes BH. Jan and Gohar [17] examined the Hawking temperature by quantum tunneling of scalars particles applying Klein-Gordon equation in WKB approximation. Kruglov [18] calculated the Hawking radiation by quantum tunneling of vector particles of BHs in 2 dimension applying Proca equation in WKB approximation. Matsumoto et al. [19] analyzed the time evolution of a thin black ring via Hawking radiation.

The different writers [20] determined the Hawking temperature by Hamilton-Jacobi equation of vector particles of Kerr and Kerr-Newman BHs by applying Proca and Lagrangian equations in WKB approximation. Corda [21] analyzed precise model of Hawking radiation from the tunneling mechanism and he found that pre-factor of the Parikh and Wilczek probability of emission depends on the BH quantum level. Anacleto [22] analyzed the GUP in the tunneling phenomena through Hamilton–Jacobi process to find the corrected temperature and entropy for three-dimensional noncommutative acoustic BHs. Anacleto et al. [23] studied the Hawking temperature by the Hamilton–Jacobi equation of spin $\frac{3}{2}$-particles of accelerating BHs, applying the Rarita–Schwinger equation in the WKB approximation. Chen and Huang [24] determined the Hawking temperature by quantum tunneling phenomena of vector particles of Vaidya BHs in applying the Proca equation in WKB approximation. Anacleto et al. [25] examined the quantum-corrected of self-dual BH entropy in tunneling phenomena with GUP. Li and Zu [26] analyzed the tunneling phenomena by the Hamilton–Jacobi equation of scalar particles of Gibbons–Maeda–Dilation BHs, applying the Klein–Gordon equation in the WKB approximation. Feng et al. [27] calculated the tunneling phenomena by the Hamilton–Jacobi equation of scalar particles of 4D and 5D BHs, applying the Proca equation in the WKB approximation. Saleh et al. [28] studied the Hawking radiation of 5D Lovelock BH with the Hamilton–Jacobi equation by using the Klein–Gordon equation.

The authors [29] analyzed the cosmology of inflation by modifying terms of gravity and inflation in F(R) gravity. In the F(R) gravity, the Starobinsky inflation is discussed with the geometry of gravitational theories to the inflationary models. Övgun and Jusufi [30] calculated the tunneling phenomena by Hamilton–Jacobi process in a Lagrangian equation of spin-1 massive particle noncommutative BHs. Jusufi and Övgun [31] examined the Hawking temperature of vector and scalar particles from 5D Myers–Perry BHs and solved the Proca and Klein–Gordon equations by using the WKB approximation and Hamilton–Jacobi process in these cases.

The cosmological solutions, BH solutions and spherically symmetric developing through F(T) gravity were discussed in different cosmic expansion eras [32]. Singh et al. [33] discussed the Hawking temperature of vector particles from Kerr–Newman BHs in the Proca equation by applying the WKB approximation in the Hamilton–Jacobi process. Jusufi and Övgun [34] examined the Hawking radiation of massive particles from rotating charged black strings. Li and Zhao [35] calculated the tunneling process of massive particles from the neutral rotating anti-de Sitter BHs using the Proca wave equation in the WKB approximation. The different authors [36,37] determined the temperature of massive vector particles from the different types BHs by using tunneling phenomena. The nutshell, bounce, late time evolution and inflation were studied through modified gravity theories [38]. The future of gravitational theories in the framework of gravitational wave in astronomy was analyzed in [39]. The charged vector particles tunneling from black ring and 5D BH [40] is studied by wave equation to calculate the tunneling phenomena for charged particles as well as Hawking temperature. In this article, the authors have calculated the tunneling probability/rate and Hawking temperature for charged boson particles tunneling from horizon.

This paper is organization as follows: in Section 2 we discuss the tunneling rate and Hawking temperature of charged vector $W^{\pm}$ boson particles for 4D gauged super-gravity BH and also calculate quantum corrected tunneling probability and Hawking temperature. Section 3 is based on the analysis of for 7D gauged super-gravity BH. In Section 4, we discuss the graphical behavior of radiation for these types of BHs and visualize the stable and unstable state of BHs. In Section 5 we explain the conclusions and discussion.

2. 4-Dimension Gauged Super-Gravity Black Holes

The super-gravity theory defined gauged theory through which the gauge boson, the super-partner of the particle is charged in some internal gauge group. Moreover, this theory is more important as compared to the ungauged case, therefore this theory has a negative cosmological constant (Λ), where Λ is stated in an anti-de Sitter BH. Now, for the study of a boson particle tunneling process form a BH in $(3+1)$ dimension theory of gauged super-gravity, we calculate the Hawking temperature of BH by tunneling phenomena at event horizon. The solution of BH occur in $D=4$ $N=8$ theory of gauged super-gravity (symmetry) [41]. The metric for such theory is given by [41]

$$ds^2 = -\left(H_1H_2H_3H_4\right)^{-\frac{1}{2}} fdt^2 + \left(H_1H_2H_3H_4\right)^{\frac{1}{2}} \left(f^{-1}dr^2 + r^2 d\Omega^2_{2,k}\right), \tag{1}$$

where $g = 1/L$ and L is related to the cosmological constant $\Lambda = -3g^2 = -3/L^2$ and the μ represent the non-extremality parameter [42]

$$f = k - \frac{\mu}{r^2} + g^2 r^2 H_1H_2H_3H_4, \quad H_i = \frac{q_i}{r^2} + 1, \quad (\text{for } i = 1,2,3,4).$$

for radius $k=1$ and $k=0$, then $d\Omega^2_{2,k}$ represents the metrics on $\mathbf{S}^2$ and $\mathbf{R}^2$ respectively. The four electric potentials A^i_μ are defined as;

$$A^i_0 = \frac{\tilde{q}_i}{r^2 + q_i} \quad (\text{for } i = 1,2,3,4),$$

where q_i and $\tilde{q}_i$ represent charges and physical charges of a BH. The line element from Equation (1) can be rewritten as

$$ds^2 = -F(r)dt^2 + L^{-1}(r)dr^2 + M(r)d\theta^2 + N(r)d\phi^2, \tag{2}$$

where

$$\begin{aligned} F(r) &= f(H_1H_2H_3H_4)^{-\frac{1}{2}} \qquad L^{-1}(r) = f^{-1}(H_1H_2H_3H_4)^{\frac{1}{2}} \\ M(r) &= r^2(H_1H_2H_3H_4)^{\frac{1}{2}} \qquad N(r) = r^2\sin^2\theta(H_1H_2H_3H_4)^{\frac{1}{2}}. \end{aligned}$$

The wave equation of motion comprises of GUP obtained from the Glashow–Weinberg–Salam model [20,43]

$$\begin{aligned} &\partial_\mu(\sqrt{-g}\Phi^{\nu\mu}) + \sqrt{-g}\frac{m^2}{\hbar^2}\Phi^\nu + \sqrt{-g}\frac{i}{\hbar}eA_\mu\Phi^{\nu\mu} + \sqrt{-g}\frac{i}{\hbar}eF^{\nu\mu}\Phi_\mu + \alpha\hbar^2\partial_0\partial_0\partial_0 \\ &(\sqrt{-g}g^{00}\Phi^{0\nu}) - \alpha\hbar^2\partial_i\partial_i\partial_i(\sqrt{-g}g^{ii}\Phi^{i\nu}) = 0, \end{aligned} \tag{3}$$

where g is a determinant coefficient matrix, $\Phi^{\mu\nu}$ is anti-symmetric tensor and m is particles mass, since

$$\begin{aligned} \Phi_{\nu\mu} &= (1-\alpha\hbar^2\partial_\nu^2)\partial_\nu\Phi_\mu - (1-\alpha\hbar^2\partial_\mu^2)\partial_\mu\Phi_\nu + (1-\alpha\hbar^2\partial_\nu^2)\frac{\iota}{\hbar}eA_\nu\Phi_\mu \\ &- (1-\alpha\hbar^2\partial_\mu^2)\frac{\iota}{\hbar}eA_\mu\Phi_\nu \end{aligned}$$

where α, A_μ, e and $\triangle_\mu$ are the quantum gravity parameter (dimensionless positive parameter), vector potential of the charged BH, the charge of the particle and covariant derivative, respectively. As the wave equations for the W^+ and W^- boson particles are alike, the tunneling actions should be alike too ($W^+ = -W^-$). We will view the W^+ boson particle case after simplification and the results of such case can be changed to multiply negative sign W^- boson particles due to the digitalization of the metric. There value of Φ^μ and $\Phi^{\nu\mu}$ are given by

$$\begin{aligned} \Phi^0 &= \frac{\Phi_0}{F(r)}, \quad \Phi^1 = \frac{\Phi_1}{L^{-1}(r)}, \quad \Phi^2 = \frac{\Phi_2}{M(r)}, \quad \Phi^3 = \frac{\Phi_3}{N(r)}, \\ \Phi^{01} &= \frac{\Phi_{01}}{F(r)L^{-1}(r)}, \quad \Phi^{02} = \frac{\Phi_{02}}{F(r)M(r)}, \quad \Phi^{03} = \frac{\Phi_{03}}{F(r)N(r)}, \\ \Phi^{12} &= \frac{\Phi_{12}}{L^{-1}(r)M(r)}, \quad \Phi^{13} = \frac{\Phi_{13}}{L^{-1}(r)N(r)}, \quad \Phi^{23} = \frac{\Phi_{23}}{M(r)N(r)}. \end{aligned}$$

The WKB approximation is given in [44], i.e.,

$$\Phi_\nu = c_\nu \exp[\frac{i}{\hbar}\oplus_0(t,r,\theta,\phi) + \sum_{i=1}^{i=n}\hbar^i\oplus_i(t,r,\theta,\phi)]. \tag{4}$$

Substituting the Equation (4) into the wave Equation (3), where $i = 1, 2, 3, ...$ neglecting the terms. We get the set of equations below:

$$
\begin{aligned}
&L(r)[c_1(\partial_0\oplus_0)(\partial_1\oplus_0) + c_1\alpha(\partial_0\oplus_0)^3(\partial_1\oplus_0) - c_0(\partial_1\oplus_0)^2 - c_0(\partial_1\oplus_0)^4\alpha \\
&+eA_0c_1(\partial_1\oplus_0) + eA_0c_1\alpha(\partial_1\oplus_0)(\partial_0\oplus_0)^2] + \frac{1}{M(r)}[c_2(\partial_0\oplus_0)(\partial_2\oplus_0) + \alpha c_2 \\
&(\partial_0\oplus_0)^3(\partial_2\oplus_0) - c_0(\partial_2\oplus_0)^2 - \alpha c_0(\partial_2\oplus_0)^4 + eA_0c_2(\partial_2\oplus_0) + \alpha eA_0c_2(\partial_0\oplus_0)^2 \\
&(\partial_2\oplus_0)] + \frac{1}{N(r)}[c_3(\partial_0\oplus_0)(\partial_3\oplus_0) + \alpha c_3(\partial_0\oplus_0)^3(\partial_3\oplus_0) + c_0(\partial_3\oplus_0)^2 \\
&+\alpha c_0(\partial_3\oplus_0)^4 + eA_0c_3(\partial_3\oplus_0) + \alpha c_3eA_0(\partial_0\oplus_0)^2(\partial_3\oplus_0)] - m^2c_0 = 0
\end{aligned} \tag{5}
$$

$$
\begin{aligned}
&\frac{-1}{F(r)}[c_0(\partial_0\oplus_0)(\partial_1\oplus_0) + c_0\alpha(\partial_0\oplus_0)(\partial_1\oplus_0)^3 - c_1(\partial_0\oplus_0)^2 - c_1\alpha(\partial_0\oplus_0)^4 \\
&-eA_0c_1(\partial_0\oplus_0) - \alpha eA_0c_1(\partial_1\oplus_0)^2(\partial_0\oplus_0)] + \frac{1}{M(r)}[c_2(\partial_1\oplus_0)(\partial_2\oplus_0) \\
&+\alpha c_2(\partial_1\oplus_0)^3(\partial_2\oplus_0) - c_1(\partial_2\oplus_0)^2 - \alpha c_1(\partial_2\oplus_0)^4] + \frac{1}{N(r)}[c_3(\partial_1\oplus_0)(\partial_3\oplus_0) \\
&+c_3\alpha(\partial_1\oplus_0)^3(\partial_3\oplus_0) - c_1(\partial_3S_0)^2 - c_1\alpha(\partial_3\oplus_0)^4] - m^2c_1 - \frac{1}{F}eA_0 \\
&[c_0(\partial_1\oplus_0) + \alpha c_0(\partial_1\oplus_0)^3 - c_1(\partial_0\oplus_0) - \alpha c_1(\partial_0\oplus_0)^3 - c_1eA_0 \\
&-eA_0\alpha c_1(\partial_1\oplus_0)^2] = 0
\end{aligned} \tag{6}
$$

$$
\begin{aligned}
&\frac{1}{F(r)}[c_0(\partial_0\oplus_0)(\partial_2\oplus_0) + \alpha c_0(\partial_0\oplus_0)(\partial_2\oplus_0)^3 - c_2(\partial_0\oplus_0)^2 - \alpha c_2(\partial_0\oplus_0)^4 \\
&-eA_0(\partial_0\oplus_0)c_2 - eA_0(\partial_0\oplus_0)^3c_2\alpha] - \frac{1}{L^{-1}(r)}[c_2(\partial_1\oplus_0)^2 + \alpha c_2(\partial_1\oplus_0)^4 \\
&-c_1(\partial_1\oplus_0)(\partial_2\oplus_0) - \alpha c_1(\partial_1\oplus_0)(\partial_2\oplus_0)^3] + \frac{1}{N(r)}[c_3(\partial_2\oplus_0)(\partial_3\oplus_0) \\
&+\alpha c_3(\partial_2\oplus_0)^3(\partial_3\oplus_0) - c_2(\partial_3\oplus_0)^2 - \alpha c_2(\partial_3\oplus_0)^4] - \frac{eA_0}{F(r)}[c_0(\partial_2\oplus_0) \\
&+\alpha c_0(\partial_2\oplus_0)^3 - c_2(\partial_0\oplus_0) - \alpha c_2(\partial_0\oplus_0)^3 + c_2eA_0 + \alpha c_2eA_0(\partial_0\oplus_0)^2] \\
&-m^2c_2 = 0
\end{aligned} \tag{7}
$$

$$
\begin{aligned}
&\frac{1}{F(r)}[c_0(\partial_0\oplus_0)(\partial_3\oplus_0) + \alpha c_0(\partial_0\oplus_0)(\partial_3\oplus_0)^3 - c_3(\partial_0\oplus_0)^2 - \alpha c_3(\partial_0\oplus_0)^4 \\
&-eA_0(\partial_0\oplus_0)c_3 - eA_0(\partial_3\oplus_0)^2(\partial_0\oplus_0)c_3\alpha] + \frac{1}{L^{-1}(r)}[c_3(\partial_1\oplus_0)^2 + \alpha c_3(\partial_1\oplus_0)^4 \\
&-c_1(\partial_3\oplus_0)(\partial_1\oplus_0) - \alpha c_1(\partial_1\oplus_0)(\partial_3\oplus_0)^3] + \frac{1}{M(r)}[c_3(\partial_2\oplus_0)^2 \\
&+\alpha c_3(\partial_2\oplus_0)^4 - c_2(\partial_2\oplus_0)(\partial_3\oplus_0) - \alpha c_2(\partial_3\oplus_0)^3(\partial_2\oplus_0)] + \frac{eA_0}{F(r)}[c_0(\partial_3\oplus_0) \\
&+\alpha c_0(\partial_3\oplus_0)^3 - c_3(\partial_0\oplus_0) - \alpha c_3(\partial_0\oplus_0)^3 - c_3eA_0 - \alpha c_3eA_0(\partial_3\oplus_0)^2] \\
&-m^2c_3 = 0.
\end{aligned} \tag{8}
$$

We can choose the separation of variables,

$$\oplus_0 = -(E - j\Omega)t + W(r) + j\phi + v(\theta), \tag{9}$$

where j, E and Ω represent angular momentum, energy and angular velocity of particle, respectively. Here, $W(r)$ and $v(\theta)$ are two arbitrary functions. The matrix equation can be obtain from the Equations (5)–(8),

$$K(c_0, c_1, c_2, c_3)^T = 0,$$

which gives "K" is a order of '4×4' matrix and its components are given by:

$$\begin{aligned}
K_{00} &= \frac{\dot{W}^2 + \alpha\dot{W}^4}{L^{-1}(r)} - \frac{j^2 + \alpha j^4}{M(r)} + \frac{\dot{v}^3 + \alpha\dot{v}^4}{N(r)} - m^2, \\
K_{01} &= -\frac{\dot{W}(E - j\Omega) + \alpha\dot{W}(E - j\Omega)^3}{L^{-1}(r)} + \frac{\dot{W}eA_0 + \alpha\dot{W}eA_0(E - j\Omega)^2}{L^{-1}(r)}, \\
K_{02} &= -\frac{(E - j\Omega)j + \alpha(E - j\Omega)j}{M(r)} + \frac{eA_0 j + \alpha(E - j\Omega)^2 eA_0 j}{M(r)}, \\
K_{03} &= -\frac{\dot{v}(E - j\Omega) + \alpha\dot{v}(E - j\Omega)^3}{N(r)} + \frac{eA_0\dot{v} + \alpha eA_0\dot{v}(E - j\Omega)^2}{N(r)}, \\
K_{10} &= \frac{(E - j\Omega)\dot{W} + \alpha(E - j\Omega)\dot{W}^3}{F(r)} - \frac{eA_0\dot{W} + \alpha eA_0\dot{W}^3}{F(r)}, \\
K_{11} &= \frac{(E - j\Omega)^2 + \alpha(E - j\Omega)^4}{F(r)} + \frac{(E - j\Omega)eA_0 - \alpha\dot{W}(E - j\Omega)eA_0}{F(r)} \\
&- \frac{j^2 - \alpha j^4}{M(r)} - \frac{\dot{v}^2 - \alpha\dot{v}^4}{N(r)} - m^2 - \frac{1}{F(r)}eA_0[(E - j\Omega) + \alpha(E - j\Omega)^3 \\
&- eA_0 - \alpha eA_0\dot{W}^2],
\end{aligned}$$

$$\begin{aligned}
K_{12} &= \frac{\dot{W}j + \alpha\dot{W}^3 j}{M(r)}, \quad K_{13} = \frac{\dot{v}\dot{W} + \dot{v}\alpha\dot{W}^3}{N(r)}, \\
K_{20} &= -\frac{j(E - j\Omega) + \alpha j^3(E - j\Omega)}{F(r)} - eA_0\frac{j + \alpha j^3}{F(r)}, \quad K_{21} = \frac{\dot{W}j + \alpha\dot{W}j^3}{L^{-1}(r)}, \\
K_{22} &= -\frac{1}{F(r)}[-(E - j\Omega)^2 - \alpha(E - j\Omega)^4 + eA_0(E - j\Omega) + eA_0\alpha(E - j\Omega)^3] \\
&- \frac{\dot{W}^2 + \alpha\dot{W}^4}{L^{-1}(r)} - \frac{\dot{v}^2 + \alpha\dot{v}^4}{N(r)} - m^2, \quad K_{23} = \frac{j\dot{v} + \alpha j^3\dot{v}}{N(r)}, \\
K_{30} &= \frac{-1}{F(r)}[(E - j\Omega)\dot{v} + \alpha(E - j\Omega)\dot{v}^3] + \frac{eA_0\dot{v} + eA_0\alpha\dot{v}^3}{F(r)}, \\
K_{31} &= \frac{-\dot{W}\dot{v} - \alpha\dot{W}\dot{v}^3}{L^{-1}(r)}, \quad K_{32} = \frac{-j\dot{v} - \alpha j\dot{v}^3}{M(r)}, \\
K_{33} &= -\frac{1}{F(r)}[(E - j\Omega)^2 + \alpha(E - j\Omega)^4 - (E - j\Omega)eA_0 - \alpha(E - j\Omega)eA_0\dot{v}^3] + \\
&\frac{\dot{W}^2 + \alpha\dot{W}^4}{L^{-1}(r)} - \frac{j^2 + \alpha j^4}{M(r)} + \frac{eA_0}{F(r)}[(E - j\Omega) + \alpha(E - j\Omega)^3 - eA_0 - \alpha eA_0\dot{v}^3] \\
&-m^2,
\end{aligned}$$

where $\dot{W} = \partial_r \oplus_0$, $\dot{v} = \partial_\theta \oplus_0$ and $j = \partial_\phi \oplus_0$. The non-trivial solution is $|\ \mathbf{K}\ | = 0$ and solving these equations yields:

$$imW^{\pm} = \pm \int \sqrt{\frac{(E - eA_0 - j\Omega)^2 + X_1(1 + \frac{X_2}{X_1}\alpha)}{L(r)}}dr, \tag{10}$$

where $-$ and $+$ denote the incoming and outgoing particles, respectively. The function 'X_1' can be defined as $X_1 = \frac{j^2}{M(r)}$ and $X_2 = \frac{\alpha(E-j\Omega)^4}{F(r)} - \frac{\alpha(E-j\Omega)eA_0\dot{v}^3}{F(r)} - \frac{\alpha\dot{W}^4}{L^{-1}(r)} + \alpha\frac{j^4}{M(r)} - \frac{eA_0}{F(r)}[\alpha(E - j\Omega)^3 -$

$\alpha e A_0 \dot{v}^3] + m^2$ represent the angular velocity at the event horizon. Integrating Equation (10) around the pole, we get

$$imW^{\pm} = \pm i\pi \frac{(E - A_0 e - j\Omega)}{2\kappa(r_+)}(1 + \Xi\alpha), \tag{11}$$

and the surface gravity of the 4D gauged super-gravity BH [41] is given by

$$\kappa(r_+) = \frac{3r_+^4 + 2r_+^3 q_1 q_2 q_3 q_4 + r_+^2(\sum_{i<j}^4 q_i q_j + 1) - q_1 q_2 q_3 q_4}{2r_+\sqrt{\prod_{i=1}^4 (r_+ + q_i)}}. \tag{12}$$

The tunneling probability $\Gamma(imW^+)$ for boson vector particles is given by

$$\begin{aligned}
\Gamma(imW^+) &= \frac{Prob[emission]}{Prob[absorption]} = \frac{\exp[-2(imW^+ + imv)]}{\exp[-2(imW^- - imv)]} = \exp[-4imW^+] \\
&= \exp\left[-\pi \frac{(E - eA_0 - j\Omega)r_+\sqrt{\prod_{i=1}^4 (r_+ + q_i)}}{3r_+^4 + 2r_+^3 q_1 q_2 q_3 q_4 + r_+^2(\sum_{i<j}^4 q_i q_j + 1) - q_1 q_2 q_3 q_4}\right] \\
&\times \; (1 + \Xi\alpha).
\end{aligned} \tag{13}$$

The particles that tunnel outside the event horizon will fall into the BH, and one has $Prob[emission] = 1$ then $imW^- - imv = 0$.

Now, we can calculate the $T_H(imW^+)$ by comparing the $\Gamma(imW^+)$ with the Boltzmann formula $\Gamma_B(imW^+) \approx e^{-(E - eA_0 - j\Omega)/T_H(imW^+)}$, we get

$$\begin{aligned}
T_H(imW^+) &= \frac{3r_+^4 + 2r_+^3 q_1 q_2 q_3 q_4 + r_+^2(\sum_{i<j}^4 q_i q_j + 1) - q_1 q_2 q_3 q_4}{4\pi r_+\sqrt{\prod_{i=1}^4 (r_+ + q_i)}} \\
&\times \; (1 + \Xi\alpha)^{-1}.
\end{aligned} \tag{14}$$

The $\Gamma(imW^+)$ depends on the radial coordinate at the outer horizon r_+, A_0 vector potentials, E energy, j angular momentum, e charge of particles, q_i charge of a 4D gauged super-gravity BHs, α quantum gravity and Ω represent the angular velocity on this horizon.

3. 5-Dimension Gauged Super-Gravity Black Holes

This BH solution occurs for $N = 8$, $D = 5$, in gauged super-gravity theory (symmetry) [41]. Now, a particular case is discussed, where the solution was developed (STU-model) for the results of $N = 2$, $D = 5$, gauged super-gravity theory wave equation of motion. The line element for 5D BH in the theory of gauged super-gravity is given as [41]

$$ds^2 = -f\,(H_3 H_2 H_1)^{-\frac{2}{3}}\,dt^2 + f^{-1}\,(H_3 H_2 H_1)^{\frac{1}{3}}\,dr^2 + (H_3 H_2 H_1)^{\frac{1}{3}}\,r^2 d\Omega_{3,k}^2, \tag{15}$$

where

$$f = g^2 r^2 H_3 H_2 H_1 - \frac{\mu}{r^2} + k, \quad H_i = 1 + \frac{q_i}{r^2},$$

here i = 1,2,3 and for radius $k = 1$ and $k = 0$, then $d\Omega_{3,k}^2$ represents the metrics on $\mathbf{S}^3$ and $\mathbf{R}^3$ respectively. It is connected to ADM mass i.e., $g = 1/L$, which indicates AdS_5's inverse radius and depends upon the cosmological constant, $\Lambda = -6/L^2 = -6g^2$, and the q_i are BH charges. The result of the wave equation is the form of the three gauge potential field A_μ^i from

$$A_0^i = \frac{\tilde{q}_i}{q_i + r^2}, \tag{16}$$

here, $i = 1, 2, 3$ and $\tilde{q}_i$ are BH physical charges. It is observed that Gauss's theorem is applicable for these charges. The corrected temperature (T'_H) can be calculated as

$$T'_H(imW^+) = \frac{(\Sigma_{i=1}^{3} q_i + 1)r_+^4 - \prod_{i=1}^{3} q_i + 2r_+^6}{2\pi r_+^2 \sqrt{\prod_{i=1}^{3}(q_i + r_+^2)}}(1 + \Xi\alpha)^{-1}. \tag{17}$$

The corrected tunneling rate depends on energy (E), potential (A_0), angular momentum (Ω_H), the outer horizon (r_+) radial coordinate, correction parameter (α) and BH charge (q_i). We notice that the corrected temperature of boson particles denoted by Equation (17) is same as ($\alpha = 0$), the 5D BH temperature in the theory of gauged super-gravity in Equation (3.12) of Reference [43]. The $T_H(imW^+)$ is related to the radial coordinate on the outer horizon r_+, α quantum gravity and charge q_i of a 4D gauged super-gravity BHs respectively.

4. 7-Dimension Black Holes in Theory of Gauged Super-Gravity

We calculate a boson particle's quantum tunneling spectrum from a BH in 7D gauged super-gravity theory and also determine the tunneling rate of boson particles and the corresponding temperature at BH outer horizon r_+. The solutions of BH occur when $D = 7$ and $N = 4$ in the gauged super-gravity theory (symmetry) [41]. Firstly, this result was developed in as a special case of solutions of cases when $D = 7$, $N = 4$ gauged super-gravity through the equations of motion. The metric of a BH in 7D gauged super-gravity theory is [41]

$$ds^2 = -(H_1H_2)^{-\frac{4}{5}} f dt^2 + (H_1H_2)^{\frac{1}{5}} \left(f^{-1}dr^2 + r^2 d\Omega_{5,k}^2\right), \tag{18}$$

where

$$f = g^2r^2H_1H_2 - \frac{\mu}{r^4} + k, \quad H_i = \frac{q_i}{r^4} + 1, \quad (\text{for } i = 1, 2)$$

where $g = 1/L = 1$ and L is related to the cosmological constant $\Lambda = -15/L^2$. The two gauge field electric potentials A^i_μ through the result ofthe wave equation of motion are given by

$$A^i_0 = \frac{\tilde{q}_i}{r^4 + q_i} \quad (\text{for } i = 1, 2).$$

The corresponding Hawking temperature at the horizon can be obtained as

$$\check{T}(imW^+) = \left[\frac{3r_+^8 + 2r_+^6 + r_+^4(q_1 + q_2) - q_1q_2}{\pi r_+^3\sqrt{(r_+^4 + q_1)(r_+^4 + q_2)}}\right](1 + \Xi\alpha)^{-1}. \tag{19}$$

The Hawking temperature depends on parameters r_0, q_2, and q_1.

5. Graphical Analysis

In this section, we describe the graphical behavior of quantum corrected Hawking temperature in Equations (14), (17) and (19) as shown in Figures 1–3, respectively, for arbitrary parameter $\Xi = 1$ and also study the stable and unstable states of BHs.

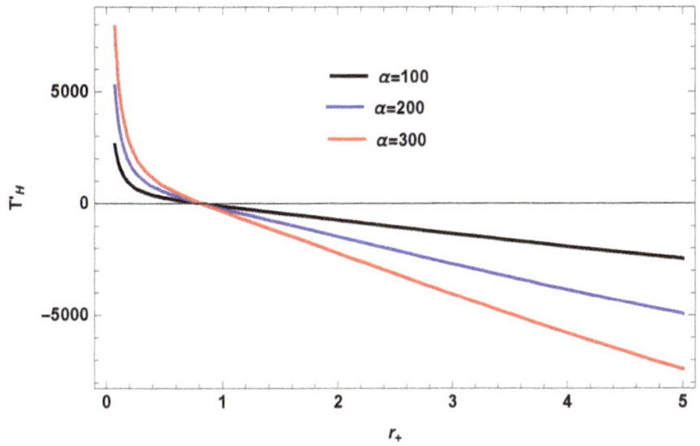

Figure 1. $T_H(imW^+)$ versus r_+ for $q_1 = q_2 = q_3 = q_4 = 0.5$ and $q_1 = q_2 = q_3 = q_4 = 5$.

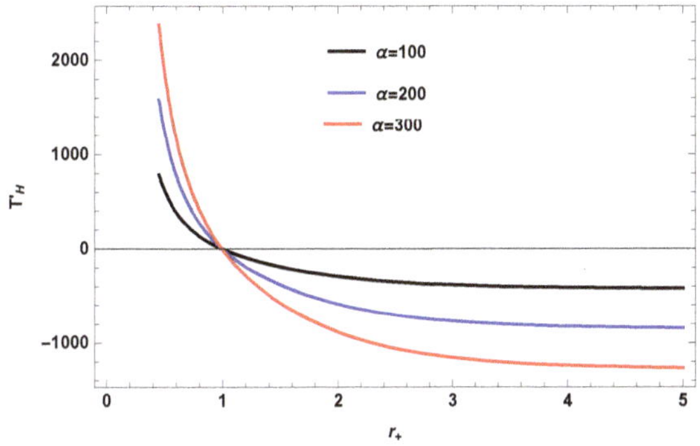

Figure 2. $T'_H(imW^+)$ versus r_+ for $q_1 = q_2 = q_3 = 0.5$ and $q_1 = q_2 = q_3 = 5$.

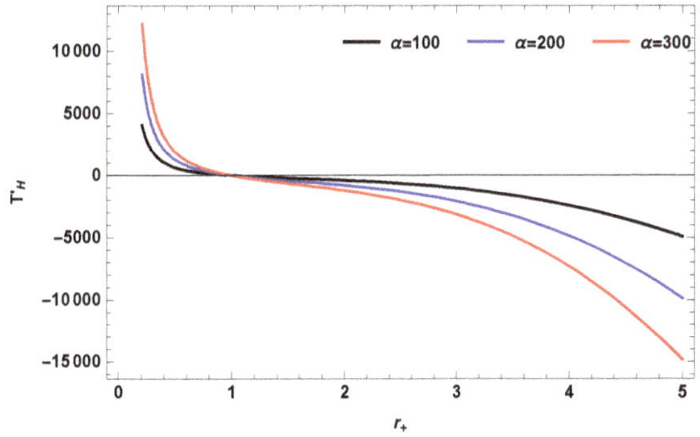

Figure 3. $\check{T}(imW^+)$ versus r_+ for $q_1 = q_2 = 0.5$ and $q_1 = q_2 = 5$.

T_H Versus r_+

In this subsection, we analyze the graphical behavior of corrected Hawking temperature T_H w.r.t the horizon r_+ for the $4D$, $5D$ and $7D$ gauged super-gravity BHs. Moreover, we study the physical significance of these graphs in the presence of correction parameter α and discuss the stable and unstable condition of corresponding BHs.

The $T_H(imW^+)$ slightly increase with increasing horizon and a slight change in the value of the correction parameter $\alpha = 1$ can cause a small increase in temperature, but the non-physical behavior identifies the unstable state of BHs.

In Figures 1–3, after initial increases in the particular range the temperature sharply increases with positive value. The non-physical behavior of the temperature increases with increasing horizon shows the instability of BH.

6. Conclusions and Discussion

In summary, applying the Hamilton–Jacobi phenomena of the tunneling formalism, we have studied the metric of the four, five and seven dimensional gauged super-gravity BHs. For this aim, we applied the Lagrangian wave equation with the setting of electromagnetism to analyze the tunneling of a massive charged boson (1-spin) particles from four, five and seven dimensional gauged super-gravity BHs having charges and physical charges. In this paper, we have extended the work of massive vector particles tunneling probability/rate for more generalized BHs in four, five and seven dimensional spaces and also observed the Hawking temperatures at which the particles tunnel through horizons. We have applied the Lagrangian equation to study the tunneling probabilty/rate of massive boson particles from four, five and seven dimensional gauged super-gravity BHs. In the Lagrangian equation, we applied the WKB approximation and which implies to the set of field wave equations, then apply separation of variables to find these wave equations.

The radial part can be obtained by applying the matrix of coefficients, whose determinant is equal to zero. We have developed the tunneling probability and temperature for these BHs at the outer horizon using surface gravity. The tunneling and temperature depend on the setting parameters of the BHs and quantum gravity. It is worth to study that the back-reaction and self-gravitating effects of boson charged particles on these BHs have been ignored, the calculated temperature are the parameters of BHs and quantum gravity.

The significance of the BHs, for the all types of particles having charged and uncharged, the tunneling rate will be change by viewing their semi-classical phenomenon and corresponding temperatures must be same for all types of charged and uncharged particles. We analyzed the part of the action which is imaginary, the tunneling probability/rate and temperature were introduced by charged massive vector particles due to gravity near the outer horizon r_+. Moreover, for the correction to the energy and tunneling rate of the massive boson particle GUP was introduced near the outer horizon r_+ in our computation. From our analysis, we have analyzed that the corrected temperature at which charged boson particles tunnel through the outer horizon r_+ is independent of the dimension of a BHs, and temperature is dependent on parameters of a metric and quantum gravity. The corrected temperature is shown to depend on the quantum gravity effect α. Both temperatures have the standard Hawking temperature limit when $(\alpha = 0)$, then the GUP effect completely vanished.

From our analysis we also concluded that the temperature at which particles tunnel through the outer horizon r_+ does not depend of the dimension of BHs in space. In particular the BH geometries, for the particles having different spin up and spin down the tunneling probabilities will be discovered to be the same by considering semi-classical phenomenon. Thus, their corresponding temperatures must be the same for all spin up and spin down particles. For these cases, we have carried out the calculations for more general BHs. Hence, the result still applies if the set BH parameters are more general.

- In the presence of charges, the BH was initially stable and attained a stability in a small domain and then becomes unstable till $r_+ \to +\infty$.
- The 4D, 5D and 7D BHs remained stable and unstable in quantum gravity minima and maxima respectively.
- The 4D, 5D and 7D BHs in the theory of gauged super-gravity remains unstable in the presence of the charge and correction parameter α.

Author Contributions: All authors contributed equally to this paper.

Funding: This research received no external funding.

Acknowledgments: The work of KB was partially supported by the JSPS KAKENHI Grant Number JP 25800136 and Competitive Research Funds for Fukushima University Faculty (18RI009).

Conflicts of Interest: The authors declare no conflict of interest.

References

1. Akhmedov, E.T.; Akhmedova, V.; Singleton, D.; Pilling, T. Thermal radiation of various gravitational backgrounds. *Int. J. Mod. Phys. A* **2007**, *22*, 1705–1715. [CrossRef]
2. Chowdhury, B.D. Problems with tunneling of thin shells from black holes. *Pramana* **2008**, *70*, 3–26. [CrossRef]
3. Jiang, Q.Q. Dirac particle tunneling from black rings. *Phys. Rev. D* **2008**, *78*, 044009. [CrossRef]
4. Akhmedov, E.T.; Pilling, T.; Singleton, D. Subtleties in the quasiclassical calculation of Hawking radiation. *Int. J. Mod. Phys. D* **2008**, *17*, 2453–2458. [CrossRef]
5. Akhmedova, V.; Pilling, T.; Gill, A.D.; Singleton, D. Temporal contribution to gravitational WKB-like calculations. *Phys. Lett. B* **2008**, *666*, 269–271. [CrossRef]
6. Akhmedova, V.; Pilling, T.; Gill, A.D.; Singleton, D. Comments on anomaly versus WKB/tunneling methods for calculating Unruh radiation. *Phys. Lett. B* **2009**, *673*, 227–231. [CrossRef]
7. Ali, A.F.; Das, S.; Vagenas, E.C. Discreteness of Space from the Generalized Uncertainty Principle. *Phys. Lett. B* **2009**, *678*, 497–499. [CrossRef]
8. Corda, C. Interferometric detection of gravitational waves: The definitive test for General Relativity. *Int. J. Mod. Phys. D* **2009**, *18*, 2275–2282. [CrossRef]
9. Singleton, D.; Vagenas, E.C.; Zhu, T.; Ren, J.R. Insights and possible resolution to the information loss paradox via the tunneling picture. *JHEP* **2010**, *1008*, 089. [CrossRef]
10. Nojiri, S.; Odintsov, S.D. Unified cosmic history in modified gravity: from F(R) theory to Lorentz non-invariant models. *Phys. Rept.* **2011**, *505*, 59–144. [CrossRef]
11. Capozziello, S.; Laurentis, M.D. Extended Theories of Gravity. *Phys. Rept.* **2011**, *509*, 167. [CrossRef]
12. Valerio, F.; Salvatore, C. *Beyond Einstein Gravity: A survey of Gravitational Theories for Cosmology and Astrophysics*; Fundamental Theories of Physics; Spring: New York, NY, USA, 2011.
13. Yale, A. Exact Hawking Radiation of Scalars, Fermions, and Bosons Using the Tunneling Method without Back-Reaction. *Phys. Lett. B* **2011**, *697*, 398–403. [CrossRef]
14. Bamba, K.; Capozziello, S.; Nojiri, S.; Odintsov, S.D. Dark energy cosmology: The equivalent description via different theoretical models and cosmography tests. *Astrophys. Space Sci.* **2012**, *342*, 155-228. [CrossRef]
15. Sharif, M.; Javed, W. Fermion tunneling for traversable wormholes. *Can. J. Phys.* **2013**, *91*, 43–47. [CrossRef]
16. Corda, C.; Hendi, S.H.; Katebi, R.; Schmidt, N.O. Effective state, Hawking radiation and quasi-normal modes for Kerr black holes. *JHEP* **2013**, *6*, 8. [CrossRef]
17. Jan, K.; Gohar, H. Hawking radiation of scalars from accelerating and rotating black hole with NUT parameter. *Astrophys. Space Sci.* **2014**, *350*, 279–284. [CrossRef]
18. Kruglov, S.I. Black hole emission of vector particles in (1+1) dimensions. *Int. J. Mod. Phys. A* **2014**, *29*, 1450118. [CrossRef]
19. Matsumoto, M.; Yoshino, H.; Kodama, H. Time evolution of a thin black ring via Hawking radiation. *Phys. Rev. D* **2014**, *89*, 044016. [CrossRef]
20. Li, X.Q.; Chen, G.R. Massive vector particles tunneling from Kerr and Kerr-Newman black holes. *Phys. Lett. B* **2015**, *751*, 34–38. [CrossRef]
21. Corda, C. Precise model of Hawking radiation from the tunnelling mechanism. *Class. Quantum Grav.* **2015**, *32*, 195007. [CrossRef]
22. Anacleto, M.A.; Brito, F.A.; Luna, G.C.; Passos, E.; Spinelly, J. Quantum-corrected finite entropy of noncommutative acoustic balck holes. *Annals Phys.* **2015**, *362*, 436. [CrossRef]
23. Lin, H.; Saifullah, K.; Yau, S.T. Accelerating black hole, spin-$\frac{3}{2}$ fields and C-metric. *Mod. Phys. Lett. A* **2015**, *30*, 1550044. [CrossRef]
24. Chen, G.R.; Huang, Y.C. Hawking radiation of vector particles as tunneling from the apparent horizon of Vaidya black holes. *Int. J. Mod. Phys. A* **2015**, *30*, 1550083. [CrossRef]
25. Anacleto, M.A; Brito, F.A.; Passos, E. Quantum-corrected self-dual black hole entropy in tunneling formalism with GUP. *Phys. Lett. B* **2015**, *749*, 181–186. [CrossRef]

26. Lin, G.; Zu, X. Scalar Particles Tunneling and Effect of Quantum Gravity. *J. Appl. Math. Phys.* **2015**, *3*, 134–139.
27. Feng, Z.; Chen, Y.; Zu, X. Hawking radiation of vector particles via tunneling from 4-dimensional and 5-dimensional black holes. *Astrophys. Space Sci.* **2015**, *359*, 48. [CrossRef]
28. Saleh, M.; Thomas, B.B.; Kofane, T.C. Hawking radiation from a five-dimensional Lovelock black hole. *Front. Phys.* **2015**, *10*, 100401. [CrossRef]
29. Bamba, K.; Odintsov, S.D. Inflationary Cosmology in Modified Gravity Theories. *Symmetry* **2015**, *7*, 220–240. [CrossRef]
30. Övgun, A.; Jusufi, K. Massive Vector Particles Tunneling From Noncommutative Charged Black Holes and its GUP-corrected Thermodynamics. *Eur. Phys. J. Plus* **2016**, *131*, 177. [CrossRef]
31. Jusufi, K.; Övgun, A. Hawking Radiation of Scalar and Vector Particles From 5D Myers-Perry Balck Holes. *Int. J. Theor. Phys.* **2017**, *56*, 1725. [CrossRef]
32. Cai, Y.F.; Capozziello, S.; Laurentis, M.D.; Saridakis, E.N. f(T) teleparallel gravity and cosmology. *Rept. Prog. Phys.* **2016**, *79*, 106901. [CrossRef] [PubMed]
33. Singh, T.I.; Meitei, I.A.; Singh, K.Y. Hawking radiation as tunneling of vector particles from Kerr-Newman black hole. *Astrophys. Space Sci.* **2016**, *361*, 103. [CrossRef]
34. Jusufi, K.; Övgun, A. Tunneling of Massive Vector Particles From Rotating Charged Black Strings. *Astrophys. Space Sci.* **2016**, *361*, 207. [CrossRef]
35. Li, R.; Zhao, J.K. Massive Vector Particles Tunneling from the Neutral Rotating Anti-di Sitter Black Holes in Conformal Gravity. *Commun. Theor. Phys.* **2016**, *65*, 469–472. [CrossRef]
36. Ghrsel, H.; Sakalli, I. Hawking Radiation of Massive Vector Particles From Warped AdS_3 Black Hole. *Can. J. Phys.* **2016**, *94*, 147–149. [CrossRef]
37. Li, X.Q. Massive vector particles tunneling from black holes influenced by the generalized uncertainty principle. *Phys. Lett. B* **2016**, *763*, 80–86. [CrossRef]
38. Nojiri, S.; Odintsov, S.D.; Oikonomou, V.K. Modified Gravity Theories on a Nutshell: Inflation, Bounce and Late-time Evolution. *Phys. Rept.* **2011**, *692*, 1–104 . [CrossRef]
39. Corda, C. The future of gravitational theories in the era of the gravitational wave astronomy. *Int. J. Mod. Phys. D* **2018**, *27*, 1850060. [CrossRef]
40. Javed, W.; Ali, R.; Abbas, G. Charged Vector Particles Tunneling From Black Ring and 5D Black Hole. *Can. J. Phys.* **2019**, *97*, 176–186. [CrossRef]
41. Cvetič, M.; Gubser, S.S. Phases of R-charged Black Holes, Spinning Branes and Strongly Coupled Gauge Theories. *JHEP* **1999**, *9904*, 024. [CrossRef]
42. Behrndt, K.; Cvetic, M.; Sabra, W.A. Non-Extreme Black Holes of Five Dimensional N=2 AdS Supergravity. *Nucl. Phys. B* **1999**, *553*, 317–332. [CrossRef]
43. Javed, W.; Abbas, G.; Ali, R.; Charged vector particle tunneling from a pair of accelerating and rotating and 5D gauged super-gravity black holes. *Eur. Phys. J. C* **2017**, *77*, 296. [CrossRef]
44. Shivalingaswamy, T.; Kagali, B.A. Eigenenergies of a Relativistic Particle in an Infinite Range Linear Potential Using WKB Method. *Eur. J. Phys. Educ.* **2011**, 2, 1309.

MDPI
St. Alban-Anlage 66
4052 Basel
Switzerland
Tel. +41 61 683 77 34
Fax +41 61 302 89 18
www.mdpi.com

Symmetry Editorial Office
E-mail: symmetry@mdpi.com
www.mdpi.com/journal/symmetry

www.ingramcontent.com/pod-product-compliance
Lightning Source LLC
Chambersburg PA
CBHW051837210326
41597CB00033B/5687
* 9 7 8 3 0 3 9 2 1 7 6 4 9 *